Frontiers in Planar Lightwave Circuit Technology

NATO Science Series

A Series presenting the results of scientific meetings supported under the NATO Science Programme.

The Series is published by IOS Press, Amsterdam, and Springer in conjunction with the NATO Public Diplomacy Division

Sub-Series

I. Life and Behavioural Sciences	IOS Press
II. Mathematics, Physics and Chemistry	Springer
III. Computer and Systems Science	IOS Press
IV. Earth and Environmental Sciences	Springer

The NATO Science Series continues the series of books published formerly as the NATO ASI Series.

The NATO Science Programme offers support for collaboration in civil science between scientists of countries of the Euro-Atlantic Partnership Council. The types of scientific meeting generally supported are “Advanced Study Institutes” and “Advanced Research Workshops”, and the NATO Science Series collects together the results of these meetings. The meetings are co-organized by scientists from NATO countries and scientists from NATO’s Partner countries –countries of the CIS and Central and Eastern Europe.

Advanced Study Institutes are high-level tutorial courses offering in-depth study of latest advances in a field.
Advanced Research Workshops are expert meetings aimed at critical assessment of a field, and identification of directions for future action.

As a consequence of the restructuring of the NATO Science Programme in 1999, the NATO Science Series was re-organized to the four sub-series noted above. Please consult the following web sites for information on previous volumes published in the Series.

http://www.nato.int/science
http://www.springer.com
http://www.iospress.nl

Series II: Mathematics, Physics and Chemistry – Vol. 216

Frontiers in Planar Lightwave Circuit Technology

Design, Simulation, and Fabrication

edited by

Siegfried Janz
National Research Council Canada

Jiří Čtyroký
Academy of Sciences of the Czech Republic

and

Stoyan Tanev
Vitesse Re-Skilling, Canada

Springer

Published in cooperation with NATO Public Diplomacy Division

Proceeding of the NATO Advanced Research Workshop on
Frontiers in Planar Lightwave Circuit Technology
Ottawa, Canada
September 21-25, 2004

A C.I.P. Catalogue record for this book is available from the Library of Congress.

ISBN-10 1-4020-4164-0 (HB)
ISBN-13 978-1-4020-4164-8 (HB)
ISBN-10 1-4020-4167-5 (e-book)
ISBN-13 978-1-4020-4167-9 (e-book)

Published by Springer,
P.O. Box 17, 3300 AA Dordrecht, The Netherlands.

www.springer.com

Printed on acid-free paper

Printed in the Netherlands.

Table of Contents

Preface

This book is the result of the NATO Advanced Research Workshop on Frontiers in Planar Lightwave Circuit Technology, which took place in Ottawa, Canada from September 21-25, 2004. Many of the world's leading experts in integrated photonic design, theory and experiment were invited to give lectures in their fields of expertise, and participate in discussions on current research and applications, as well as the new directions planar lightwave circuit technology is evolving towards. The sum of their contributions to this book constitutes an excellent record of many key issues and scientific problems in planar lightwave circuit research at the time of writing. In this volume the reader will find detailed overviews of experimental and theoretical work in high index contrast waveguide systems, micro-optical resonators, nonlinear optics, and advanced optical simulation methods, as well as articles describing emerging applications of integrated optics for medical and biological applications.

Acknowledgements

We would like to acknowledge and express our gratitude to the several organizations whose support and involvement made this NATO Advanced Research Workshop possible. This workshop was of course made possible because of the initial financial grant awarded by the NATO Security through Science Programme in Brussels, Belgium. We also acknowledge the Canadian Institute for Photonic Innovation (CIPI), the Canadian Photonics Fabrication Center and the National Research Council Canada, the Center for Research in Photonics at the University of Ottawa, Carleton University, Optiwave Systems Inc. and the Canadian Microelectronic Center for their generous support and/or cooperation. This workshop would not have been possible without the vision, expertise and commitment of the project management team of Vitesse Re-Skilling™ Canada. Finally, the editors wish to thank their respective institutions for supporting and encouraging their involvement in the organization of this workshop.

MICROPHOTONICS
Current challenges and applications

S. Janz, P. Cheben, A. Delâge, B. Lamontagne, M.-J. Picard, D.-X. Xu, K.P. Yap, W.N. Ye
Institute for Microstructural Sciences, National Research Council Canada

Abstract: This chapter gives an overview of the current challenges encountered in implementing devices based on high index contrast microphotonic waveguides, along with some new applications. An input coupler based on graded index (GRIN) waveguides is described. Theoretical and experimental results on using of cladding stress to eliminate the polarization dependence in SOI waveguide devices are reviewed. Recent work on output coupling of data on many output waveguides using waveguide to free space coupler array is also described. Design rules are presented for increasing bandwidth and resolution of integrated waveguide microspectrometers, to address applications in spectroscopic sensing and analysis. Finally, the potential for high index contrast microphotonic waveguides in evanescent field sensing is explored.

Key words: Microphotonics, silicon-on-insulator, integrated optics, birefringence, silicon photonics, wavelength division multiplexing, evanescent field sensors, couplers, waveguides, waveguide loss

1. INTRODUCTION

Optical waveguides and planar lightwave circuits have reshaped the field of optics over the last two decades. In classical free space optics light can only be delivered to a given point in space and time by line of sight propagation. The propagation direction is modified through refraction and reflection of freely propagating beams using bulk optical elements such as lenses and mirrors. In contrast, optical amplitude and phase information can be delivered through a waveguide to any specified location over an arbitrary path. Waveguides can be used to split, route, and recombine light on

S. Janz et al. (eds.), Frontiers in Planar Lightwave Circuit Technology, 1–38.

integrated planar waveguide circuit (PLC) of a few centimeters in diameter, thereby achieving almost any optical function. Modern optical telecommunications systems depend on waveguide technology for light generation by the waveguide based semiconductor lasers, transmission through optical fibre waveguides, and more recently for optical processing and routing using, for example, arrayed waveguide grating demultiplexers. As existing planar waveguide technologies evolve, waveguide optics will play an increasing role in telecommunications and penetrate the closely related technologies for optical interconnects in sensor and control networks, and between or within chips inside computers.

The challenge for the coming decade is to extend waveguide technology to new areas including chemical sensing, spectroscopy, medical instrumentation, biology, and space-based sensing. The functionality and small size that waveguide optics provides can be used to great advantage in these areas, but the existing telecommunication devices do not meet the requirements of many of the latter applications. For example most analytical spectroscopy applications involve optical wavelengths well outside the traditional C-band ($\lambda \sim 1525 - 1565$ nm), in the ultra-violet, visible and infrared. New advances, particularly in the areas in microphotonics and high refractive index contrast materials now make it possible to adapt and extend the capabilities of integrated optics to address these applications.

Microphotonics begins where the structures that confine and guide light are on the same scale or smaller than the wavelength of light. Microphotonic devices rely on the use of high index contrast material systems with a large refractive index step between the confining waveguide core and the surrounding cladding. In optical devices the wavelength sets the length scale for all interference and diffraction phenomena. The wavelength of light inside a medium, for a given vacuum wavelength λ_0, scales inversely with the index of refraction so that $\lambda = \lambda_0/n$. Any optical device that relies on optical interference or diffraction can therefore also be scaled down by the same factor 1/n, yet give identical performance. The larger index step between the waveguide core and cladding also makes it possible to confine light into a smaller volume. Hence high index contrast material systems allow more reduction in size than the simple 1/n scaling rule would suggest. For example, while conventional glass waveguide with a core cladding index step of Δn=0.01 have cross sectional dimensions of approximately 5 µm, Si waveguides with an air cladding (Δn= 2.5) or SiO_2 cladding (Δn =2.0) can be as small as 250 nm in cross section and still confine most of the optical mode within the silicon layer. As a result, waveguide devices fabricated in high refractive index materials such as silicon can be much smaller than their equivalents in glass or polymer.

The optical design rules for high index contrast materials are clearly very different from those for traditional glass waveguides, as suggested by the comparison in Table I. A unique feature in high index contrast materials is that the interaction of light with optical structures is non-perturbative, due to the large field discontinuities occurring at the interfaces. This allows much more scope for the design of optical components, but increases the difficulty in accurately designing and modeling waveguide structures since established perturbative solutions do not apply.

The most common examples of high index contrast waveguides are silicon-on-insulator (SOI) waveguides, and silicon oxynitride

Table 1. Comparison of normal and high index contrast material systems

Waveguide	Material	Bend Radius	Waveguide size	Interaction
Low Δn	Glass, Polymer, III-V semiconductors	> 1 mm	> 1.0 μm	Perturbative
High Δn	SOI, SiO_xN_y	> 1 μm	> 0.2 μm	Non-perturbative

waveguides. This combination of strong optical confinement and ability to manipulate light at the sub-wavelength scale give the optical designer unprecedented ability to design devices with high performance, small footprint, and to open up new application areas for optics. However this new capability in optical engineering has a cost. With decreasing device scale and higher index contrast, design tolerances rapidly become very difficult to meet. Polarization sensitivity and waveguide loss are difficult to control, and the coupling of light into and out of the devices is a major engineering challenge. Given these difficulties, this chapter will try to answer two fundamental questions; does microphotonic technology make sense given the associated technical hurdles, and can cost-effective manufacturing processes can be developed for microphotonic components?

2. MICROPHOTONIC WAVEGUIDE SPECTROMETERS

Waveguide based spectrometers came to maturity with the advent of optical telecommunications and wavelength division multiplexing (WDM) systems. Spectrometers of various forms are the essential component for WDM networks, where they are used to multiplex several wavelength channels together onto a single fibre, demultiplex or separate the different wavelength channels at the end of the fibre link, and also to allow regeneration, processing, and routing of single channels or groups of channels at various points in the network. State-of-the-art waveguide demultiplexers have reached very high levels of performance. The most common wavelength channel spacing is $\Delta\lambda$ = 0.8 nm (100 GHz), and the crosstalk between adjacent channels is often below -40 dB (a factor of 10000). Polarization sensitivity can be reduced to the level where the peak wavelength of any given output channel shifts by less than 0.01 nm for arbitrary changes in the input polarization. Finally the total insertion loss of telecommunications demultiplexers can be significantly less than 3 dB, including fibre to waveguide coupling loss.

Nevertheless, spectrometers or demultiplexers for telecommunications are specialized devices designed to function over a very narrow range of operating parameters. The operating wavelength range is usually just the bandwidth of the Er^+-doped fiber amplifier (EDFA) from $\lambda = 1525$ nm to $\lambda = 1565$ nm. For many applications in spectroscopy it may useful to have a free-spectral range over a much wider wavelength range or have much finer wavelength channel spacing than 0.8 nm, e.g. for high-resolution spectroscopy. Although some devices with much denser channel spacing have been reported[1-3], little work has been done to extend the performance of these devices to compete directly with laboratory spectrometers based on free space optics.

The operation of an echelle grating based waveguide spectrometer[1,3], shown schematically in Figure 1 is perhaps easiest to visualize since it is essentially a grating spectrometer fabricated in the two-dimension waveguide plane. Light is coupled to an input waveguide which terminates at a slab waveguide section, through which the light propagates freely within the waveguide plane to illuminate a curved grating etched vertically into the slab waveguide layers. Light is diffracted from the grating and focused into a spot on the output focal plane (or more accurately, the focal line in a two dimensional optical system), as in a free space grating spectrometer. The output waveguides, each one positioned to accept diffracted light at a specific wavelength, are arrayed at appropriate spacing along the focal line.

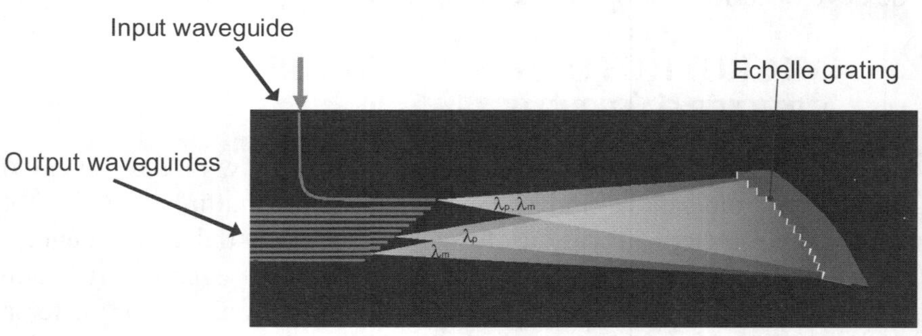

Figure 1. Schematic layout of an echelle grating waveguide spectrometer.

Each output waveguide leads to either an output fiber or photodetector, depending on the function of the spectrometer chip. The wavelength resolution is determined by several factors. The first is the change in diffraction angle θ with wavelength, dθ/dλ, which determines the rate at which the diffracted beam sweeps past a given output waveguide on the focal line as the wavelength changes. The angular dispersion is directly proportional to the order of the grating, M; the path length difference between rays diffracted from adjacent grating facets expressed in units of optical wavelength. The grating order also sets the operating wavelength range of the spectrometer, since the free spectral range is given approximately by λ/M. The second factor affecting resolution is the width of the input waveguide mode. The spectrometer creates an image of the input waveguide mode at the output focal plane – hence the smaller the input mode size the better the spectrometer resolution Δλ. Thus the mode size plays the same role as the slit width in a conventional free space spectrometer. Finally, the resolution is limited by the spot size of the diffracted beam at the focal plane. In telecommunications applications, the latter two parameters are coupled, since waveguide mode width and the focal spot size must be matched to achieve maximum coupling into the output waveguide for specific wavelength channels.

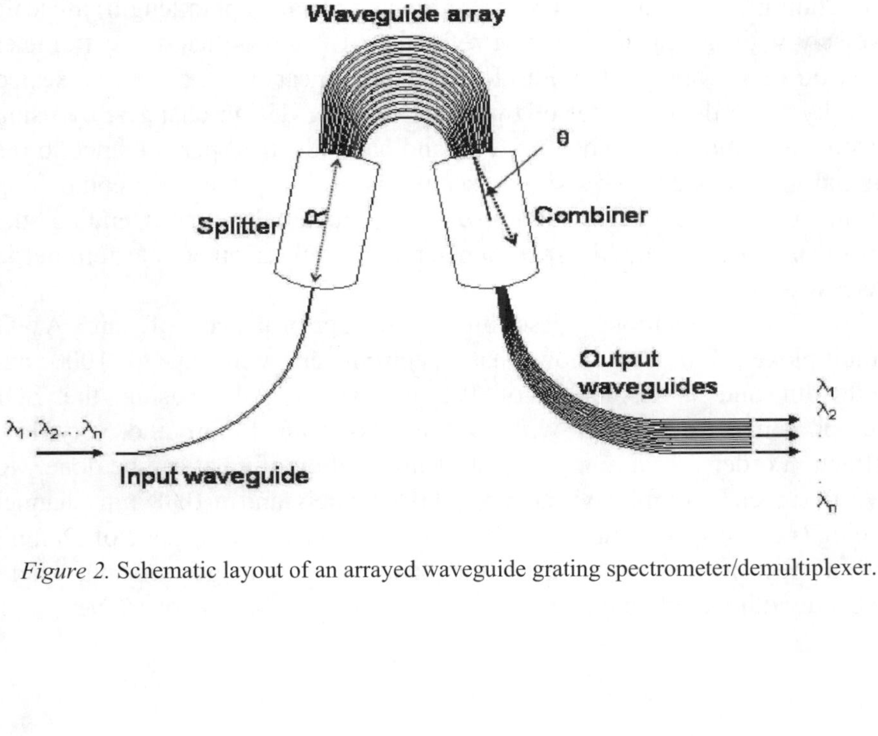

Figure 2. Schematic layout of an arrayed waveguide grating spectrometer/demultiplexer.

Despite recent developments in waveguide echelle grating technology, most telecommunication spectrometers and demultiplexers are still based on arrayed waveguide grating (AWG) designs, as shown in Figure 2. In an AWG, the grating has been replaced by an array of waveguides, in which each adjacent waveguide is a fixed increment longer than the next waveguide. This length increment, when expressed in wavelengths, is just the order M of the AWG. Thus each waveguide in the array plays the same role as a single grating tooth in the echelle grating device, and the physical principles of operation of the two designs are otherwise identical.

In adapting waveguide spectrometers to new applications, the immediate challenge is to either improve the resolution of the spectrometer, or to extend the spectrometer bandwidth beyond the 40 nm range typically found in telecommunications devices. As discussed in the previous section, the size of the spectrometer and all the parameters that set spectrometer resolution are dependent upon the wavelength of light in the material. The grating order M is the physical path length difference, in wavelengths, between light rays diffracted from adjacent grating teeth or propagating through adjacent waveguides in an AWG. The waveguide mode size is dependent on the wavelength of light, and the smallest achievable focal spot size is of course limited by diffraction and thus depends on wavelength as well. This suggests that spectrometer performance can be enhanced simply by using materials with high index of refraction. Waveguide mode size, the focal spot diffraction limit, and grating order (for a given physical path length) all scale inversely with increasing index of refraction. It follows that a spectrometer based on glass waveguide with effective index near $n = 1.5$, can be scaled down by more than a factor of two with no other design changes, by using silicon waveguide with index $n = 3.5$, and have identical performance to the original glass device. This simple scaling is probably the most compelling reason to consider high refractive index materials for extending the performance of waveguide spectrometers, as well as other interferometric devices.

From the equations describing the performance of an AWG demultiplexer, one can show that spectrometers with up to 1000 nm bandwidth and a resolution of 0.1 nm are possible using the SOI microphotonic waveguides with widths less than 1 μm and very low diffraction orders. As a more modest demonstration of what can be done, we have designed a simple AWG with 100 channels and a 0.08 nm channel spacing (i.e. the spectrometer resolution) and a free spectral range of 20 nm[4]. Such a spectrometer can be used in applications such as gas sensing, where the bandwidth would be targeted at a specific set of vibration overtone

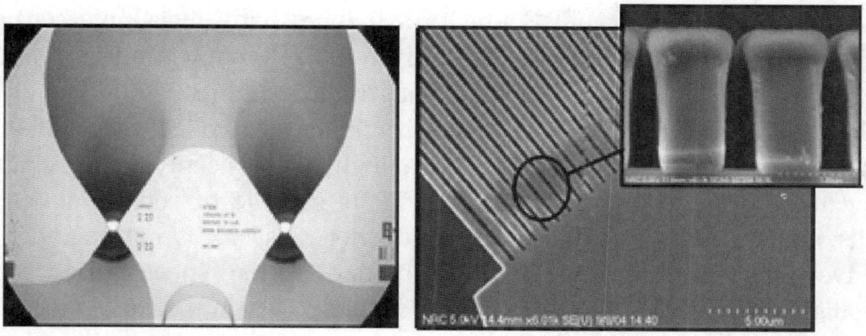

(a) (b)

Figure 3. (a) Plan view image of a SOI waveguide microspectrometer with 0.08 nm channel spacing and 20 nm free spectral range. (b) Plan view and a cross-section scanning electron microscope view of the waveguide apertures at the combiner output.

absorption lines[5], or for measuring small frequency shifts in such lines for Doppler velocimetry. A plan view of the first such spectrometer is shown in Figure 3(a). This spectrometer has an order M = 80 waveguide grating, with 250 waveguides in the array section. The overall device size is approximately 5×5 mm, small enough that it can be incorporated in a wide variety of analytical instruments with very little impact on instrument footprint.

A key design feature of this spectrometer is to narrow the input and output waveguide apertures at the combiner sections, as shown in Figure 3 (b). These apertures play the same role as the input and exit slits in a conventional free space grating monochromator, where the instrument wavelength resolution $\Delta\lambda$ is proportional to the input and out put slit widths. By reducing the waveguide mode width at the entrance of the combiner slab waveguide section, it is possible to improve the spectrometer resolution by a similar scaling factor, provided lithography and fabrication tools of sufficient spatial resolution are available and the fundamental limit set by diffraction is not reached. Using SOI waveguides with a very high index contrast between the Si waveguide core and the SiO_2 cladding, it becomes possible to design and fabricate waveguides with dimensions as small as a few hundred nanometers[6,7]. Such manipulation of the waveguide mode width is only possible when using high index contrast materials such as SOI. In conventional glass waveguides with an index step of the order of 1%, the waveguide mode width cannot be compressed to less than a few microns. If the glass waveguide core is decreased further the optical confinement is insufficient to prevent the waveguide mode to expand again. In our spectrometer design the waveguide is tapered down from a conventional SOI ridge waveguide formed in a 1.5 um thick SOI layer, with ridge width of 1.5

um and an etch depth of 0.8 um, to a 0.7 um wide rectangular channel waveguide that is etched down to the oxide layer, as shown in Fig. 4. These narrow apertures are the key to achieving high resolution. They launch the light into the spectrometer at the input combiner, and collect the light at the output side of the output combiner. Although such a deeply etched channel waveguide provides a narrow aperture, it is capable of supporting higher order modes. To ensure that the spectrometer crosstalk is not degraded by inadvertently exciting higher order modes at the spectrometer input, adiabatic converters are used to couple all the power in the ridge mode into the fundamental mode of the rectangular aperture waveguide. This is accomplished using the two level taper mode converter design shown in Figure 4. The converter design consists of a deep etched lateral taper that first transforms the waveguide from a ridge waveguide to a channel waveguide of the same width, and then tapers down the channel waveguide to the desired 0.7 μm aperture. Figure 5 shows simulations of the mode coupling efficiency from fundamental mode of the ridge waveguide to the fundamental mode of the deeply etched channel waveguide. The results show that when the two sections of the taper approach 100 μm in length, mode conversion to higher modes is less than -30 dB.

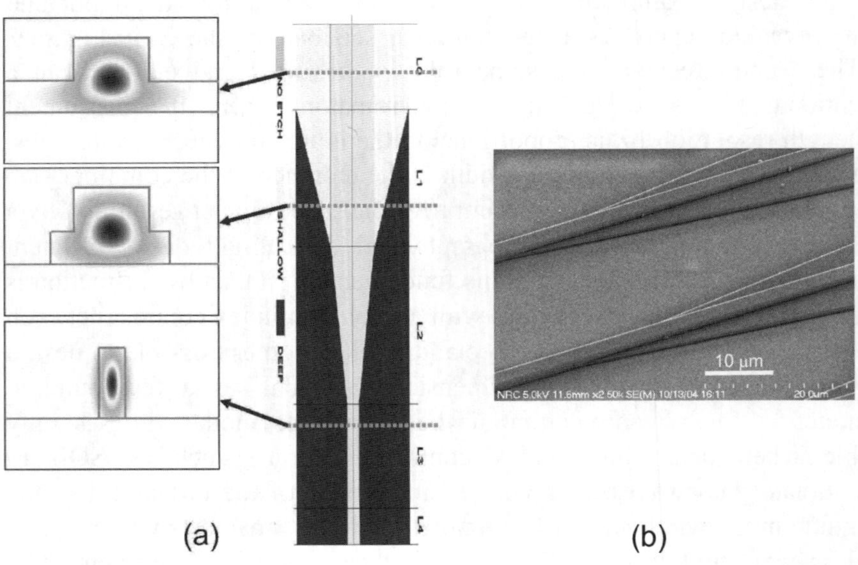

Figure 4. (a) Design of an SOI two level taper leading to a waveguide aperture in the microspectrometer, and the corresponding mode size at the indicated positions. (b) An SEM view of the fabricated mode converters.

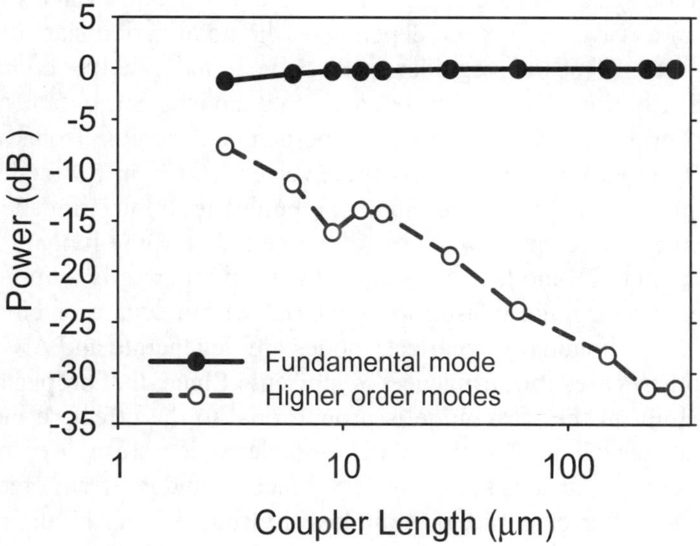

Figure 5. BPM simulations of the power coupling efficiency from the fundamental mode of the initial 1.5 μm wide ridge waveguide to the fundamental and cumulative higher order modes at the end of the 0.7 μm aperture waveguide.

Insertion loss measurements were performed on a series of fabricated mode converters of different lengths. Series of two, four and six converters of identical length were fabricated end to end. By measuring the total waveguide insertion loss with increasing number of converters, an indication of the converter excess loss could be obtained. As in the case of the microspectrometer aperture waveguides, the ridge waveguides were formed in a 1.5 um thick SOI, with a 1.5 um wide ridge width. The mode converter tapers down to a 0.7 μm wide channel waveguide. The measured insertion loss for converters that are 50, 100, 200 and 300 μm long are independent of converter length. These results indicate that the additional insertion loss of a single converter is less than 1 dB, and since the ridge waveguides are single mode, most of the power is maintained in the local fundamental mode.

In summary, the use of a narrowed aperture waveguide can be employed to increase spectrometer resolution. Here we have described a simple mode converter that can be used to couple light from a ridge waveguide to aperture waveguide (i.e. a narrow rectangular channel waveguide etched to the buried oxide layer) with negligible higher order mode generation.

3. POLARIZATION

The waveguide effective index N, and hence the propagation wave vector $\beta = N\,(2\pi/\lambda_0)$ of a waveguide mode depends on the polarization state of the waveguide mode. Some waveguides will have modes with complex polarization distributions, but in most cases the mode electric fields are predominantly oriented along either the perpendicular axis (transverse magnetic or TM mode) or parallel axis (transverse electric or TE mode) to the substrate plane. In this discussion, the polarization birefringence is assumed to be the difference between the effective index of the fundamental TM and fundamental TE modes, $\Delta N = N_{TM} - N_{TE}$. In a perfectly symmetric waveguide (i.e. a waveguide of isotropic material with a square or circular cross-section) the orthogonally polarized modes are degenerate and $\Delta N = 0$, but in most other cases birefringence is present. Since the propagation wavevector of light in the waveguide is proportional to the effective index, birefringence causes an unwanted polarization dependent response in all interferometric devices such as spectrometers, Mach-Zehnder interferometers and ring resonators. For example, if the index birefringence is ΔN, the peak wavelength of a waveguide spectrometer channel will differ by $\Delta\lambda \sim \lambda_0\, \Delta N/N$ for the two orthogonally polarized modes. Similarly, the phase of TE and TM polarized light after propagating through a waveguide of length L will differ by $\Delta\phi \sim (2\pi\Delta N/\lambda_0)\cdot L$.

The source of polarization anisotropy may be the intrinsic birefringence of the waveguide material itself, as is the case of $LiNbO_3$, or may be induced by anisotropic stress fields in the waveguide layer[8]. The latter is often the case in waveguides formed by film deposition or epitaxial growth. Finally, there is a geometrical birefringence determined by the shape of the waveguide and the boundary conditions for the electromagnetic field across the waveguide core-cladding interfaces. Since the tangential electric field is continuous across a boundary, while the normal component of the displacement $D = \varepsilon E$ must be continuous, the field distributions near the interface for orthogonally polarized modes can be very different for TE and TM modes, as for the Si channel waveguide modes shown in Figure 6. The difference in field distribution is naturally reflected in the difference in effective index between these modes. Since this geometrical birefringence is determined by the field amplitude at the interface, and by the dielectric constant discontinuity at the waveguide cladding interfaces, ΔN increases as the waveguide shrinks and as the index contrast between core and cladding becomes larger.

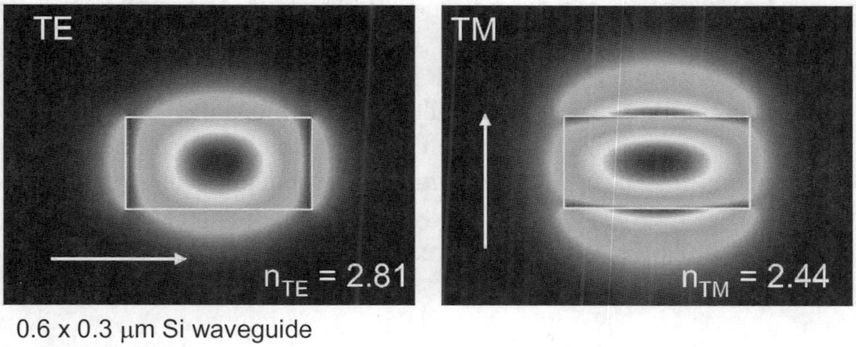

Figure 6. TE and TM polarized waveguide modes for a 0.6 x 0.3 μm rectangular waveguide embedded in an SiO_2 cladding. The arrows indicate the dominant electric field polarization.

In a typical high index contrast SOI ridge waveguide with width and height approaching one micron, ΔN can easily be larger than 10^{-3}. In applications such as wavelength demultiplexing and high-resolution spectroscopy, the birefringence should be less than $\Delta N \sim 10^{-5}$ for acceptable polarization insensitive performance. The birefringence problem may be circumvented by adopting a polarization diversity strategy, wherein the two orthogonally polarized components of an incoming optical signal are separated and directed through two waveguide circuits each optimized to process a single polarization state correctly. The polarization diversity approach requires polarization splitters, filters, and converters, all of which can introduce complexity and undesirable compromises in overall system performance. The difficulty encountered by coherent communications systems, which rely on polarization diversity, to be accepted by the telecommunications industry in the past suggests that polarization diversity is unlikely to be commercially viable.

In the case of SOI ridge waveguides, a specific aspect ratio (e.g. the ratio of ridge width and ridge height) can often be found for which the geometrical birefringence is zero[9]. While this approach can be effective in controlling birefringence, for small waveguides it becomes increasingly difficult to fabricate waveguides with dimensions within the tolerances required to reduce birefringence to an acceptable level. Also, by adopting this approach the waveguide ridge aspect ratio is fixed by the birefringence constraint, and it is difficult to design the waveguide to meet other performance criteria such as mode confinement, bend loss, and higher order mode suppression.

Recently a new approach has been proposed and demonstrated by Xu et al.[10], in which the geometrical birefringence in an SOI waveguide is balanced by the stress birefringence created by a cladding layer. By applying claddings of different thickness and internal stress, the total waveguide birefringence can be precisely controlled. Furthermore the birefringence of waveguides of arbitrary shape can be corrected, thus allowing considerably more freedom to design waveguide independently of birefringence constraints.

Through the photoelastic effect, stress will modify the refractive index of silicon according to the relations

$$n_x - n_0 = -C_1\sigma_x - C_2(\sigma_y - \sigma_z)$$
$$n_y - n_0 = -C_1\sigma_y - C_2(\sigma_x - \sigma_z) \qquad (1)$$

where z is along the waveguide propagation direction, and y points along the surface normal. The local stress field at any point in the waveguide is described by the three stress tensor components σ_x, σ_y and σ_z , and C_1 and C_2 are the stress optic constants given in Table 2. Here n_x and n_y are the effective refractive indices for light polarized along the x- and y-axes respectively, while n_0 is the unperturbed index of refraction.

The stress tensor elements σ_i are related to the strain field ε_k in the material by the equations

$$\varepsilon_x = \frac{1}{E}\left[\sigma_x - \nu(\sigma_y + \sigma_z)\right] + \alpha\,\Delta T$$
$$\varepsilon_y = \frac{1}{E}\left[\sigma_y - \nu(\sigma_x + \sigma_z)\right] + \alpha\,\Delta T \qquad (2)$$
$$\varepsilon_z = \frac{1}{E}\left[\sigma_z - \nu(\sigma_x + \sigma_y)\right] + \alpha\,\Delta T$$

where *E* is Young's modulus, ν is the Poisson ratio, and α is the thermal expansion coefficient. The last term in each of these equations represents the strain in the material that would arise from thermal expansion in the absence of external forces, for a temperature change ΔT.

Table 2. Mechanical and optical constants of Si and SiO_2 (refractive index N is given at λ=1550 nm)

Material	E (GPa)	ν	α (K^{-1})	N	C_1 (GPa^{-1})	C_2 (GPa^{-1})
Si	130	0.27	3.6×10^{-6}	3.476	-17.79×10^{-3}	5.63×10^{-3}
SiO_2	76.7	0.186	5.4×10^{-7}	1.444	0.65×10^{-3}	4.5×10^{-3}

The thermal expansion coefficient of silicon is approximately seven times larger than that of SiO_2, as given in Table 2. When SiO_2 is deposited, typically at temperatures of several hundred degrees C, an in-plane compressive stress develops in the oxide layer as the Si wafer is cooled by ΔT to room temperature. For a uniform 2-dimensional thin film deposited on a substrate, the in-plane stress obtained from Equation 2 is:

$$\sigma_{film} = \sigma_x = \sigma_y = \frac{E}{1-\nu}(\alpha_s - \alpha)\Delta T \tag{3}$$

where x and z are taken to be in the wafer plane, and y is normal to the surface. The coefficients E and ν are the Young's modulus and the Poisson ratio of the SiO_2 film, as given in Table 2. In a uniform film, the film stress σ_{film} depends on the difference between in thermal expansion coefficients, α_s and α, of the silicon substrate and SiO_2 layer, respectively. Assuming the system is cooled by ΔT = 400 ºC, the calculated film stress is σ_{film} = -115 MPa. The out-of-plane stress σ_y is zero, since no force acts perpendicular to the surface. However, in response to the in-plane stress, a uniform SiO_2 film will expand along the surface normal with a tensile strain given by

$$\varepsilon_y = -\frac{2\nu}{E}\sigma_{film} \tag{4}$$

In real films, the deposition conditions, composition, and subsequent annealing also determine the stress. Stresses are usually larger than predicted by the simple thermal model, with values in the range from σ_{film} = -200 to -300 MPa for PECVD grown SiO_2 films. This thermal stress is the dominant source of birefringence in glass waveguides, and hence its control and elimination are among the fundamental issues in silica waveguide device design. Surprisingly, this same stress mechanism provides a simple and elegant method of eliminating the birefringence in SOI waveguides.

As an illustrative model, consider a rectangular silicon-on-insulator channel waveguide embedded in an oxide cladding, as shown in Figure 7.

$$\varepsilon_y = -\frac{2\nu}{E}\sigma_{film}$$

$$\varepsilon_x = \alpha_s \Delta T$$

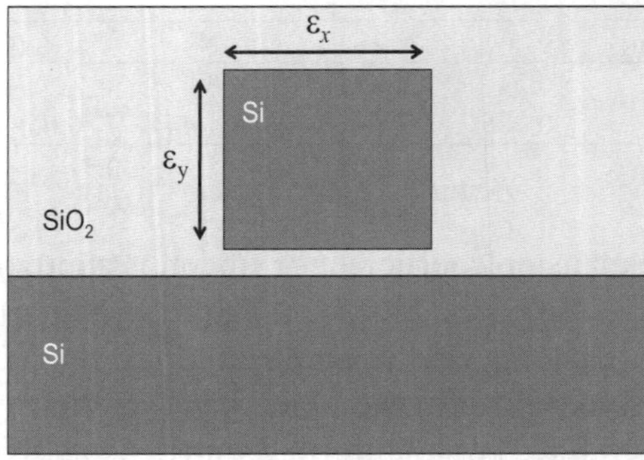

Figure 7. Schematic diagram of a Si channel waveguide, buried in a SiO_2 cladding under compressive stress $\sigma_x = \sigma_{film}$ in the horizontal plane.The equations indicate the approximat strains ε_x and ε_y estimated according to the simple model described in the text.

The in-plane strain ε_x inside the Si is expected to be dominated by intrinsic thermal expansion since the Si is effectively lattice matched to the substrate with only a thin intervening layer of SiO_2. However, the Si channel will be stretched along the surface normal by the oxide, with approximately the same tensile strain as in the oxide given by Equation 2. In reality the Si waveguide will deform in a more complex way, but this simple picture shows how a strongly anisotropic strain field is created within an embedded Si waveguide by a uniform cladding. The approximate magnitude of the resulting birefringence can be estimated using the strain given by Equation 2 to estimate the stress anisotropy in the Si waveguide[11]. For typical SiO_2 film stresses (σ_{film} = -200 to -400 MPa) the stress birefringence turns out to be comparable in magnitude (i.e. $|\Delta N| \sim 10^{-3}$) and more importantly, opposite in sign to the geometrical birefringence of a typical SOI ridge waveguide. As a result of this fortuitous circumstance, careful manipulation of the cladding film stress can be used to eliminate the total waveguide birefringence.

Designing an SOI ridge waveguide to have zero birefringence requires a numerical model that calculates both the stress distribution within the SiO_2 cladding and the Si waveguide core, as well as the optical field distribution of the TE and TM waveguide modes. Well established mode solving methods can be used to evaluate the electromagnetic fields, but the stress calculation can be problematic. The model system must extend far enough that stress artifacts originating at the edge of the calculation window boundary do not perturb the calculated stress in the vicinity of the waveguide ridge. The dimensions required for the calculation window may be as large as 100 μm. At the same time, the spatial resolution of the calculation must be fine enough to accurately model the local stress distribution at the sub-micron scale in a ridge waveguide that is only a few microns in size.

The finite element method (FEM) is used by Xu et al.[10,12] for both the mechanical problem of determining the strain distribution, and the electromagnetic problem of solving for the waveguide modes. The model system consists of an SOI wafer with the top Si layer patterned to form the ridge waveguide, and the desired SiO_2 cladding placed over the entire wafer and ridge waveguide. The width and height of the model system, i.e. the calculation window, were chosen to be large enough that edge effects would be negligible, as discussed below. Initially the layers are in equilibrium and there is no stress in any of the model layers. The stress field is generated by a simulated cooling of the waveguide, where the temperature change ΔT is chosen to induce an oxide film stress σ_{film} equivalent to that measured on experimental wafers. This procedure should reproduce the stress distribution due to thermal expansion mismatch. However, as noted above, intrinsic

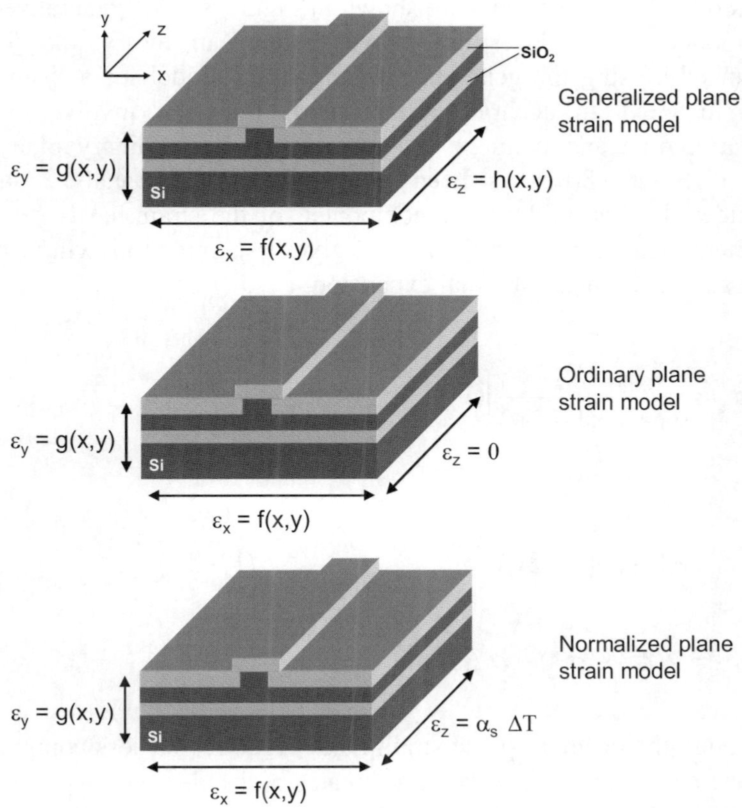

Figure 8. The generalized plane strain model, ordinary plane strain model, and the normalized plane strain model as applied to a single SOI ridge waveguide.

material stress will also contribute to the total stress in the waveguide-cladding system, but there is no reliable quantitative model for predicting this intrinsic stress. Implicit in this work is the assumption the stress distribution created by the simulated cooling of the model system is an accurate representation of the total stress distribution in the corresponding real waveguide cladding system. The excellent agreement eventually achieved between experimental measurements and theory suggests that the assumption is a reasonable one.

A variety of models and approximations have been used to analyze stress in waveguides[13,14]. In plane strain models, the strain distribution is allowed to vary in the x and y directions, but remains constant along the z-direction. The problem is thereby reduced from three to two dimensions. Plane strain models are appropriate for systems that do not change in cross sectional profile along one direction, and are very long compared with relevant structural features, as in the case for ridge waveguides. Three different versions of plane strain model are shown in Figure 8. The generalized plane strain model allows the strain to vary freely in the x and y plane. Unfortunately, using the generalized plane strain model for a system large enough to yield an accurate stress distribution can involve excessive computation time and memory usage. The simpler ordinary plane strain model of Figure 8(b) has been used previously to analyze stress in waveguides. In this model the z-component of the strain, ε_z, is set to zero everywhere. Setting $\varepsilon_z = 0$ in Equation 2 gives rise to the following equations relating stress, strain and thermal expansion.

$$\sigma_x = \frac{E}{(1+\nu)(1-2\nu)}\left[(1-\nu)\varepsilon_x + \nu\varepsilon_y\right] - \frac{E}{(1-2\nu)}\alpha\,\Delta T$$

$$\sigma_y = \frac{E}{(1+\nu)(1-2\nu)}\left[\nu\varepsilon_x + (1-\nu)\varepsilon_y\right] - \frac{E}{(1-2\nu)}\alpha\,\Delta T \qquad (5)$$

$$\sigma_z = \nu(\sigma_x + \sigma_y) - \alpha\,\Delta T$$

Although the ordinary plane strain model requires far less computational resources, constraining ε_z to be zero creates a sizable spurious stress field. Physically, this artifact arises as the model system is cooled by ΔT. Contraction can only occur in x and y, since the system length is locked along z. The presence of the stress artifact will i) shift the average refractive index, and ii) cause an error in the calculated stress anisotropy in the x-y plane (and hence birefringence). The anisotropy error is due to changes in the effective thermal expansion along x and y when the substrate is under tensile stress along the z-axis. The shift in calculated birefringence is of the order of 10% for SOI waveguides[12], but can be larger when the Poisson ratios of the layers in the structure are sufficiently different.

The accuracy of stress and birefringence calculations can be improved by an extension of the ordinary plane strain model, proposed by Ye et al. [12], in which the strain along ε_z is again fixed but now set equal to the thermal strain of the substrate, $\varepsilon_z = \alpha_S \Delta T$, after simulated cooling. If the substrate is much thicker than the layers composing the waveguide and cladding (i.e. substrate bending is negligible) this model should give equivalent results to the generalized plane strain model. The equations for this normalized plane strain model can be written in a form where σ_x and σ_y are again independent of σ_Z, and are formally identical to the Equation 5 defining the ordinary plane strain model.

$$\sigma'_x = \frac{E}{(1+\nu)(1-2\nu)}\left[(1-\nu)\varepsilon_x + \nu\varepsilon_y\right] - \frac{E}{(1-2\nu)}(\alpha - \frac{\nu}{1+\nu}\alpha_s)\Delta T$$

$$\sigma'_y = \frac{E}{(1+\nu)(1-2\nu)}\left[\nu\varepsilon_x + (1-\nu)\varepsilon_y\right] - \frac{E}{(1-2\nu)}(\alpha - \frac{\nu}{1+\nu}\alpha_s)\Delta T$$

$$\sigma'_z = \nu(\sigma'_x + \sigma'_y) - (\alpha - \alpha_s)\Delta T \tag{6}$$

Comparing Equations 5 and 6, it is evident that the stress distribution in the x-y plane can be calculated simply by using the ordinary plane strain model with the thermal expansion coefficient α in Equation 5 replaced by a renormalized expansion coefficient

$$\alpha' = \alpha + \frac{\nu}{1+\nu}\alpha_s \tag{7}$$

Proceeding by solving Equation 5 with α' in this manner will yield an incorrect value for σ'_z, but since σ'_z is by assumption independent of x and y, the correct value is obtained simply adding a constant to the σ_Z from equation 4:

$$\sigma'_z = \sigma_z + \frac{E}{1+\nu}\alpha_s\,\Delta T \tag{8}$$

The normalized plane strain procedure therefore requires the same modest computational resources as the ordinary plane strain model. Furthermore, the same algorithms used for the ordinary plane strain model can be employed, since the revised model simply requires the substitutions defined by Equations 7 and 8.

Finally, neither the normalized plane strain model or the ordinary plane strain model can correctly calculate the effect of bending of the entire waveguide/substrate system by the stress in the SiO_2 layers, since by assumption the strain along z is independent of x and y. The modification of film stress in the Si waveguide layer and SiO_2 layers can be estimated using known relations[12,15]. While this correction may be significant when the substrate thickness is comparable to the total SiO_2 thickness, it is negligible for typical Si substrate thicknesses 500 μm or more. In any case, after packaging of devices, the bending stress is can be comparable to local stresses induced by the bonding the chip to a carrier.

One further approximation greatly improves the efficiency of stress distribution calculations. The wafer bending is directly proportional to the stiffness, or Young's modulus E_s, of the substrate wafer. If bending can be neglected (e.g. when the substrate is much thicker than the oxide film), one can set E_s to be arbitrarily large[12]. The film stresses in the oxide and Si waveguide layers are determined mainly by the thermal expansion of the substrate, not by the substrate elastic constants, so this approximation will have little effect on the calculated stress distributions in Si waveguide or SiO_2 layers. The advantage of assuming a stiff substrate is a substantial reduction in edge effects for the numerical stress calculation. Using the large E_s approximation, the calculation window for a typical SOI waveguide can be reduced from 100 x 100 μm to 40 μm wide by 5 μm deep, with no change in the calculated stress field in the vicinity of the waveguide ridge. Combining the large E_s approximation with the normalized plane strain procedure, stress distributions in SOI waveguides can be obtained that are equivalent to those using the full generalized plane strain model, but with several orders of magnitude reduction in computation time and resources.

Using the model described here, the anisotropic stress field inside the silicon ridge waveguides shown in Figure 9 were calculated. Applying Equation 1 to evaluate the anisotropic index of refraction, the FEM mode solver was used to determine the effective index values N_{TE} and N_{TM} for the

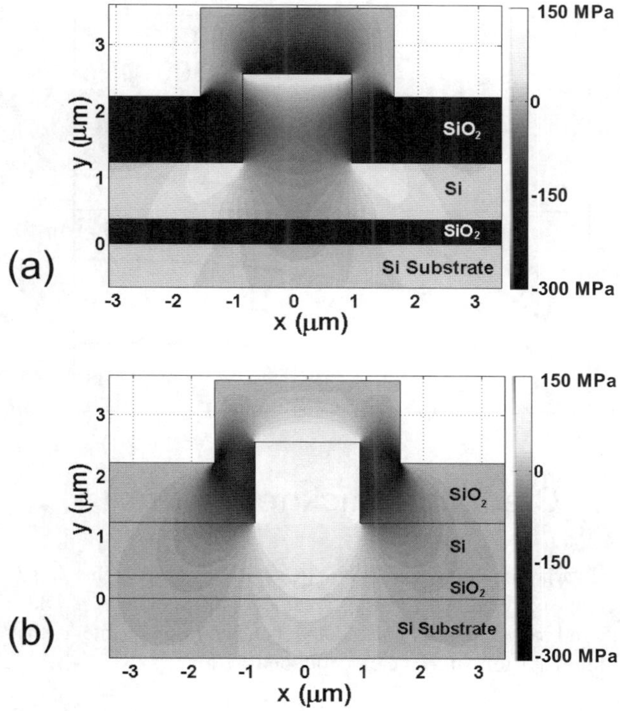

Figure 9. (a) Calculated in-plane (σ_x), and (b)calculated out-of-plane (σ_y) stress fields inside a Si ridge waveguide

TE and TM polarized fundamental modes. The resulting birefringence value $\Delta N = N_{TM}\text{-}N_{TE}$ incorporates the contribution of the geometric birefringence and the stress induced birefringence.

The geometrical birefringence of SOI can be compensated in waveguides over a range of waveguide dimensions and cross-sectional shapes, using SiO_2 films with modest film thickness (e.g. < 1 μm), and film stresses between σ_{film} = -200 to -400 MPa, as commonly found in oxides deposited by conventional PECVD. For example, Figure 10 shows the calculated birefringence in a 1.5 μm wide trapezoidal ridge waveguide, with a ridge depth of 1.5 μm etched into a 2.2 μm silicon layer. The variation of the total birefringence with SiO_2 cladding thickness is shown for films with stresses ranging from σ_{film} = 0 to -300 MPa. With no cladding stress, the birefringence is approximately $\Delta N \sim 10^{-3}$, already unacceptably large for most applications in telecommunications and spectroscopy. However, the

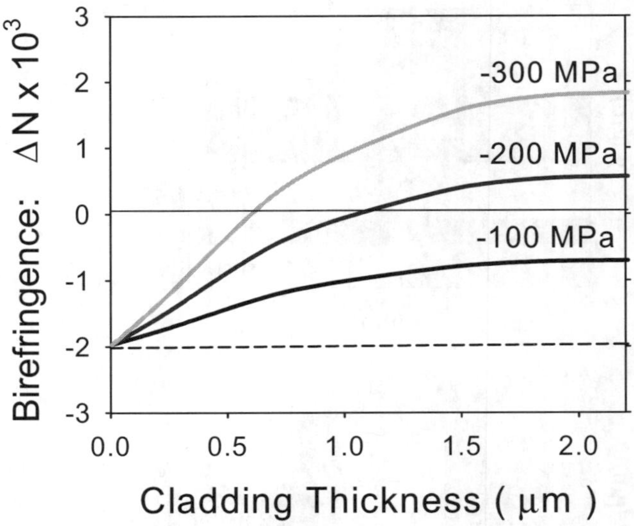

Figure 10. Calculated variation with cladding thickness of the birefringence in a SOI ridge waveguide, for different SiO_2 cladding film stress values. The model waveguide is formed in a 2.2 μm thick Si layer and has a typical trapezoidal wet etched ridge profile, with a base width of 3.8 μm, a top width of 1.1 μm, and an etch depth or 1.47 μm.

stress field created by SiO_2 film with σ_{film} = -300 MPa and a thickness of 0.6 μm fully cancels this large geometrical birefringence, resulting in a polarization independent waveguide.

Waveguide birefringence can be measured using an arrayed waveguide grating (AWG) demultiplexer, in which the waveguide array is composed of the waveguides under test. The birefringence will induce a shift in the wavelength, $\Delta\lambda = \lambda\cdot(\Delta N/N)$ for each output channel, as the input polarization is varied between TE and TM. In our experiment a nine wavelength channel AWG with a 1.6 nm (200 GHz) channel spacing was used. The device used to obtain the data in Figure 11 had trapezoidal ridge waveguides of width 1.1 μm, formed in a 2.2 μm thick Si layer using a 1.5 μm deep etch. Figure 11 shows the calculated and measured wavelength shifts $\Delta\lambda$ for increasing SiO_2 cladding thicknesses. Remarkable agreement is achieved between theory and experiment. Furthermore, these results demonstrate experimentally that a polarization independent AWG demultiplexer can be fabricated using SOI waveguides with a SiO_2 cladding less than 1 μm thick and a typical film stress of -320 MPa. Figure 12 shows the TE and TM polarized spectra of 9 output channels of a similar AWG demultiplexer before and after deposition

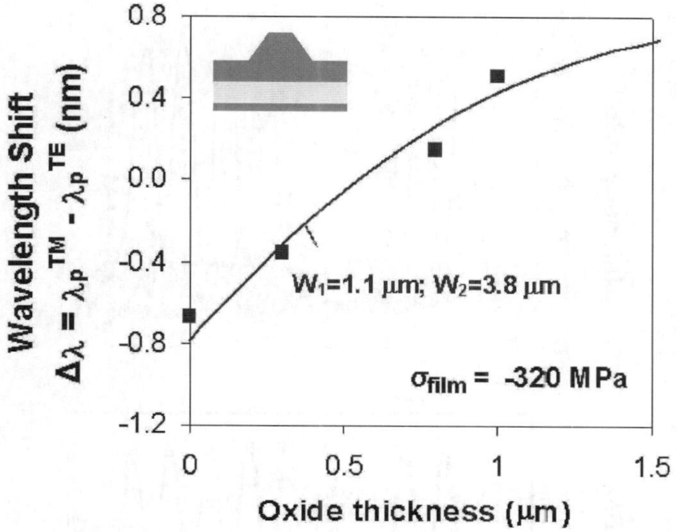

Figure 11. Comparison of calculated and measured birefringence for different SiO_2 cladding thicknesses for a trapezoidal waveguide formed in a 2 μm thick Si layer as shown in the inset. The ridge base width is 3.8 μm, the top width is 1.1 μm, and the etch depth is 1.47 μm.

of a birefringence compensating 0.6 μm thick SiO_2 layer. In this case the cladding induced stress is reduced from $\Delta\lambda \sim 0.6$ nm to achieve a fully birefringence compensated device with a polarization dependent wavelength shift of less than 0.05 nm.

The use of cladding stress to correct birefringence of a range of waveguide sizes allows a considerable degree of freedom in designing SOI waveguides to meet other performance criteria such as higher order mode suppression and reduced loss at waveguide bends. The birefringence can then be corrected as a final step using deposition of the appropriate cladding. This technique also allows the waveguide shape and dimensions to be chosen for relaxed dimensional tolerances, again with polarization correction applied as a last step. The cladding correction technique is also amenable to post fabrication birefringence tuning by annealing to modify the film stress, or adding or removing a small amount of cladding [11].

Since the effect of a SiO_2 cladding on mode shape is negligible, there is no mode mismatch loss or PDL associated with the junction between waveguide sections with and without claddings. Therefore it is possible to apply tailored cladding patches at discrete locations in a planar waveguide

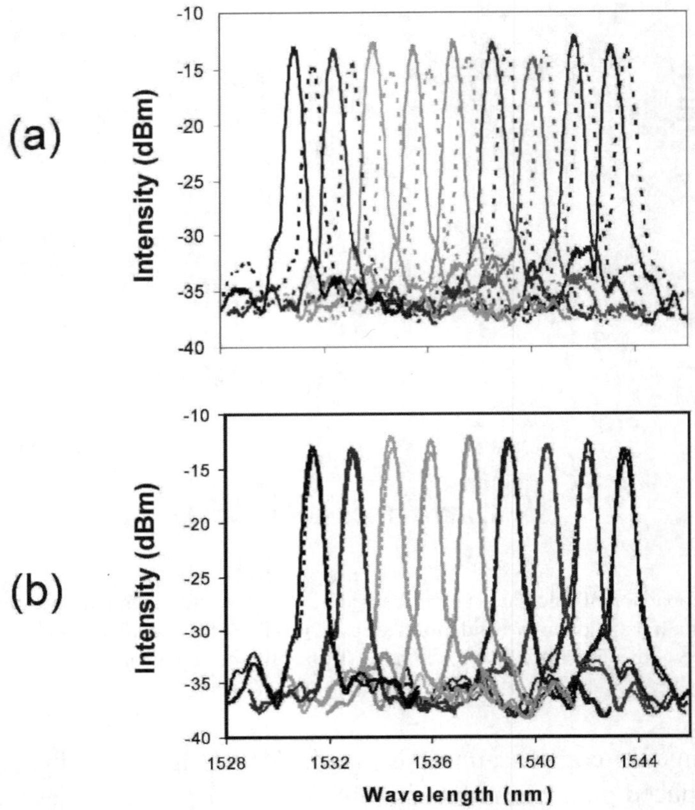

Figure 12. SOI AWG channel spectra (a) before and (b) after cladding stress compensation.

circuit with no insertion loss penalty. Although the initial experimental demonstration of this technique involved correction of an AWG demultiplexer, oxide stress can be applied in many situations including polarization correction of ring resonators, waveguide grating filters, and Mach-Zehnder interferometers. Cladding stress can also be used to create a specific non-zero birefringence for polarization control devices including polarization filters and splitters.

4. INPUT COUPLING

The potential advantages of Si microphotonic components can be offset by the difficulty in coupling light into and out of waveguides with micron or submicron dimensions. In most applications, light is coupled from an optical fibre or free space beam with a beam waist of at least several microns. Conceptually, the simplest approach for coupling light from a large to a small waveguide is to use an adiabatic mode converter, in which the waveguide size in continuously tapered down over a long enough distance that all the light remains in the local fundamental mode. This approach is easily implemented in two-dimensions, i.e. in the wafer plane, by tapering the waveguide width using conventional lithography and etching. However, three-dimensional fabrication methods are required to taper both the width and height of the waveguide. Such three dimensional waveguide tapers have been fabricated using gray scale lithography to vary the etch rate linearly along a waveguide. In one such coupler, a taper varying in thickness from 10 to 0.5 μm[16] has been produced, with measured coupling efficiencies of 60% from fibre to waveguide. By using a two-dimensional inverted taper coupler, the need for true three-dimensional fabrication is avoided. The waveguide width is tapered adiabatically in the opposite sense, from its nominal on chip width to an extremely narrow waveguide[17]. As the waveguide narrows down below a critical value, the mode begins to expand not only laterally but also in the vertical direction. By adjusting the final width of the waveguide the mode size can be closely matched to the much larger fibre mode. Although the need for three-dimensional fabrication is avoided, high-resolution lithography (e.g. using e-beam) is still required to produce tip diameters of the order of 100 nm in size. Coupling efficiencies of -3.3 dB have been measured. Other approaches include the use of gratings to couple light from an initial large input waveguide to successively smaller waveguides[18], or from free space beams. All the above approaches require high-resolution e-beam or deep UV lithography, and correspondingly accurate fabrication procedures.

We have proposed an alternative coupling scheme that uses a graded-index (GRIN) waveguide lens monolithically integrated on the waveguide chip[19]. A conventional cylindrical GRIN lens has a refractive index that decreases quadratically with radius from the central axis of the cylinder. Light launched along the cylinder axis comes to a focus in a length f, after which the beam repeatedly diverges and converges. Such bulk optic GRIN lenses are used in the packaging and fiber pig-tailing of optoelectronic components. Our proposed GRIN waveguide coupler is shown schematically in Figure 13 for an SOI waveguide. The guiding layer, in this example, is formed by a thin silicon layer on a SiO_2 buried oxide layer cladding. The much thicker GRIN coupler layer is deposited on top of this waveguide. Ideally, the refractive index N(x) of this GRIN layer should have a quadratic dependence on distance x from the base of the layer,

$$N(x) = N_0(1 + ax^2) \tag{9}$$

where N_0 is the refractive index of the waveguide layer and a is a constant. By analogy with a conventional cylindrical GRIN lens, one may expect that light launched into the composite structure formed by the Si waveguide and GRIN layer in figure 3 should also periodically come to a focus and then diverge, provided the bottom edge of the waveguide core acts as a perfect mirror. The coupler is then formed by etching away the GRIN layer beyond the first focal point of the lens so that light remains confined in the Si waveguide beyond that point. In fact, the bottom surface of the waveguide is never a perfect mirror, since even for total internal reflection there will be a finite phase difference in the reflected beam. Nevertheless, calculations using the beam propagation method (BPM) and also exact modal expansion propagation suggest the analogy is almost exact for a quadratic index profile.

Silicon-on-insulator (SOI) and silicon oxynitride (SiO_xN_y) on silica are the most common waveguide material systems for high index contrast microphotonics. For the SiO_xN_y system, layers with index of refraction ranging from n = 1.46 (SiO_2) and n = 2.0 (Si_3N_4) can be grown using PECVD. Coupling to SOI waveguides requires GRIN layers of much higher refractive index to match the index of Si (n = 3.47 at λ = 1550 nm). Materials ranging in composition from amorphous silicon to silicon dioxide can also be grown by PECVD with the index of refraction ranging from n = 1.46 (SiO_2) to n ~ 3.4 (a-Si)[19]. Figure 14 shows the optical intensity distribution for light at wavelength λ =1550 nm propagating through a SiO_xN_y GRIN coupler structure composed of a 0.5 μm Si_3N_4 waveguide with a 3.5 μm GRIN layer.

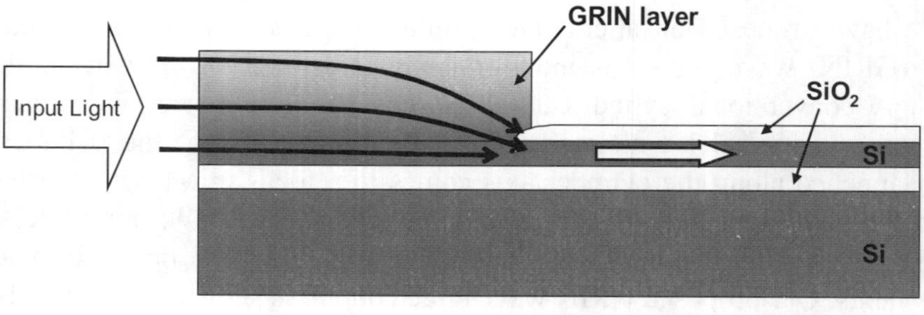

Figure 13. GRIN waveguide lens on SOI, consisting of a thick graded index a-Si_xO_y layer deposited on a thin Si waveguide.

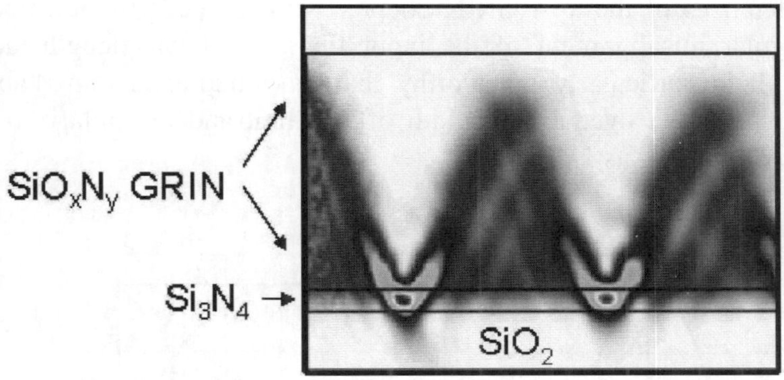

Figure 14. Intensity distribution for light at wavelength λ = 1550 nm propagating through a SiO_xN_y GRIN coupler structure composed of a 0.5 μm Si_3N_4 waveguide with a 3.5 μm SiO_xN_y GRIN layer.

The index changes from n = 1.5 at the top surface to n = 1.9 at the boundary between the waveguide and GRIN structure. This coupler achieves a 90% coupling efficiency from a 4 μm top hat intensity profile to the Si_3N_4 waveguide. Similar calculations show that effective couplers can be fabricated using a step index profile with as few as two or three discrete layers[19]. Figure 15 (a) shows refractive index profiles for continuous, two layer, and single layer a-Si GRIN couplers as depicted schematically in Figure 13. Figure 15 (b gives the corresponding coupling efficiency into

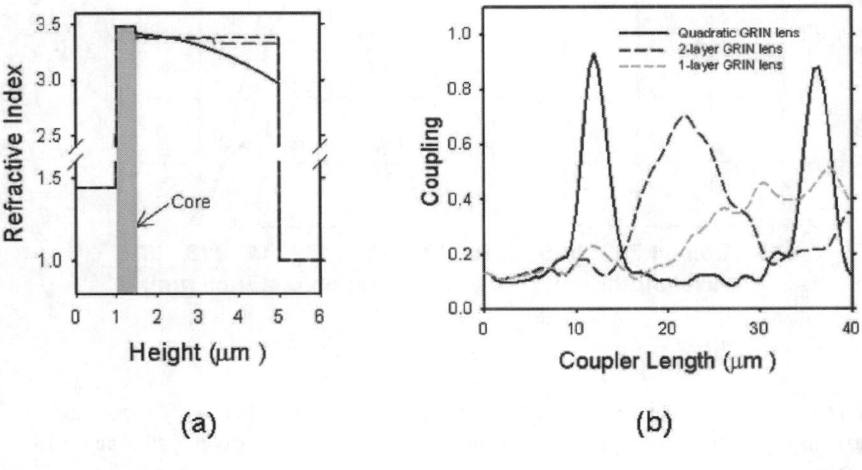

Figure 15. (a) Index profile for one, two and three layer GRIN waveguide coupler structures and (b) corresponding coupling efficiencies.

the 0.5 um SOI waveguide, as the coupler length is varied. Figures 16 a)-d) show the tolerances of the a-Si GRIN coupler with respect to input mode position, angular misalignment of the input fiber, input wavelength, and polarization. It is particularly noteworthy that this coupler is remarkably wavelength insensitive over a bandwidth of 200 nm, and the polarization dependent loss is as low as 1 dB.

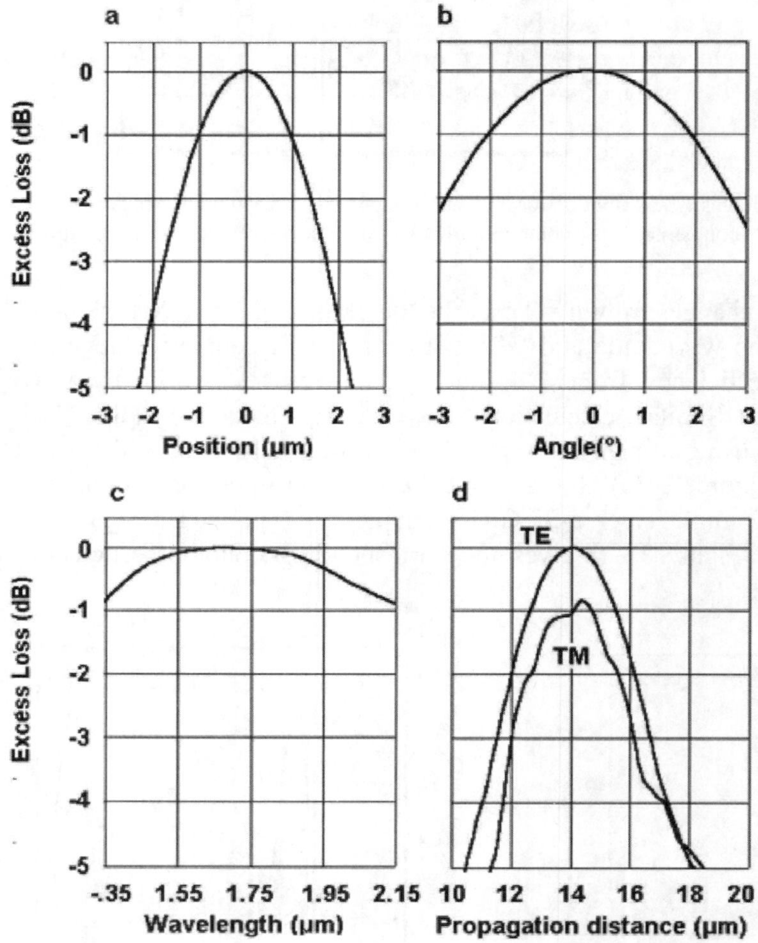

Figure 16. Tolerances of the a-Si Ọ ÇGRIN coupler efficiency with respect to (a) input mode angular misalignment, (b) input mode displacement, (c) input wavelength, and (d) polarization.

The fabrication of GRIN waveguide coupler can be much simpler than of other coupler structures. The basic GRIN structure can be grown with a well calibrated PECVD machine. Patterning the GRIN coupler requires lithography with a ±1 μm resolution. The GRIN coupling scheme does require that the input facet position relative to the end of the coupler section is within ±1 μm of the focal point. The input/output facets of glass and SOI waveguide die are often diced and polished, with a resulting uncertainty in final facet position of several tens of microns. To address this problem, we have developed a process that forms a lithographically defined coupling facet for an SOI waveguide. The facet position is therefore determined to within the spatial accuracy of the lithography and processing. Although essential for implementation of a GRIN coupler, lithographically defined coupling facets considerably simplify device fabrication, since labor intensive mechanical polishing is no longer required. Figure 17 shows a scanning electron microscope (SEM) image of a typical waveguide facet produced by deep etching. The facet was etched using an ICP etcher at -120°C using O_2 + SF_6 gas mixture. Waveguide to fiber insertion loss measurements results for waveguides with etched facets were better than or comparable to those obtained by polishing.

Figure 17. A lithographically defined ridge waveguide facet waveguide formed by ICP etching.

5. OUTPUT COUPLING OF MULTICHANNEL DEVICES

Applications of integrated optics to other areas such as chemical sensing and spectroscopy may involve the simultaneous acquisition of optical signals in large number of optical waveguides. For example, monitoring a full spectrum in real time using a high resolution waveguide spectrometer may involve the continuous measurement and processing of information (i.e. the optical intensity) on hundreds of output channels. Existing packaging technologies used in telecommunications applications rely on the alignment and bonding of fibre arrays to the edge of the waveguide chip. Alternatively, the waveguide chip may be aligned with linear photodetector array. As the number of output waveguides increases, the cost and effort of fibre or detector array alignment and attachment grows quickly. Furthermore, if the output waveguide spacing is set by the fibre diameter (e.g. 125 μm) or the photodetector array pitch, the chip size can become excessively large. An alternative approach to this packaging problem is to terminate each output waveguide with a coupler that redirects the guided light into a free space beam propagating approximately normal to the chip surface. By arranging the output couplers in an appropriate square grid pattern, for example that

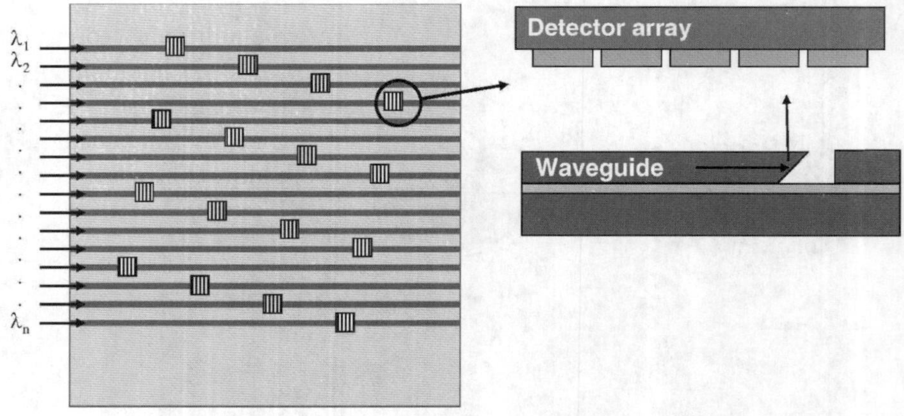

Figure 18. Array of waveguide to free space couplers (a) in top view and (b) in a side view showing imaging detector array used for channel intensity acquisition.

shown in Figure 18, the resulting array of free space beams can be monitored using a two dimensional imaging array such as a CCD or CMOS arrays for wavelengths less than $\lambda = 1.0$ μm, or commercially available IR detector arrays for further in the infra-red. As an example, an image of the output coupling array in Figure 18(a) can be captured by the imaging detector array as shown in Figure 18 (b). Precise alignment becomes unnecessary in such an output coupling scheme. The pixels responding to a given waveguide can be identified during a calibration procedure, and simply allocated to that channel through the controlling software. Imaging arrays are available at relatively low cost with array dimension well over 1000×1000 pixels, and pixel sizes of the order of a few microns. The only draw-back to this approach is that the speed of data acquisition is limited to the read-out time of the imaging array used. The output coupling array arrangement can lead to a significant reduction in packaging cost for high channel count devices, wherever the spectrum acquisition time need not be shorter than the typical array read-out times.

As is clear from the preceding discussion, a manufacturable waveguide to free space output coupler is another essential component that would allow integrated optics to penetrate into fields of application beyond telecommunications. The simplest form of waveguide to vertical beam coupler is a total internal reflection mirror etched into the waveguide, as shown in Figure 19. This figure presents a cross-section diagram of such a structure designed for a 2 μm SOI waveguide. Mirrors of this form can be fabricated using a directional dry etch to create an undercut facet at approximately 45° to the vertical. Figure 19(b) shows an example of such an

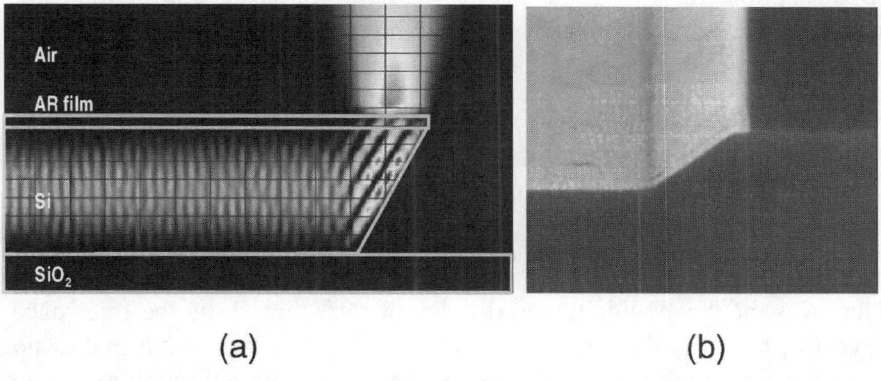

Figure 19. (a) Cross-sectional view of the calculated field distribution in a total internal reflection (TIR) coupler in a 2 μm SOI waveguide, and (b) an etched TIR structure in SOI.

etched structure in SOI. As a final step, the surface of the Si waveguide can be coated with an antireflection coating to minimize back reflections into the waveguide circuit, and increase the efficiency of the coupler. Finite difference time domain calculations of light propagating through the coupler shown in Figure 19(a) predict a 96.5% coupling efficiency from the waveguide mode to the free space beam, the latter with a 56° divergence angle at full width half maximum (FWHM) of the optical intensity. An important advantage of this simple TIR coupler is that it is both wavelength and polarization independent, unlike couplers based on second-order gratings or photonic crystals structures. To be useful for applications such as a multichannel waveguide spectrometer, the output couplers must be fabricated at numerous specified locations on a wafer. Preliminary work suggests that structures similar to that shown in Fig. 18 can be fabricated using a single angle dry etch through appropriately placed windows in an etch mask. Other alternative forms of output coupler include gratings of integer M wavelength pitch (i.e. M-λ gratings), and photonic crystal or waveguide holograms. As discussed above, since such structures depend on the relative wavevector difference of the grating and the optical mode, they will always be inherently polarization and wavelength dependent.

6. CLADDING AND INTERFACE EFFECTS

The interaction of light with cladding materials comes to the fore as the waveguide size decreases. In the case of Si waveguides, as waveguide width shrinks below 1 um, an increasing fraction of the optical mode propagates in the cladding material. At the same time, the effective index of the waveguide becomes increasingly dependent on the refractive index of the cladding, and more susceptible to waveguide loss caused by scattering due to interface roughness. Using a well known expression developed by Payne and Lacey[21], the maximum attenuation coefficient due to scattering loss expected from a waveguide with root-mean-square (rms) sidewall roughness σ is

$$\alpha = \frac{\sigma^2}{nk_0 d^4} \tag{10}$$

Here n is the waveguide material index of refraction, k_0 is the free space wavevector, and d is the waveguide core width. This expression makes no reference to the length scale of the roughness, for example through inclusion of the correlation length. Hence it is only an estimate of the maximum attenuation that would occur if the roughness correlation length was comparable to the wavelength of light. An important prediction of this expression is the inverse fourth power dependence of the scattering loss on waveguide dimension. This dependence on waveguide width has been confirmed by experimental measurements on SOI waveguides[22]. Using a more detailed version of scattering loss theory from Reference 21, Figure 20 shows the calculated insertion loss for an SOI slab waveguide with rms surface roughness of $\sigma = 10$ nm, and a roughness correlation length of $L_c =$ 500 nm. For waveguides smaller than 2 um, the predicted propagation loss rapidly diverges. In the case of square SOI channel waveguides, the loss increases with decreasing width as for the slab waveguide, but Grillot et al.[23] show that as the dimensions decrease below 300 nm, the interface losses begin to decrease as the mode expands into the cladding. It is clear that viable microphotonic technology will depend on fabrication techniques that can eliminate roughness between the waveguide core and cladding.

The various approaches to removing surface roughness include thermal oxidation to preferentially consume asperities on the waveguide surface, resulting in loss improvement from 32 dB/cm to 0.8 dB/cm for a 0.5 um wide SOI waveguide[24]. The residual rms roughness after oxidation smoothing was 2 nm. Another method is to use a wet etch after conventional dry etching to form Si ridge waveguides. The resulting rms roughness is also approximately 2 nm. Recent developments in lithography are also having an

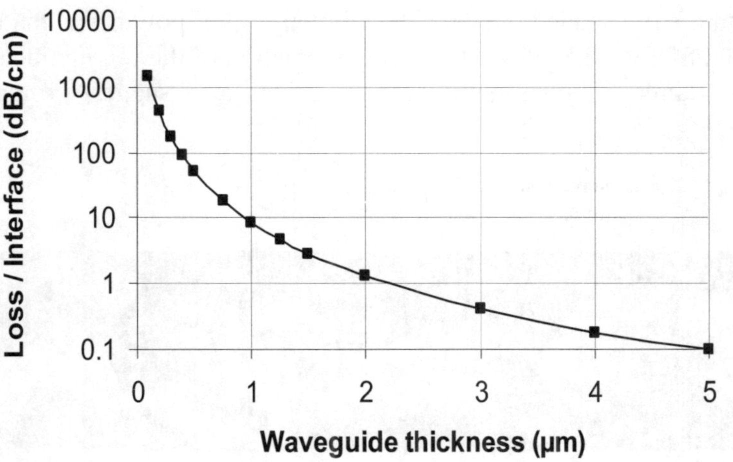

Figure 20. Calculated insertion loss for an SOI slab waveguide with surface roughness parameterized by $\sigma_{rms} = 10$ nm and $L_c = 500$ nm.

impact on achievable waveguide loss. For example, the use of deep UV lithography and dry etching has yielded 0.5 x 0.33 μm^2 waveguides with losses as low as 2.4 dB/cm[25] .

As waveguides shrink, the waveguide mode will shrink along with the waveguide cross-section, until the index step between the core and the cladding is insufficient to confine the mode. If the waveguide core size is decreased further, the mode begins to expand into the cladding. For glass waveguides, this dimension threshold is near 4 μm cross-section, while for Si waveguides this threshold is about one order of magnitude smaller, near 0.4 μm. Figure 21 shows the mode size evolution for square waveguides of different sizes composed of glass and silicon. This behaviour is the basis of the inverse taper mode coupler described in Section 4, but also has other useful implications. For example, a microphotonic waveguide where a significant fraction of the mode field propagates in the cladding can be considered a composite system, the properties of which are determined by both the core material and the cladding material. It is therefore possible to combine silicon waveguides with electro-optic or magneto-optic materials to create devices not possible in silicon alone. This has been exploited by Osgoode et al.[26] to create optical isolators based on SOI waveguides clad with magnetic garnet films. Potentially a practical solution to the quest for a high-speed modulator may be a combination of Si and an electro-optic polymer cladding. Because of the strong waveguide more cladding interaction, it may also be possible to fabricate optical structures such as gratings and strip loaded waveguides directly into polymeric materials deposited on SOI waveguides. The strong interaction of the waveguide mode with the cladding in microphotonic waveguides also suggests important

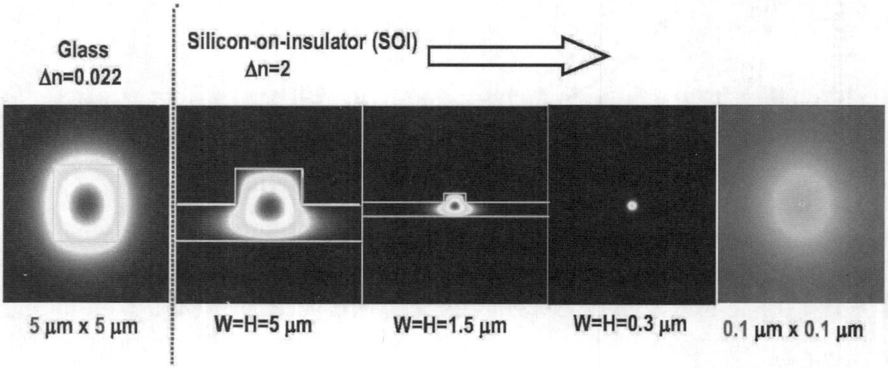

Figure 20. Variation of mode size with core cladding index step and waveguide dimensions for glass and SOI waveguides. W is the width and H is the height of the waveguide.

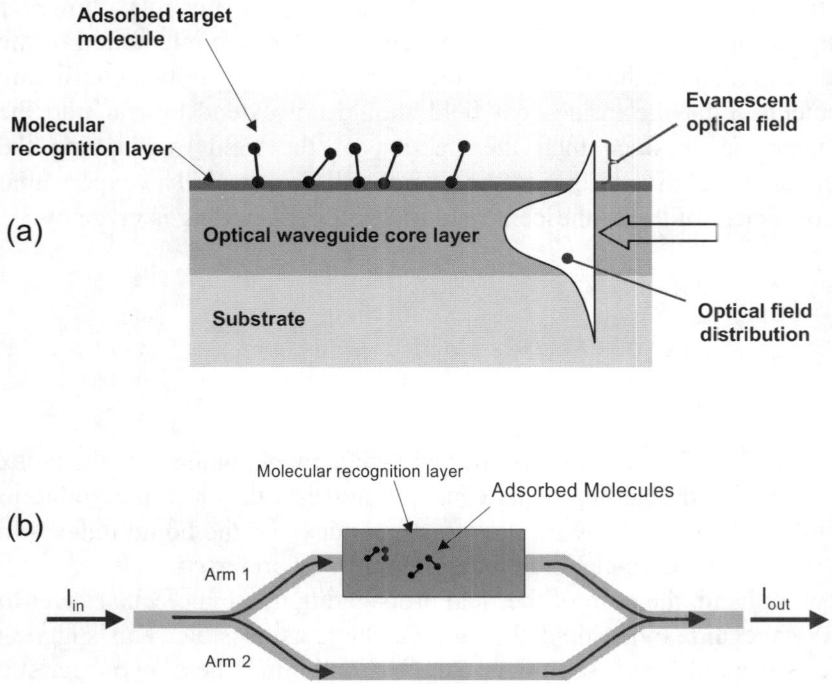

Figure 21. Schematic diagram of (a) an evanescent field sensor, and (b) a simple waveguide circuit where the sensor is incorporated into a Mach-Zehnder interferometer.

applications in sensing and detection using evanescent field waveguide sensors.

A schematic of an evanescent field sensor is shown in Figure 22. The light is guided by the waveguide core layer, but the evanescent tail of the mode extends into the cladding. In Figure 22 the upper cladding consists of a thin molecular recognition layer (e.g. antibody receptors), an adsorbed layer of molecules attached to the receptor molecules, and finally a liquid medium containing the molecules to be detected. The effective index of the guided mode is modified by the molecules adsorbed on the surface of the waveguide. The effective index can be measured accurately, for example by placing the evanescent field sensor in a Mach-Zehnder interferometer (as shown in Figure 22) or a ring-resonator, and the number of adsorbed molecules can be inferred from changes in an optical signal.

The concept of evanescent field sensing is well known and has been exploited for applications such as chemical and biological sensing. The application of microphotonic SOI waveguides to this area may however bring significant increases in sensitivity. To maximize the interaction of the waveguide mode with the adsorbed molecular layer, there are two conflicting requirements. First the evanescent field should not extend too far into the liquid upper layer since then the fraction of the field overlapping the molecule layer, relative to that in the liquid, will decrease. The exponential decay coefficient of the evanescent field in the upper cladding is given by

$$d = k_0 (N_{eff}^2 - n_1^2) \tag{11}$$

where N_{eff} is the effective index of the waveguide mode, while n_1 is the index of the upper liquid cladding. This equation suggests that it is preferable to have a large difference between the effective index and the liquid index – in other words a large core-cladding index step may be preferred.
On the other hand, the ratio of the field propagating in the molecular layer to the total waveguide mode field should be as large as possible. This suggests that the waveguide core should be small. Combining these two heuristic considerations suggests that very small microphotonic waveguides in high index contrast materials as the best approach for making evanescent field sensors. This conclusion is illustrated by the calculations shown in Figure 23, showing the relative index change induced by the addition of a 4 nm layer with index n = 1.5 deposited on top of SOI and glass waveguide layers respectively.

The sensitivity of the SOI waveguides is two orders of magnitude greater than that of glass. As the silicon waveguide layer is decreased from 0.5 to 0.2 μm in thickness, the index change increases by more than one order of magnitude both for TE and TM polarized light, and there is a large polarization anisotropy in the induced index shift. This last effect suggests a potential for polarimetric detection using SOI evanescent field sensors. Despite the promise of SOI waveguide for this application, little work has been done to date in sensors using this material system. However, given the increasing interest in optical techniques for genomics research, biological and chemical sensing, and pathogen detection, this seems a very promising and natural application of microphotonics technology. The use of silicon for this type of sensor is also particularly suitable given the large amount of research being carried out on methods for the functionalization of silicon and SiO_2 surfaces.

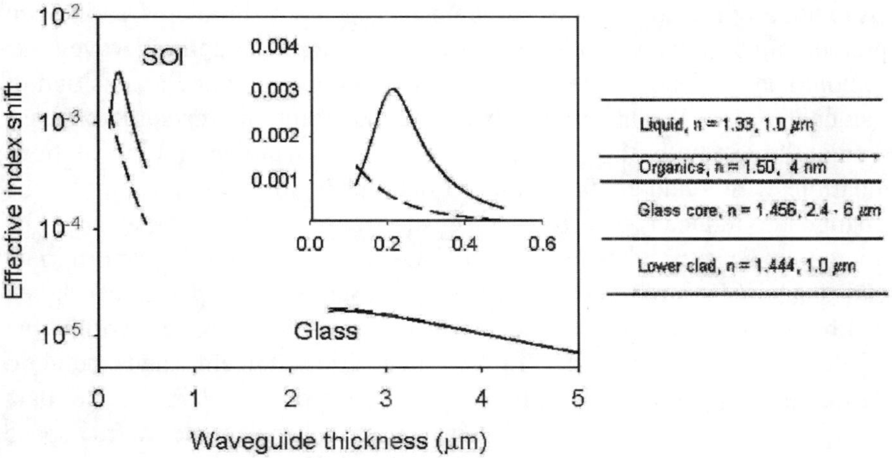

Figure 22. Calculated waveguide effective index shift due to the presence of a 4 nm thick layer with refractive index n = 1.5 on top of the waveguide core. This layer is intended to emulate the the adsorption of a layer of large organic molecules. The index shift is shown for SOI and glass waveguides of varying core layer thickness. The uppermost layer is assumed to be water (n = 1.3). The inset shows an expanded view of the SOI data.

7. SUMMARY

This chapter has reviewed some of the important themes encountered in implementing technologies based on high index contrast microphotonic waveguides. The applications driving research in this area divide into two broad areas. One direction is to extend existing telecommunications technology further into the local area network and into the computer itself, where it is anticipated that optics will eventually take over the task of transporting data from board to board, chip to chip, and possibly even within chips. The other direction involves extending the capabilities of integrated optics beyond the narrow specifications of the telecommunications world, to address applications in spectroscopic measurement and analysis, and chemical and biological sensing. The two specific device examples highlighted here waveguide microspectrometers and evanescent field sensors. In both cases it is clear that the capabilities of microphotonics in high index contrast materials such as SOI can be used to great advantage in improving and extending the performance of optical devices.

As with any new technology, a number of technical problems arise when implementing integrated optics in a new area: input coupling, waveguide loss, polarization dependence, and finally output coupling and acquisition of the signal or data. We have described potential solutions for some of these perennial issues with the emphasis on simple, manufacturable solutions compatible with standard fabrication tools and facilities. For example, the birefringence induced polarization sensitivity of Si waveguides can be corrected using the stress of a thin SiO_2 cladding with the appropriate thickness and film stress. An input coupler based on a simple graded index structure has been proposed, the fabrication of which is well within the capabilities of a calibrated PECVD deposition system, and standard photolithography. Output coupling and packaging for large scale data acquisition may be best addressed using arrays of waveguide to free space beam couplers, combined with commercially available imaging arrays. The combination of simple solutions to outstanding problems as described here, with the new design possibilities microphotonic technology offers, will lead to the expansion of integrated optics as the technologies originally created by the telecommunications industry mature and new applications in many fields are realized.

REFERENCES

1. S. Janz, A. Balakrishnan, S. Charbonneau, P. Cheben, M. Cloutier, A. Delâge, K. Dossou, L. Erickson, M. Gao, P.A. Krug, B. Lamontagne, M. Packirisamy, M. Pearson, and D.-X. Xu, "Planar waveguide echelle gratings in silica-on-silicon," *Photon. Technol. Lett.* **16**, 503-505 (2004).
2. Y. Hibino, "Recent Advances in high density and large scale AWG multi-demultiplexers with higher index contrast silica based PLCs," IEEE J. Sel. Top. Quant. Electron. **17**, 1090-1101 (2002).
3. A. Delâge, S. Bidnyk, P. Cheben, K. Dossou, S. Janz, B. Lamontagne, M. Packirisamy, and D.-X. Xu, "Recent developments in integrated spectrometers," *Proceedings of the 6th International Conference on Transparent Optical Networks*, Vol. 2, 78-83 (2004).
4. P. Cheben, A. Bogdanov, A. Delâge, S. Janz, B. Lamontagne, M.-J. Picard, E. Post, and D.-X. Xu, "A 100 channel near-infra-red SOI waveguide microspectrometer: Design and fabrication challenges," *Proceedings of the SPIE*, Vol. 5644, 103-110 (SPIE, Bellingham, MA, 2005).
5. J.C. Petersen, J. Henningsen, "Molecules as absolute standards for optical telecommunications requirements and characterization," *Proceedings of the International*

Union of Radio Science 27th General Assembly, Maastricht, The Netherlands, CD ROM (2002).
6. C. Manolatou, S.G. Johnson, S. Fan, P.R. Villeneuve, H.A. Haus, and J.D. Joannopoulos, "High density integrated optics," *J. Lightwave Technol.* **17**, 1682-1692 (1999).
7. Q. Xu, V.R. Almeida, R.R. Panepucci, and M. Lipson, "Experimental demonstration of guiding and confining light in nanometer-size low-refractive index material," *Opt. Lett.* **29**, 1626-1628 (2004).
8. S. Janz. P. Cheben, H. Dayan, and R. Deakos, "Measurement of birefringence in thin-film waveguides by Rayleigh scattering," *Opt. Lett.* **28**, 1778-1780 (2003).
9. L. Vivien, S. Laval, B. Dumont, S. Lardenois, A. Koster, and E. Kassan, "Polarization independent single mode rib waveguides on silicon-on-insulator for telecommunications wavelengths," *Opt. Commun.* **210**, 43-49 (2002).
10. D.-X. Xu, P. Cheben, D. Dalacu, A. Delâge, S. Janz, B. Lamontagne, M.-J. Picard, and W.N. Ye, "Eliminating the birefringence in silicon-on-insulator ridge waveguides by use of cladding stress," *Opt. Lett.* **29**, 2384-2386 (2004).
11. D.-X. Xu, P. Cheben, S. Janz, D. Dalacu, "Control of SOI waveguide polarization properties for microphotonic applications," *Proc. of CLEO/Pacific RIM 2003*, CD-ROM (IEEE, Piscataway, NJ, 2003).
12. W.N. Ye, D.-X. Xu, S. Janz, P. Cheben, M.-J. Picard, B. Lamontagne, and N.G.T. Tarr, "Birefringence control using stress engineering in silicon-on-insulator (SOI) waveguides," *IEEE J. Lightwave Technol.* (in press, 2005).
13. M. Huang, "Stress effects on the performance of optical waveguides," *Intnl. J. of Sol. and Structures* **40**, 1615-1632 (2003).
14. X. Zhao, C. Li, and Y.Z. Xu, "Stress induced birefringence control in optical planar waveguides," *Opt. Lett.* **28**, 564-566 (2003).
15. K. Röll, "Analysis of stress and strain distribution in thin films and substrates," *J. Appl. Phys.* **47**, 3224-3229 (1976).
16. A. Sure, T. Dillon, J. Murakowski, C. Lin, D. Pustai, and D.W. Prather, "Fabrication and characterization of three-dimensional silicon tapers," *Optics Express* **11**, 3555-3561 (2003).
17. V.R. Almeida, R.R. Panepucci, and M. Lipson, "Nanotaper for compact mode conversion," *Opt. Lett.* **28**, 1302-1304 (2003).
18. G. Masanovic, V.M.N. Passaro, and G.T. Reed, "Dual grating-assisted directional coupling between fibers and thin semiconductor waveguides, *IEEE Phot. Technol. Lett.* **15**, 1395-1397 (2003).
19. A. Delâge, S. Janz, D.-X. Xu, D. Dalacu, B. Lamontagne, and A. Bogdanov, Graded-index coupler for microphotonic SOI waveguides," *SPIE Proc. Vol.* 5577**,** Photonics North 2004 (in press, 2004).
20. K. Shiraishi, C.S. Tsai, H. Yoda, and K. Minagawa, *Proc. of CLEO/Pacific RIM 2003*, CD-ROM (IEEE, Piscataway, NJ, 2003).
21. F.P. Payne and J.P.R. Lacey, "A theoretical analysis of scattering loss from planar optical waveguides," *Opt. Quantum Electron.* **26**, 977-986 (1994).
22. K.K. Lee, D.R. Lim, H.-C. Luan, A. Agarawal, J. Foresi, and L.C Kimmerling, "Effect of size and roughness on light transmission in a Si/SiO_2 waveguide: Experiments and model," *Appl. Phys. Lett.* **77**, 1617-1619 (2000).
23. F. Grillot, L. Vivien, S. Laval, D. Pascal, E. Cassan, "Size influence on the propagation loss induced by sidewall roughness in ultrasmall SOI waveguides," *IEEE Phot. Technol. Lett.* **16**, 1661-1663 (2004).

24. K.K. Lee, D.R. Lim, L.C. Kimmerling, J. Shin, and F. Cerrina, "Fabrication of ultra-low loss Si/SiO2 waveguides by roughness reduction," *Opt. Lett.* **26**, 1888-1890 (2001).
25. P. Dumon, W. Bogaerts, V. Wiaux, J. Wouters, S. Beckx, J. van Campenhout, D. Taillaert, B. Luyssaert, P. Bientman, D. van Thourhout, and R. Baets, "Low loss photonic wires and ring resonators fabricated with deep UV lithography," *IEEE Phot. Technol. Lett.* **16**, 1328-1330 (2004).
26. R.L. Espinola, T. Izuhara, M.-C. Tsai, R.M. Osgoode, and H. Dotsch, "Magneto-optical non-reciprocal phase shift in garnet/silicon-on-insulator waveguides," *Opt. Lett.* **29**, 941-943 (2004).

MICRO-OPTICAL RESONATORS FOR MICROLASERS AND INTEGRATED OPTOELECTRONICS

Recent advances and future challenges

Trevor M. Benson[1], Svetlana V. Boriskina[1], Phillip Sewell[1], Ana Vukovic[1], Stephen C. Greedy[1] and Alexander I. Nosich[1,2]

[1]*George Green Institute for Electromagnetics Research, University of Nottingham, Nottingham NG7 2RD, UK;* [2]*Institute of Radio Physics and Electronics NASU, Kharkov 61085, Ukraine*

Abstract: Optical microcavities trap light in compact volumes by the mechanisms of almost total internal reflection or distributed Bragg reflection, enable light amplification, and select out specific (resonant) frequencies of light that can be emitted or coupled into optical guides, and lower the thresholds of lasing. Such resonators have radii from 1 to 100 μm and can be fabricated in a wide range of materials. Devices based on optical resonators are essential for cavity-quantum-electro-dynamic experiments, frequency stabilization, optical filtering and switching, light generation, biosensing, and nonlinear optics.

Key words: optical resonators; photonic crystal defect cavities; whispering gallery modes; integrated optics; add/drop filters; photonic biosensors; spontaneous emission control; quantum wells/dots; semiconductor microdisk lasers; optical device fabrication.

1. INTRODUCTION

Optoelectronic devices based on optical microresonators that strongly confine photons and electrons form a basis for next-generation compact-size, low-power and high-speed photonic circuits. By tailoring the resonator shape, size or material composition, the microresonator can be tuned to support a spectrum of optical (i.e., electromagnetic) modes with required polarization, frequency and emission patterns. This offers the potential for

S. Janz et al. (eds.), Frontiers in Planar Lightwave Circuit Technology, 39–70.

developing new types of photonic devices such as light emitting diodes, low-threshold microlasers, ultra-small optical filters and switches for wavelength-division-multiplexed (WDM) networks, colour displays, etc. Furthermore, novel designs of microresonators open up very challenging fundamental-science applications beyond optoelectronic device technologies. The interaction of active or reactive material with the modal fields of optical microresonators provides key physical models for basic research such as cavity quantum electrodynamics (QED) experiments, spontaneous emission control, nonlinear optics, bio-chemical sensing and quantum information processing (Yokoyama, 1992, 1995; Yamamoto, 1993; Chang, 1996; Vahala, 2003). We shall briefly review the state-of-the-art in microresonator design tools, fabrication technologies and observed optical phenomena, and outline the challenges for future research.

2. MECHANISMS OF LIGHT CONFINEMENT AND RESONATOR BASIC FEATURES

Optical resonators can be fabricated by exploiting either (almost) total internal reflection (ATIR) of light at the interface between a dielectric (e.g. semiconductor) material and the surrounding air or distributed Bragg reflection (DBR) from periodical structures such as multilayered structures or arrays of holes. The spectra of optical modes supported by microresonators are shape and size dependent. The large variety of resonator geometries that can be realized by using either or combining both of the light confinement mechanisms opens up a wide field of research and device opportunities.

For various applications it is often critical to realize a microresonator with compact size (small modal volume, V), high mode quality factor, Q, and large free spectral range (FSR). Ultra-compact microresonators enable large-scale integration and single-mode operation for a broad range of wavelengths. Q-factor is a measure of the resonator capacity to circulate and store light, and is usually defined as the ratio of the energy stored to the energy dissipated in the microresonator. In practical device applications, the high Q of the microresonator mode translates into a narrow resonance linewidth, long decay time, and high optical intensity. Wide FSR (the spacing between neighbouring high-Q resonances) is often required to accommodate many WDM channels within the erbium amplifier communications window. The ratio Q/V determines the strength of various light-matter interactions in the microresonator, e.g., enhancement of the

Table 1. Types and characteristics of optical microresonators.

Resonator type	**Light confinement** **Dominant modes** **Features**	**Resonator type**	**Light confinement** **Dominant modes** **Features**
Microsphere 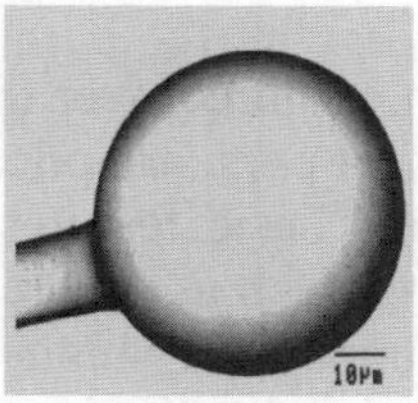 Lefevre-Seguin, V., 1999, *Opt. Mater.* **11**(2-3):153-165. © 1999 Elsevier B.V.	ATIR WGMs Ultra-high Q-factors (10^7 - 9 x 10^9); large mode volumes; dense modal spectrum (all modes are degenerate); challenging on-chip integration	Microtorus  Armani, D.K. et al, 2003, *Nature* **421**:905-908.	ATIR WGMs Mode volumes lower than for spheres; very high Q-factors (5 x 10^8); reduced azimuthal-mode spectrum; suitable for on-chip integration
Microdisk (microring)  Baba, T. et al, 1997, *IEEE Photon. Technol. Lett.* **9**(7):878-880. © 1997 IEEE	ATIR WGMs Small mode volumes; high Q-factors (10^4 – 10^5); higher-radial-order WG modes are eliminated in the ring resonators; suitable for planar integration	Quadrupolar (racetrack) microresonator 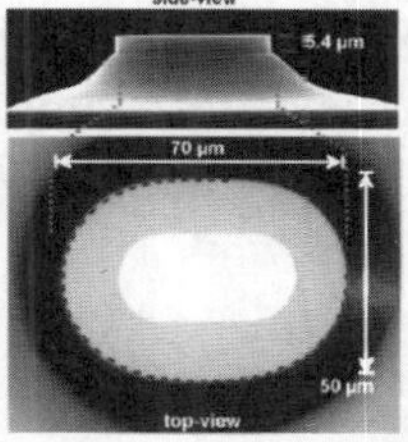 Gmachl, C. et al, 1998, *Science* **280**:1556-1564. © 1998 AAAS	ATIR WGMs, bow-tie Relatively low Q-factors (850-1500), highly directional emission and high FSR of the bow-tie modes; WGM Q-factors lower than in circular microdisks, efficient coupling to planar waveguides
Micropost/pillar  Solomon, G.S. et al, 2001, *Phys. Rev. Lett.* **86**(17):3903-3906. © 2001American Physical Society	DBR in vertical direction ATIR in horizontal direction Fabry-Perot oscillations Small mode volumes; relatively high Q-factors (1300-2000); easy coupling to fibres	Photonic crystal defect microcavity 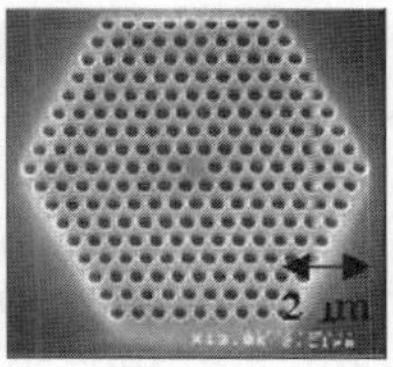 Painter, O.J. et al, 1999, *J. Lightwave Technol.* **17**(11):2082-2088. © 1999 IEEE	ATIR in vertical direction DBR in horizontal direction Symmetry-dependent spectrum of defect modes The smallest mode volumes; high Q-factors (4.5 x 10^4); suitable for planar integration

spontaneous emission rate, and should be maximised for microlaser applications and QED experiments. However, high-Q microresonators of optical-wavelength size are difficult to fabricate, as the WG-mode Q-factors decreases exponentially with the cavity size, and thus in general the demands for a high Q-factor and compactness (large FSR, small V) are contradictory.

A very wide range of microresonator shapes has been explored over the years for various applications (Table 1 lists some of the most popular optical microresonator types as well as their dominant modes and basic features). The most widely used are rotationally symmetric structures such as spheres, cylinders, toroids, and disks, which have been shown to support very high-Q whispering-gallery (WG) modes whose modal field intensity distribution is concentrated near the dielectric-air interface (Fig. 2). Silica microspheres exhibit the highest (nearly 9 billion) Q-factors (Braginsky, 1989; Gorodetsky, 1996; Vernooy, 1998; Laine, 2001; Lefevre-Seguin, 1997), yet have a very dense spectrum of multiple-degenerate WG modes, which complicates their application for spectral analysis or laser stabilization. Recently proposed micro-toroidal resonators (Ilchenko, 2001; Armani, 2003; Vahala, 2003; Polman, 2004) not only demonstrate very high WG-mode Q-factors approaching those of microspheres but also enable reduction of WG-mode volume, increase of resonator FSR, and on-chip integration with other components.

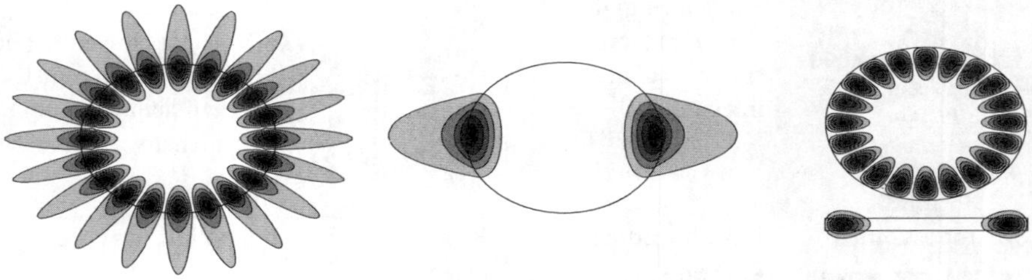

(a) (b)

Figure 1. Near-field intensity portraits (16.6% contours) of the high-Q $WG_{10,1}$ modes supported by a silica microsphere (a) and a GaAs microdisk (b).

Circular high-index-contrast microring and microdisk resonators based on planar waveguide technology with diameters as small as 1-10 μm are able to support strongly-confined WG modes with typical Q-factors of 10^4-10^5 and are widely used as microlaser cavities (Levi, 1993; Baba, 1997, 1999; Cao, 2000; Zhang, 1996) and add/drop filters for WDM networks (Hagness, 1997; Little, 1997, 1999; Chin, 1999). Recently, record Q-factors have been demonstrated in wedge-edge microdisk resonators (Q in excess of 1 million, Kippenberg, 2003) and in polished crystalline microcavities ($Q>10^{10}$, Savchenkov, 2004).

Along with circular microdisk resonators, cavities of elliptical (Noeckel, 1994; Backes, 1999; Boriskina, 2003; Kim, 2004), quadrupolar (Noeckel, 1994; Gmachl, 1998; Fukushima, 2004; Gianordoli, 2000; Chin, 1999) and square (Poon, 2001; Ling, 2003; Manolatou, 1999; Hammer, 2002; Fong, 2003; Guo, 2003; Boriskina, 2004, 2005) shapes have attracted much interest. Depending on their size and degree of deformation, these microresonators can support several types of optical modes with significantly different Q-factors, near-field intensity distributions, and emission patterns (WG-like modes in square resonators (Fig. 2a), bow-tie modes in quadrupoles (Fig. 2b), distorted WG modes, volume and two-bounce oscillations, etc.). Such resonators offer advantage for various filter and laser applications as they provide splitting of the double-degenerate WG modes, directional light emission and more efficient microresonator-to-straight-waveguide coupling.

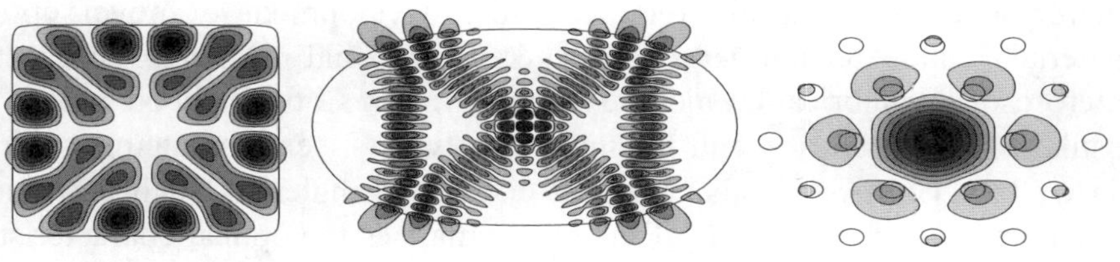
(a) (b) (c)

Figure 2. Near-field intensity portraits of a WG-like mode in (a) a square microdisk resonator and (b) a bow-tie mode in a quadrupolar (stadium) resonator and (c) a monopole mode in a hexagonal photonic crystal defect cavity (Boriskina, 2005).

Planar photonic crystal (PC) microcavities (Fig. 2c), formed, e.g., as arrays of air holes etched into a slab, have been demonstrated to simultaneously exhibit high Q-factors and ultra-small, wavelength-scale modal volumes (Foresi, 1997; Benisty, 1999; Chow, 2000; Yoshie, 2001; Kim, 2002; Akahane, 2003; Srinivasan, 2004; Noda, 2000; Painter, 1999; Boroditsky, 1999). In these cavities, the photonic-bandgap effect (fulfilment of the Bragg reflection conditions for all the propagation directions in a certain frequency range) is used for strong light confinement in the cavity plane, and TIR, for light confinement at the air-slab interface (Russel, 1996; Krauss, 1999). Modern fabrication technologies enable precise control of the PC cavity geometry, and the inherent flexibility in hole shape, size, and pattern makes fine-tuning of the defect mode wavelengths, Q-factors, and emission patterns possible (Painter, 2001; Coccioli, 1998).

In micropost and micropillar resonators (Gayral, 1998; Solomon, 2001; Pelton, 2002; Benyoucef, 2004; Santori, 2004), the transverse mode confinement is due to TIR at the semiconductor-air interface, while confinement in the vertical direction is provided by a pair of distributed Bragg reflectors. Thus, these microresonator structures can be seen as 1-D PC defect cavities in a fibre. They support Fabry-Perot-type modes with relatively high Q-factors and small modal volumes, which makes them promising candidates for microlaser applications and the observation of cavity-QED phenomena.

Other types of optical microcavities employing the DBR mechanism of light confinement include planar annular Bragg resonators (Scheuer, 2005), based on a radial defect surrounded by Bragg reflectors, and their 3-D equivalent, spherical Bragg "onion" resonators (Liang, 2004).

3. RESONATOR MATERIAL SYSTEMS

Progress in the design and fabrication of high-quality optical microresonators is closely related to the development of novel optical materials and technologies. The key material systems used for microresonator fabrication include silica, silica on silicon, silicon, silicon on insulator, silicon nitride and oxynitride, polymers, semiconductors such as GaAs, InP, GaInAsP, GaN, etc, and crystalline materials such as lithium niobate and calcium fluoride. Table 2 summarises the optical characteristics of these materials (see Eldada, 2000, 2001; Hillmer, 2003; Poulsen, 2003 for more detail).

Silica glass is a widely used resonator material, which combines the advantage of a very low intrinsic material loss, a large transparency window, and compatibility with standard fibre-optic technologies, e.g., efficient coupling to optical fibres (Sandoghdar, 1996; Gorodetsky, 1996; Laine, 2001; Cai, 2000; Ilchenko, 2001). The silica-on-silicon technology that involves growing silica layers on silicon substrates is one the most widely used technologies to fabricate planar microresonator devices (Eldada, 2001; Kippenberg, 2003; Armani, 2003; Polman, 2004), although microring resonators made of compound glasses (Ta_2O_5-SiO_2) have also been demonstrated (Little, 1999). A glass platform is relatively inexpensive, and as the index contrasts are smaller than for semiconductors, larger-size single-mode resonators can be fabricated by using cheaper lithographic techniques. However, wavelength-scale resonators with wide FSR in high-index-contrast silicon or silicon-on-insulator material systems (Little, 1998; Akahane, 2003; Dumon, 2004) are crucial for ultra-compact integration of photonic integrated circuits. Another relatively new planar resonator platform is based on using silicon nitride and oxynitride (Krioukov, 2002; Melloni, 2003; Barwicz, 2004). It enables the index contrast to be adjusted by as much as 30% by changing the resonator material composition in between that of SiO_2 and Si_3N_4.

Table 2. Properties of some key optical microresonator material systems.

Material system	**λ emission**	**Refractive index at 1.55μm**
Fused silica (SiO_2); Silica on silicon		1.44-1.47
Silicon on insulator (SOI)		3.4757
Silicon oxynitride (SiO_xN_y)		1.44-1.99
Polymers		1.3-1.7
Gallium Arsenide (GaAs)	0.8-1.0 μm	3.3737
Indium Phosphide (InP)	1.3-1.7 μm	3.1
Gallium Nitride (GaN)	0.3-0.6 μm	2.31
Lithium Niobate ($LiNbO_3$)		2.21(2.14)

Of course, GaAs, InP, and GaN- based semiconductors have attracted much attention as the materials for microresonator device fabrication (Baba, 1997; Hagness, 1997; Solomon, 2001; Gianordoli, 2000; Kim, 2002; Painter, 1999; Grover, 2001, 2004; Zhang, 1996; Kneissl, 2004), as they enable very compact resonator structures that can perform passive (coupling, splitting and multiplexing) and active (light generation, amplification, detection, modulation) functions in the same material system. Most semiconductor microdisk and planar PC resonators consist of thin heterostructure layers. Such structures enable local modification of the energy-band structure of the semiconductor (quantum wells) and thus control of the material emission wavelength (Einspruch, 1994). Recently, there has been a lot of interest in semiconductor resonators with artificially engineered material band structure based on low-dimensional heterostructures, such as one-dimensional quantum wires and zero-dimensional quantum boxes/dots (Arakawa, 1986).

Polymer materials (Eldada, 2000; Rabiei, 2002, 2003; Chao, 2003, 2004) offer great potential for use in advanced optoelectronic systems as they offer low material costs, up to 35% tunability of the refractive index contrast, excellent mechanical properties, very low optical loss, large negative thermo-optic coefficient and low-cost high-speed processing. Other attractive materials for high-Q microresonators fabrication, which combine a wide optical transparency window, a good electro-optic coefficient and nonlinearity, are crystalline materials such as calcium fluoride and lithium niobate (Ilchenko, 2002; Savchenkov, 2004, 2005).

In addition to semiconductor materials discussed above, there are other material systems in which electroluminescence has already been demonstrated, even in the visible wavelength range. These materials include:

microporous silicon fabricated by electrochemical etching of crystalline silicon wafers in a hydrofluoric acid (Chan, 2001), Erbium-doped silicon (Polman, 2004; Hillmer, 2003; Gardner, 2005), phosphate glass (Cai, 2000) and organic materials, e.g., para-phenylene-vinylene (Krauss, 1999; Hillmer, 2003). Furthermore, polymers that are not intrinsically functional can be doped with organic laser dyes, rare-earth light-amplifying complexes, and electro-optic dyes. Such materials can be used to fabricate laser microresonators and have an advantage over III-V semiconductors in either compatibility with silicon microelectronics or cost and ease of fabrication.

4. FABRICATION TECHNOLOGY

Recent advances in nanofabrication technology offer the possibility of manufacturing novel optical microresonator devices with dimensions of the order of the optical wavelength in a variety of natural and artificial material systems. Integrated micro-ring and micro-disk resonators as well as PC defect microcavities are usually micro-fabricated on wafer substrates using well-developed integrated-circuit deposition, lithographic and etching techniques (Elliott, 1989). During the fabrication process, wafers are typically cycled through three main steps: wafer preparation, lithography, and etching. Some of these steps may be repeated at various stages of the fabrication process. A schematic drawing of a typical micro-resonator fabrication sequence is presented in Fig. 3.

The first stage of the fabrication process (wafer preparation) usually involves the growth or deposition of material layers of different composition to create vertical layered structures able to support guided waves. The layered structure might be formed using various wafer growth techniques, such as molecular beam epitaxy, chemical or physical vapour deposition, or wafer bonding, depending on the material system. The choice of the growth technique depends on the desired resonator structure, material system, and required precision. For example, semiconductor materials usually require several successive growths to make the vertical structure, while polymer layers can simply be sequentially spun onto a substrate.

Molecular beam epitaxy (MBE) is an expensive yet widely used technique for producing epitaxial layers of metals, insulators and III-V and II-VI based semiconductors, both at the research and the industrial production level (Herman, 1996). It consists of deposition of 'molecular beams' of atoms or clusters of atoms, which are produced by heating up a solid source, onto a heated crystalline substrate in ultra-high vacuum. MBE is characterized by low growth temperatures and low growth rates and thus enables producing high-precision epitaxial structures with monolayer control. Other wafer growth techniques include chemical vapour deposition (Foord, 1997) of reactant gases broken down at high temperatures, which can be used to deposit GaN, GaInAsP, Si, SiO_2, Si_3N_4, and polymer materials; sputter (or physical vapour) deposition, which is a process of controlled deposition of energetic ionized particles of SiO_2, Si_3N_4, metals, and alloys (Little, 1999); and wafer bonding, or fusing of two materials (e.g., Si to glass, Si to Si, III-V semiconductors to SiN or SiO_2) at the atomic level (Tishinin, 1999).

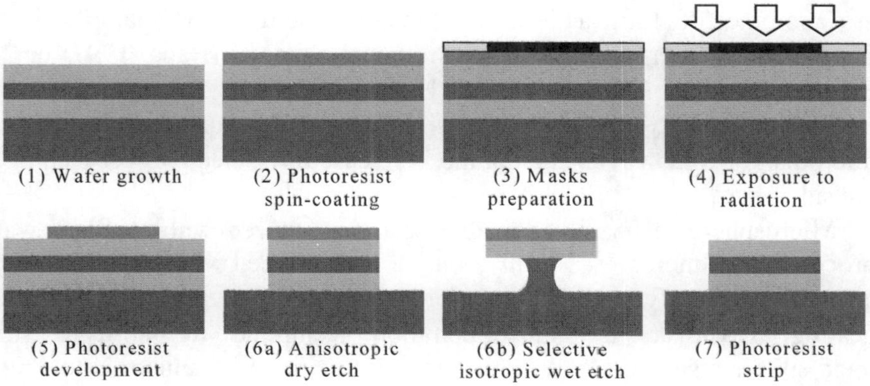

Figure 3. Schematic procedure of microdisk resonator fabrication.

A resonator structure is then imaged onto a wafer through a multi-step lithography process. To make the wafer (usually covered by a thin oxide film) sensitive to an image, a photoresist is spread on the wafer by a process called spin coating. Then, a photomask (a glass emulsion plate with a resonator pattern) is placed on top of the wafer. Light is projected through the voids in the photomask and images the mask pattern on the wafer. Several types of lithography with various resolutions can be used depending on the resonator type, size, and fabrication tolerances: optical (λ=160-430 nm; feature size ~ 0.13 μm), extreme ultraviolet (λ=13 nm; feature size ~ 45 nm), X-ray (λ=0.4-4 nm; feature size ~ 25 nm), or electron (ion) beam (λ=0.03 Å; feature size ~ 10-20 nm). When exposed to light, the resist either polymerizes (hardens) (if a negative resist is used) or un-polymerizes (if a positive resist is used). After exposure, the wafer is developed in a solution to dissolve the excess resist.

Once the resist has been patterned, the selected regions of material not protected by photoresist are removed by specially designed etchants, creating the resonator pattern in the wafer. Several isotropic (etching occurs in all directions at the same etch rate) and anisotropic (directional) etching processes are available. Wet chemical etching is an isotropic process that uses acid solutions to selectively dissolve the exposed layer of silicon dioxide. However, to produce vertical resonator sidewalls, anisotropic dry etching is required. In the most commonly used dry etching technique, plasma etching, ions reacting with atoms of the wafer are created by generating plasma by rf discharge. Dry etching techniques also include sputter etching, ion milling, reactive etching, and reactive ion beam etching. Another type of anisotropic etching technique is anodic etching - wet etching with an applied electric field. After the etching process, the remaining photoresist is removed. In most cases, several etching techniques are combined to form a desired resonator structure. For example, in the fabrication of microdisk resonators mounted on posts (Baba, 1997) or 2-D PC microcavities (Painter, 1999; Hennesy, 2003), the pattern is first transferred into a heterostructure by an anisotropic dry etch and then a microdisk or a PC cavity membrane is released by selective wet etching of the substrate.

Microsphere resonators can also be manufactured with standard wafer processing technologies. Usually, silicon posts topped with silica blocks are created on the wafer and then are molten into spherical shapes by controlled heating. However, the most common technique to fabricate silica microsphere resonators of several hundred microns in diameter is to simply melt a tip of an optical fibre by hydrogen flame or electric arc heating (Braginsky, 1989; Sandoghdar, 1996; Gorodetsky, 1996; Laine, 2001). To fabricate microspheres of smaller diameters, the fibre can first be thinned by tapering or etching (Cai, 2000). In the heating process, once the silica temperature passes the melting point, the surface tension forces shape the fused silica into a near-perfect spherical form. After the sphere is removed from the flame, solidification occurs almost instantaneously. The resulting microsphere has extremely smooth surface with Angstrom-scale surface deformations resulting in very high Q-factors of the WG modes.

Similar technique can also be used to fabricate silica microtorus resonators by compressing a small sphere of low-melting silica glass between cleaved fibre tips (Ilchenko, 2001). The combined action of surface tension and axial compression results in the desired toroidal geometry. Recently, a process for producing silica toroidal-shaped microresonators-on-a-chip with Q factors in excess of 100 million by using a combination of lithography, dry etching and a selective reflow process have been demonstrated (Armani, 2003). By selectively heating and reflowing a patterned and undercut microdisk with the use of a CO_2 laser a toroidal resonator with the atomically smooth surface was obtained.

The thermal-reflow process (or hot embossing) has also been used to reduce significantly sidewall surface roughness in polymer (polystyrene) microring resonators (Chao, 2004). This is a very useful technique because it enables post-fabrication fine-tuning of microresonator performance. However, for the fabrication of complex, multilayer devices this procedure is less useful as the high temperatures needed to reflow the polymer film could also disturb lower cladding layers and possibly alter optically active dopant molecules.

A need for submicron-scale patterning as well as stringent etching tolerances of conventional fabrication techniques often result in either very high fabrication costs or high scattering losses and frequency detuning in microresonator-based devices. An emerging lithographic technology that can achieve sub-10 nm pattern resolutions beyond the limitations set by the light diffractions or beam scattering in the conventional methods is a nanoimprint technique (Chou, 1996; Guo, 2004). Based on the mechanical embossing principle, nanoimprint technology is used not only to create resist patterns as in lithography but also to directly pattern microresonator structures in polymers. Nanoimprinting utilizes a hard mould (which plays the same role as the photomask in photolithography) with predefined nanoscale features to mechanically imprint into a heated polymer film. The created thickness contrast pattern is `frozen' into the polymer during a cooling cycle.

Recently, as processing techniques for the synthesis of monodispersed nanoparticles of various shapes have matured, considerable effort has been directed to study the mechanisms of their self-assembly as an inexpensive and fast method of fabrication of photonic structures. Electrostatic self-assembly is an emerging technology that can create microresonator structures with tailored optical properties. For example, self-assembled arrays of polystyrene microspheres coupled to each other and to rib waveguides are expected to or already find application in wavelength selection (Tai, 2004), waveguiding (Astratov, 2004), optical sensing, and optical delays. Finally, recent advances in polishing techniques have allowed the fabrication of large (~5 mm-diam & 100-micron thick) disk and toroidal microresonators of crystalline materials, such as lithium niobate and calcium fluoride (Ilchenko, 2002; Savchenkov, 2005) that find use as high-efficiency microwave and millimetre-wave electro-optical modulators.

5. DEVICES AND APPLICATIONS

It is difficult to overestimate the growing importance of optical microresonators in both fundamental and applied research, and thus it is hardly possible to cover in detail the progress in all the areas of their application. In the following sections, we shall briefly review some of the major existing and emerging microresonator applications and outline the challenges calling for novel resonator geometrical designs, spectral characteristics, material properties, or robust fabrication procedures.

5.1 Integration with other components

One of the most difficult challenges in the design and fabrication of integrated microresonator-based photonic devices and systems is the efficient coupling of light into and out of a microresonator without compromising its narrow resonance linewidth. Furthermore, the cost and robustness of fabrication, simplicity of the microresonator-to-coupler alignment as well as ability to provide on-chip integration are very important factors in the development of advanced microresonator couplers. The most widely used microresonator coupling devices are evanescent-field couplers of various geometries such as prisms, tapered fibres, planar and PC waveguides, etc. (see Fig. 4).

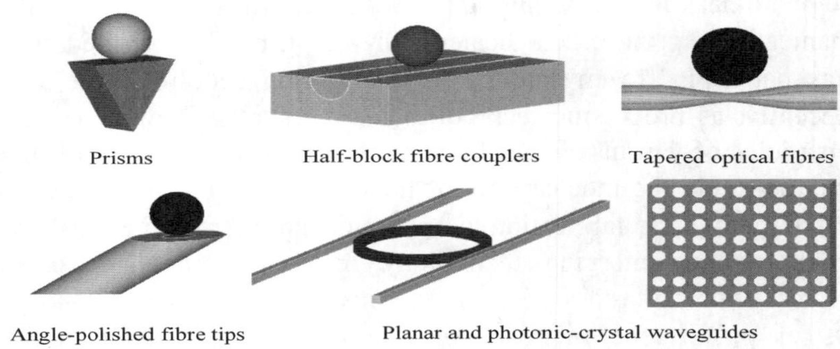

Figure 4. Various types of microresonator coupling devices.

The efficiency of optical power transfer in/out of a microresonator can be controlled by manipulating the overlap between the resonator and coupler mode fields, matching of the mode propagation constants, or changing the length of the evanescent-field coupling region. Prisms have traditionally been used to couple light into microspheres (Gorodetsky, 1996; Vernooy, 1998) and more recently, to square resonators (Pan, 2003), although they require bulk optics for focusing and alignment of the light source. Coupling to microsphere (Serpenguzel, 1995; Cai, 2000; Spillane, 2002) and microtoroid resonators (Kippenberg, 2004) through tapered fibres has several distinct advantages such as the built-in coupler alignment, relatively simple fabrication, possible on-chip integration, and control of the coupling efficiency by the change of the fibre thickness. However, integrated, wafer-fabricated microresonators are usually coupled into a photonic circuit through either planar (rib) waveguides (Hagness, 1997; Little, 1997, 1999; Chin, 1999; Grover, 2004; Zhang, 1996; Melloni, 2003) or PC waveguides (Noda, 2000).

The very small size and strong optical confinement of integrated planar microdisk and microring resonators, which make them promising candidates for large scale integration, also make them very sensitive to fabrication errors that can drastically spoil coupling efficiency. For example, if a circular microdisk is laterally coupled to a waveguide via a submicron-width air gap, the coupling efficiency strongly depends on the gap width. Accurate and repeatable fabrication of such narrow gaps by lithographic and etching techniques is a rather challenging task. The smaller the microdisk, the more difficult it is to control the coupling coefficient. For 2–5 μm-sized microdisks, the resolution of the fabrication method is often not high enough to achieve the desired narrow air gaps (<0.1 μm), and to enhance coupling, microdisks are fused to bus waveguides (Grover, 2004). Efficient lateral coupling across wider air gaps can be achieved by increasing the coupling interaction length by either curving the adjacent waveguide along the microdisk (Zhang, 1996; Hagness, 1997; Chin, 1998) or using elliptical, racetrack or square microresonators (Chin, 1999; Boriskina, 2003; Manolatou, 1999; Hammer, 2002; Fong, 2003). It has been also predicted that resonator-waveguide coupling may be enhanced by exploiting the higher field concentration at the increased-curvature portion of an elliptical microdisk resonator (Boriskina, 2003).

5.2 Wavelength-selective components for WDM systems

Microdisk, microring, and PC defect resonators are versatile building blocks for very large scale integrated photonic circuits, as they are ultra compact (10^5 devices/cm^2) and can perform a wide range of optical signal processing functions such as filtering, splitting and combining of light, switching of channels in the space domain, as well as multiplexing and demultiplexing of channels in the wavelength domain.

Wavelength-selective bandstop (Fig. 5a) or add/drop (Fig. 5b) filters that can combine or separate different wavelengths of light carrying different information are essential components for controlling and manipulating light in optical transmission systems. High-Q microdisks/microrings evanescently coupled to bus waveguides have been extensively explored for WDM channel dropping due to their ability to select a single channel with a very narrow linewidth (Hagness, 1997; Little, 1997, 1999; Chin, 1999; Boriskina, 1999; Grover, 2004). To compromise between the requirements for a narrow channel linewidth, structure compactness, and wide FSR, the radii of the microresonators used in filters are usually in between 5 and 30 micron, resulting in filters with a FSR of 20-30 nm. By using PC defect microcavities, further miniaturization of optical filters can be achieved (Fan, 1998). PC-microcavity filters can be realized by either introducing a single defect in the vicinity of the PC waveguide (Noda, 2000) or by integrating PC defect microcavities directly into a sub-micrometer-scale silicon waveguide (Foresi, 1997; Chow, 2004).

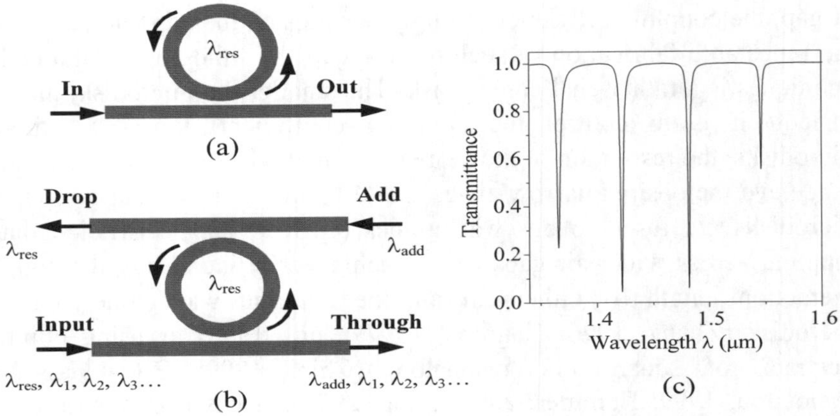

Figure 5. Schematics of (a) a bandstop (all-pass) and (b) add/drop filter designs, and (c) a transmission characteristic of a 3.0-μm-diameter GaAs microring resonator coupled to a bus waveguide.

An ideal filter should have a box-like spectral response with a flat passband, sharp roll-off from passband to stop band, and large out-of-band rejection. However, the transmission characteristic of a single microresonator is a series of Lorentzian-shape sharp resonance peaks at wavelengths corresponding to the excitation of the high-Q modes in the microresonator (Fig. 5c). For closely spaced channels, however, a Lorentzian response may not provide adequate roll-off to minimize cross-talk between different channels. To overcome this limitation, higher-order filters formed by cascading multiple resonators have been introduced (Little, 1997; Hryniewicz, 2000; Grover, 2002; Melloni, 2003; Savchenkov, 2003, 2005). Such filters have flat responses around resonances, much faster roll-off, and larger out-of-band signal rejection. Vernier filter tuning by combining microring resonators of different radii has also been successfully used to suppress non-synchronous resonances of different microrings and extend the resulting filter FSR (Yanagadze, 2002).

Optical characteristics of active microresonators (and thus spectral properties of the resonator-based filters) can be dynamically tuned by changing their refractive indices, for example, by using the thermo-optic or electro-optic effects. Tunable microring resonators have been used to demonstrate optical modulation using the electro-optic effect in polymer (Rabiei, 2002), all-optical switching using free-carrier injection in GaAs–AlGaAs (Ibrahim, 2003), and absorption-induced wavelength switching and routing (Little, 1998). Furthermore, significant enhancement of non-linear effects in microresonators made possible the demonstration of Kerr nonlinear phase shift (Heebner, 2004) and optical wavelength switching and conversion (Absil, 2000; Melloni, 2003) in microresonator structures.

Linear arrays of optical microresonators evanescently coupled to each other can also be used for optical power transfer (Fig. 6). This type of coupled-resonator optical waveguide (CROW) has recently been proposed (Yariv, 1999) and then demonstrated and studied in a variety of material and geometrical configurations, such as sequences of planar microrings (Poon, 2004), arrays of coupled microspheres (Astratov, 2004), and chains of photonic crystal defect cavities (Olivier, 2001).

Figure 6. Near-field pattern inside a short chain of evanescently coupled microdisks excited by a directional beam grazing a rim of the left-hand-side resonator (Boriskina, 2005).

Among the advantages offered by CROWs is the possibility of making reflection-less waveguide bends as well as a significant slowing of light pulses. Reduction of the light group velocity in CROWs can be explored in a variety of optical applications such as delay lines, optical memory elements, and components for nonlinear optical frequency conversion, second-harmonic generation and four-wave mixing (Poon, 2004; Melloni, 2003; Heebner, 2004; Mookherjea, 2002).

5.3 Biochemical sensors

Microresonators supporting high-Q modes also have great potential in the development of inexpensive, ultra-compact, highly sensitive and robust bio- and chemical sensors on a chip. As compared to linear optical waveguide biosensors, microresonator-based devices benefit from much smaller size (several μm rather than a few centimetres) and higher sensitivity. Several phenomena can be used for detection. For example, cylindrical microresonators, operating on the WG modes, have been shown to enhance the intensity of the fluorescence emitted by biological materials (Blair, 2001). However, most microresonator sensors rely on the measurement of transmission or scattering characteristics of a microresonator exited by an optical waveguide mode in the presence of biological material on the resonator surface or in the surrounding solution. These sensors can detect the resonance frequency shift caused by the change of the resonator effective refractive index or increased absorption (Boyd, 2001; Krioukov, 2002; Vollmer, 2003; Chao, 2003). Alternatively, the sensors can measure the change in phase and/or intensity of the light at the output of the waveguide in the forward direction at a fixed wavelength (Rosenblit, 2004).

High Q-factors of microresonator modes are crucial for achieving high sensitivity of the sensors. Indeed, the higher the Q-factor, the steeper the slope between zero and unity in the transmission characteristic of a microresonator, resulting in higher sensor sensitivity. Another way to enhance the slope between the zero and the unity transmission (and thus improve the sensor sensitivity) is to use a microresonator structure that generates a sharp asymmetric Fano-resonance line shape (Chao, 2003).

5.4 Spontaneous emission control, novel light sources and cavity QED

Another major application for microresonators is in development and fabrication of novel light sources such as resonant-cavity-enhanced light-emitting diodes (LEDs), low-threshold microlasers, and colour flat-panel displays. In wavelength-sized microresonator structures, semiconductor material luminescence can be either suppressed or enhanced, and they also enable narrowing of the spectral linewidth of the emitted light (Haroche, 1989; Yokoyama, 1992; Yamamoto, 1993; Krauss, 1999; Vahala, 2003).

Since 1946, when it was first proposed that the spontaneous emission from an excited state of an emitter can be significantly altered if it is placed into low-loss wavelength-scale cavity (Purcell,1946), various microresonator designs for efficient control of spontaneous emission have been explored

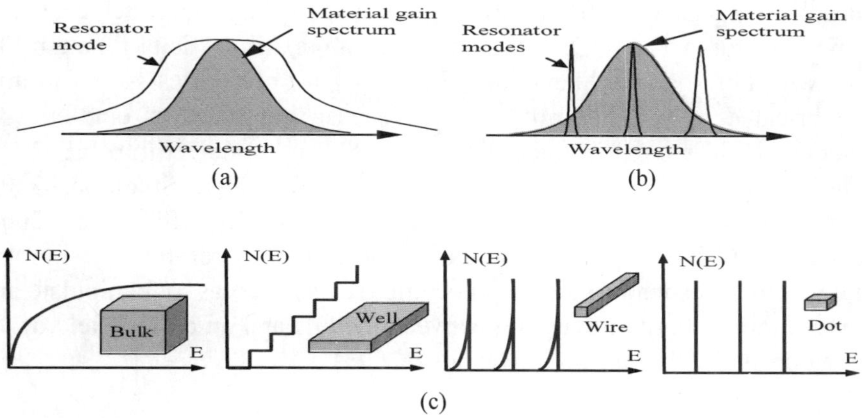

Figure 7. Difference in the spontaneous emission enhancement in a LED (a) and a microcavity laser (b); Density of electronic states in bulk semiconductor material and low-dimensional semiconductor heterostructures (c).

including microdisk (Levi, 1993; Baba, 1997, 1999; Backes, 1999; Cao, 2000; Fujita, 1999, 2001, 2002; Zhang, 1996), microsphere (Cai, 2000; Shopova, 2004; Rakovich, 2003) and micropost (Pelton, 2002; Reithmaier, 2004; Santori, 2004; Solomon, 2001; Gayral, 1998) resonators as well as PC defect cavities (Painter, 1999; Boroditsky, 1999; Gayral, 1999). Depending on the Q-factors (resonance linewidths) of the modes supported by the microresonator in the spontaneous emission range of the resonator material, the two situations illustrated in Fig. 7 can be realized. In a cavity-enhanced LED (Fig. 7a), the material gain bandwidth is smaller than the cavity resonance bandwidth, and emission of the whole material gain spectrum is enhanced. In a semiconductor laser (Fig. 7b) with a laser microcavity supporting high-Q modes, the emission at one (or several) of the modal wavelengths is strongly enhanced (being predominantly stimulated emission), while the emission at all other wavelengths is suppressed. The amount by which the spontaneous emission rate is enhanced for an emitter on resonance with a cavity mode is characterized by the Purcell factor, which is proportional to the mode Q-factor and inversely proportional to the mode volume.

Clearly, to increase the enhancement factor, it is necessary to design and fabricate high-Q, small-V microresonators. However, cavity-enhanced LEDs based on the microresonators with high-Q modes must have equally narrow material spontaneous emission linewidths (Fig. 7a), which are not easily realized in bulk or heterostructure quantum-well microresonators. The recently proposed concept of an active material system, semiconductor quantum dots (QDs) (Arakawa, 2002) combines the narrow linewidth normally associated with atomic emitters and the high gain achievable in semiconductors. The QD emission spectrum exhibits delta-function-like lines with ultra-narrow linewidths (Fig. 7c). QDs of various sizes can now be fabricated by self-assembly, and have been integrated as emitters (so-called artificial atoms) in microdisk (Cao, 2000), PC defect (Yoshie, 2001; Hennesy, 2003), micropost/micropillar (Santori, 2004; Solomon, 2001; Gayral, 1998; Benyoucef, 2004), and microsphere (Shopova, 2004; Rakovich, 2003) resonators. The challenge is then to carefully design and tune the microresonator modal and geometrical properties to manipulate and increase the strength of coupling between the QD and an optical field of the resonator mode (Pelton, 2002; Hennesy, 2003).

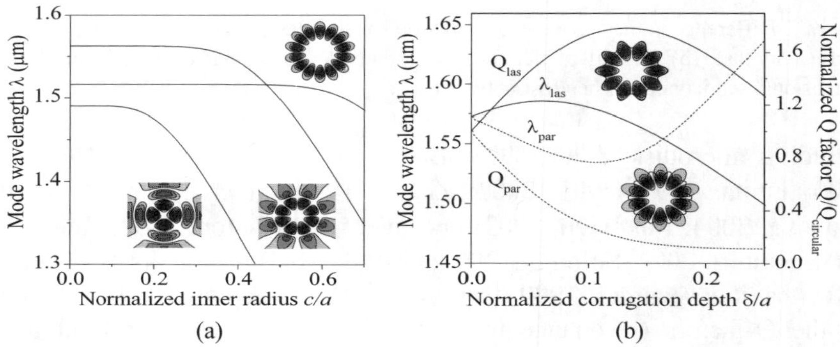

Figure 8. (a) Detuning of the higher-radial-order WG modes in a microring resonator of radius *a* with an increase of the inner radius *c* (Boriskina, 1999; 2002); and (b) splitting of a double-degenerate WG mode in a microgear resonator with simultaneous enhancement of the Q-factor of the working mode and suppression of the parasitic mode (Boriskina, 2004).

In the microcavity lasers, only a small portion of the spontaneous emission couples into a single (or several) optical modes (Fig. 7b). It should also be noted that optical modes of rotationally symmetrical microresonators are either multiple (microspheres) or double (microdisks, microrings, circular micropillars, etc.) degenerate. This often leads to the appearance of closely located doublets in their lasing spectra due to fabrication errors (sidewall roughness and shape imperfections, etc.) and thus causes spectral noise, mode hopping and polarization instabilities (Gayral, 1998; Fujita, 1999; Boriskina, 2004). Among recently proposed modified resonator designs that remove the mode degeneracy and help separate two closely located modes are elliptical (Gayral, 1998; Boriskina, 2003; Kim, 2004), square (Guo, 2003; Boriskina, 2005), microgear (Fujita, 2001, 2002; Boriskina, 2004), and notched (Backes, 1998; Boriskina, 2005) microresonators.

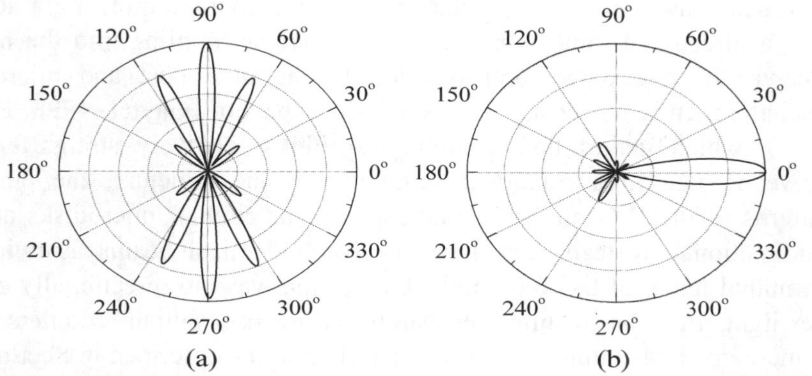

Figure 9. Directional far-field emission patterns of the WG modes in (a) an elliptical (Boriskina, 2003) and (b) a notched (Boriskina, 2005) microdisk resonator.

If, as a result of optical or electrical pumping, the optical gain in the resonator active material begins to dominate optical loss, stimulated emission in the microresonator begins to dominate spontaneous emission. To improve the microlaser efficiency, it is desirable to reduce the threshold input energy at which lasing (stimulated emission) begins to occur. The laser threshold depends on the size of a laser microresonator, the lasing mode optical confinement, the gain/absorption balance of the resonator material and the presence of several competing modes within a material spontaneous emission range. Reducing the number of available modes in the spontaneous emission range would therefore both reduce the threshold and improve the laser noise characteristics. This can be achieved by either detuning the wavelengths of all the modes but one against the spontaneous emission peak, or decreasing their Q-factors and thus reducing spontaneous emission into these modes. For example, all the higher-radial-order WG modes in microdisk resonators, with the fields occupying large areas inside the disk, can be suppressed by removing material from the resonator interior (Fig. 8a) by piercing holes (Backes, 1999) or forming microring resonators (Hagness, 1997; Boriskina, 1999, 2002). Enhancement of the working mode Q factor in a microgear (circular microdisk with a periodically corrugated rim) laser has also been achieved (Fujita, 2002), together with the suppression of the Q factor of the other mode of a doublet appearing due to splitting of a double-degenerate first-radial-order WG mode of a circular microdisk (Fujita, 2001; Boriskina, 2004; Fig. 8b). Another way to reduce the threshold in a microcavity laser is to increase the fraction of spontaneous emission into the lasing mode by improving the spectral alignment between the material gain peak and the microcavity mode wavelength (Fujita, 2001; Cao, 2003).

Finally, most practical optoelectronic applications require light sources with a directional light output, e.g., for efficient coupling into the narrow acceptance cone of an optical fibre. In the micropost and micropillar resonators, emission occurs in a single-lobe pattern (Gayral, 1998; Pelton, 2002), which makes their coupling to fibres relatively straightforward. However, due to a rotational symmetry of the structure, the emission patterns of the WG modes in microspheres or circular microdisks are not unidirectional. Instead, they have as many identical beams as twice the azimuthal index of the WG mode. One of the ways to directionally extract the light from such microresonators is to use output couplers, e.g., evanescent-field couplers of various configurations described in Section 5.1. Alternatively, microresonators with non-circularly-symmetric geometrical shapes can be designed to obtain directional emission patterns. Directional light output has been observed from deformed spherical resonators; quadrupolar, elliptical (Noeckel, 1994; Gmachl, 1998; Backes, 1999; Lee, 2002; Boriskina, 2002; Kim, 2004; Fig. 9a), and egg-shaped (Shima, 2001; Boriskina, 2002) microdisks; microdisks with patterned gratings and tabs (Levi, 1992); micropost and disk resonators with spiral cross-sections (Chern, 2003; Kneissl, 2004); and microdisks with notches and openings (Chu, 1994; Backes, 1999; Boriskina, 2006; Fig. 9b).

6. SIMULATION METHODS

All the above-discussed technological advancements in microresonator fabrication and functional applications call for innovative computational methods for solving Maxwell's equations, which govern the electromagnetic fields in the resonator structures. Development of rigorous yet efficient simulation methods and CAD tools is necessary to fully exploit the potential of the new generation of microresonators, substantially reducing the cost and time of the design stage, to create novel devices, and to study new optical phenomena in microcavities.

Unfortunately, Maxwell's equations can be solved analytically for only a few simple canonical resonator structures, such as spheres (Stratton, 1997) and infinitely long cylinders of circular cross-sections (Jones, 1964). For arbitrary-shape microresonators, numerical solution is required, even in the 2-D formulation. Most 2-D methods and algorithms for the simulation of microresonator properties rely on the Effective Index (EI) method to account for the planar microresonator finite thickness (Chin, 1994). The EI method enables reducing the original 3-D problem to a pair of 2-D problems for transverse-electric and transverse-magnetic polarized modes and perform numerical calculations in the plane of the resonator. Here, the effective refractive index of the 2-D structure is taken as the normalized propagation constant of the fundamental guided mode in an equivalent planar waveguide with the same thickness and material as the microresonator.

The finite-difference time-domain (FDTD) method (Chu, 1989) has been a main workhorse in modelling and design of optical microresonators (both in 2-D and 3-D) over the past few decades, due to its simplicity and flexibility (Fujita, 2001; Fong, 2003, 2004; Ryu, 2004; Painter, 2001; Pelton, 2002; Hagness, 1997; Vuckovic, 1999; Rosenblit, 2004). However, the FDTD method suffers from relatively large dispersive error, staircasing errors at resonator boundaries, and reduction of accuracy to the first order at material interfaces. Furthermore, optical microresonators are often placed into complex open space domains, in which case discretization of the problem can lead to errors caused by non-physical backreflections from the edges of the computational window. For large and complex domains, and especially in full-vectorial 3-D simulations, development of faster algorithms becomes imperative to reduce the cost of computations, as the grid size required by using the FDTD method becomes prohibitively expensive for even modern advanced computers. For example, CPU times for the FDTD computations of the characteristics of a 2.25 μm-diameter semiconductor microring resonator evanescently coupled to a straight semiconductor waveguide reached 35 hours on a 1-GHz, 4-GB RAM platform (Fujii, 2003).

By using fast algorithms based on approximate techniques such as geometrical optics, "billiard theory", paraxial approximation, etc. useful insight into the ray dynamics within optically large microresonators can be obtained (Chowdhury, 1992; Noeckel, 1994, 2000; Jiang, 1999; Poon, 2001; Guo 2003; Huang 2001; Fong, 2004). However, the accuracy of such methods is clearly not adequate for studying the modal spectra of microresonators whose dimensions are of the order of an optical wavelength. Among other popular first-order numerical tools for the microresonator design are: algorithms based on the Coupled Mode Theory (Yariv, 1973; Little, 1997; Manolatou, 1999; Chin, 1998, Hammer, 2002), which are able to provide fast initial designs of waveguide-coupled circular and square microdisk resonators; and the Spectral Index method (Greedy, 2000), which has been successfully applied to study the WG mode characteristics of high-index-contrast circular microdisk resonators mounted on substrates (Fig. 10).

Accurate yet computationally efficient techniques that have been already demonstrated to yield solutions to a variety of design problems in optics and photonics are based on the formulation of the problem in terms of either

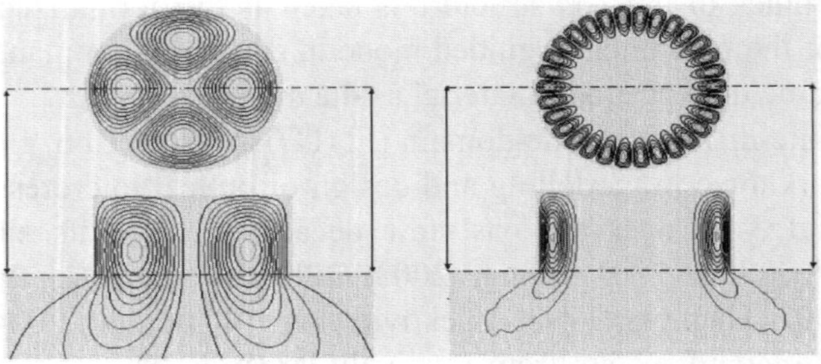

Figure 10. Field distributions for two WG modes in a microdisk resonator on a substrate predicted by the Spectral Index method (Greedy, 2000).

volume or surface integral equations (IEs). A clear advantage of such formulations is that the problem size is reduced by moving it from the open infinite domain to a finite one (a volume or a surface of the microresonator), which reduces the required computational effort. Furthermore, the use of artificial absorbing conditions is completely avoided, together with the danger of unwanted non-physical back-reflections, and microresonators located in the layered media can be modelled without loss of accuracy (Boriskina, 1999, 2002, 2003; Chremmos, 2004). The IE formulation of Maxwell's equations, in its continuous form, is exact provided that the full equivalence takes place. By using various Galerkin-type discretization schemes with low- and high-order basis functions (Harrington, 1968), numerical algorithms with various convergence rates (and thus various computational costs to reach a desired accuracy) can be constructed, with higher-order schemes providing the most computationally efficient solutions (Atkinson, 1997; Nosich, 1999). Due to the flexibility of the method, it has been successfully applied to study modal characteristics of symmetrical and asymmetrical microresonator structures in 2-D, such as circular, elliptical, hexagonal, stadium, and notched microdisk resonators (Wiersig, 2003; Lee, 2004; Boriskina, 1999-2005; Kottmann, 2000, Chremmos, 2004). In 3-D, IE techniques have been developed mainly to analyse dielectric resonators of electrically small size typical for microwave applications (Kucharski, 2000; Glisson, 1983; Umashankar, 1986; Liu, 2004).

It should be noted, however, that the Q factors of open microcavities do not characterise directly the threshold gain values of the corresponding semiconductor lasers. To overcome this difficulty a new lasing eigenvalue problem (LEP) was introduced recently (Smotrova, 2004). The LEP enables one to quantify accurately the lasing frequencies, thresholds, and near- and far-field patterns separately for various WG modes in semiconductor laser resonators. However, the threshold of a lasing mode depends on other factors, including presence of several competing modes within a material spontaneous emission range and detuning of the lasing wavelength against the spontaneous emission peak. To account for these effects, a spontaneous emission coupling factor (β factor) of a given mode can be calculated, which is defined as the ratio of the spontaneous emission rate into that mode and the spontaneous emission rate into all modes (Bjork, 1991). Various approaches to calculate the β factor have been described in the literature, including classical and quantum mechanical methods as well as FDTD (Bjork, 1991; Chin, 1994; Vuckovic, 1999; Xu, 2000).

7. CONCLUSIONS

Recent advances in the nanofabrication technology offer the possibility of manufacturing novel optical microresonator devices with unprecedented control in a variety of natural and artificial material systems. High-Q micro- and nano-scale resonators have already proven to be excellent candidates for optical signal processing in low-cost high-density photonic integrated circuits, significant enhancement of detection sensitivity in biosensors, and compact and efficient laser sources with enhanced functionality. Continued development of new materials with specially designed properties such as quantum wires and dots as well as 3-D photonic crystal structures is expected to yield more ideal high-performance optical resonators. These novel resonator designs will fuel future growth of broadband all-optical networks and will impact basic research in quantum physics and quantum information science. However, emerging resonator-based devices and applications as well as growing industrial competition impose strict requirements on the accuracy and performance of the existing resonator design and simulation tools calling for the development of advanced methods and algorithms.

ACKNOWLEDGEMENT

The authors would like to express their gratitude to the UK Engineering and Physical Sciences Research Council (EPSRC) and the Royal Society for financial support.

REFERENCES

Absil, P.P., Hryniewicz, J.V., Little, B.E., Cho, P.S., Wilson, R.A., Joneckis, L.G., Ho, P.-T., 2000, Wavelength conversion in GaAs micro-ring resonators, *Opt. Lett.* **25**(8):554-556.

Akahane, Y., Asano, T., Song, B.-S., and Noda, S., 2003, High-*Q* photonic nanocavity in a two-dimensional photonic crystal, *Nature* **425**: 944-947.

Arakawa, Y., Yariv, A., 1986, Quantum-well lasers gain, spectra, dynamics, *IEEE J. Quantum Electron.* **22**(9):1887-1899.

Arakawa, Y., 2002, Connecting the dots, *SPIE's OE Magazine* **1**:18-20.

Armani, D. K., Kippenberg, T. J., Spillane, S. M., and Vahala, K.J., 2003, Ultra-high-Q toroid microcavity on a chip, *Nature* **421**:905-908.

Astratov, V.N., Franchak, J.P., and Ashili, S.P., 2004, Optical coupling and transport phenomena in chains of spherical dielectric microresonators with size disorder, *Appl. Phys. Lett.* **85**(23):5508-5510.

Atkinson, K.E., 1997, *The numerical solution of boundary integral equations*, Cambridge University Press, Cambridge.

Baba, T., 1997, Photonic crystals and microdisk cavities based on GaInAsP-InP system, *IEEE J. Select. Topics Quantum Electron.* **3**(3):808-828.

Baba, T., Fujita, M., Sakai, A., Kihara, M., and Watanabe, R., 1997, Lasing characteristics of GaInAsP/InP strained quantum-well microdisk injection lasers with diameter of 2-10 μm, *IEEE Photon. Technol. Lett.* **9**(7):878-880.

Backes S.A. and Cleaver, J.R.A., 1998, Microdisk laser structures for mode control and directional emission, *J. Vac. Sci. Technol. B* **16**(6):3817-3820.

Backes, S.A., Cleaver, J.R.A., Heberle, A.P., Baumberg, J.J., and Köhler, K., 1999, Threshold reduction in pierced microdisk lasers, *Appl. Phys. Lett.* **74**(2):176-178.

Barwicz, T., Popovic, M.A., Rakich, P.T., Watts, M.P. Haus, H.A., Ippen, E.P., and Smith, H.I, 2004, Microring-resonator-based add-drop filters in SiN: fabrication and analysis, *Opt. Express*, **12**(7):1437-1442.

Benisty, H., Weisbuch, C., Labilloy, D., Rattier, M., Smith, C. J. M., Krauss, T. F., De La Rue, R.M., Houdre, R., Oesterle, U., Jouanin, C., and Cassagne D., 1999, Optical and confinement properties of two-dimensional photonic crystals, *J. Lightwave Technol.* **17**(11):2063-2077.

Benyoucef, M., Ulrich, S.M., Michler, P., Wiersig, J., Jahnke, F., and Forchel, A., 2004, Enhanced correlated photon pair emission from a pillar microcavity, *New J. Physics* **6**: 91.

Berman, P.R. (ed.), 1994, *Cavity Quantum Electrodynamics*, Acad. Press, San Diego, CA.

Bjork, G. and Yamamoto, Y., 1991, Analysis of semiconductor microcavity lasers using rate equations, *IEEE J. Quantum. Electron.*, **27**(11):2386-2396.

Blair, S. and Chen, Y., 2001, Resonant-enhanced evanescent-wave fluorescence biosensing with cylindrical optical cavities, *Appl. Opt.* **40**:570–582.

Blom, F.C., vanDijk, D.R., Hoekstra, H.J.W.M., Driessen, A., and Popma, T.J.A., 1997, Experimental study of integrated-optics microcavity resonators: Toward an all-optical switching device, *Appl. Phys. Lett.* **71**(6):747-749.

Boriskina, S.V., Nosich, A.I., 1999, Radiation and absorption losses of the whispering-gallery-mode dielectric resonators excited by a dielectric waveguide, *IEEE Trans. Microwave Theory Tech.* **47**:224-231.

Boriskina, S.V., Benson, T.M., Sewell, P., and Nosich, A.I., 2002, Effect of a layered environment on the complex natural frequencies of 2D WG-mode dielectric-ring resonators, *J. Lightwave Technol.* **20**: 1563-1572.

Boriskina, S.V., Benson, T.M., Sewell, P., Nosich, A.I., 2003, Tuning of elliptic whispering-gallery-mode microdisk waveguide filters, *J. Lightwave Technol.* **21**(9):1987-1995.

Boriskina, S.V., Benson, T.M., Sewell, P., and Nosich, A.I., 2003, Highly efficient design of spectrally engineered WG-mode laser resonators, *Opt. Quantum Electron.* **35**:545-559.

Boriskina, S.V., Sewell, P. Benson, T.M., and Nosich A.I., 2004, Accurate simulation of 2D optical microcavities with uniquely solvable BIEs and trigonometric-Galerkin discretisation, *J. Opt. Soc. Am. A* **21**(3):393-402.

Boriskina, S.V., Benson, T.M., Sewell, P., and Nosich, A.I., 2004, Spectral shift and Q-change of circular and square-shaped optical microcavity modes due to periodical sidewall surface roughness, *J. Opt. Soc. Am. B*, **10**:1792-1796.

Boriskina, S.V., Benson, T.M., Sewell, P., and Nosich, A.I., 2005, Optical modes in imperfect 2-D square and triangular microcavities, *IEEE J. Quantum Electron.*, **41**(6):857-862.

Boriskina, S.V., 2005, Efficient simulation and design of coupled optical resonator clusters and waveguides, ***Proc. ICTON Int. Conf.***, Barcelona, Spain.

Boriskina, S.V., 2005, Symmetry, degeneracy and optical confinement of modes in coupled microdisk resonators and photonic crystal cavities, submitted to *J. Quantum Electron.*

Boriskina, S.V., Benson, T.M., Sewell, P., and Nosich, A.I., 2006, Q-factor and emission pattern control of the WG modes in notched microdisk resonators, to appear in *IEEE J. Select. Topics Quantum. Electron.*.

Boroditsky, M., Vrijen, R., Krauss, T.F., Coccioli, R., Bhat, R., and Yablonovitch, E., 1999, Spontaneous emission extraction and Purcell enhancement from thin-film 2-D photonic crystals, *J. Lightwave Technol.* **17**(11):2096-2112.

Boyd, R.W. and Heebner, J.E., 2001, Sensitive disk resonator photonic biosensor, *Appl. Opt.* **40**(31):5742-5747.

Braginsky, V. B., Gorodetsky, M. L., and Ilchenko, V. S., 1989, Quality-factor and nonlinear properties of optical whispering-gallery modes, *Phys. Lett. A* **137**(7-8):393-397.

Buck, J.R. and Kimble, H.J., 2003, Optimal sizes of dielectric microspheres for cavity QED with strong coupling, *Phys. Rev. A* **67**(3):033806.

Cai, M., Painter, O., and Vahala, K.J., 2000, Observation of critical coupling in a fiber taper to a silica-microsphere whispering-gallery mode system, *Phys. Rev .Lett.* **85**(1):74-77.

Cai, M., Painter, O., Vahala, K.J., and Sercel, P.C., 2000, Fiber-coupled microsphere laser, *Opt. Lett.* **25**(19):1430-1432.

Campillo, A.J., Eversole, J.D., and Lin, H.-B., 1991, Cavity quantum electrodynamic enhancement of stimulated emission in microdroplets, *Phys. Rev. Lett.* **67**(4):437–440.

Cao, H., Xu, J.Y., Xiang, W.H., Ma, Y., Chang, S.-H., Ho, S.T., and Solomon, G.S., 2000, Optically pumped InAs quantum dot microdisk lasers, *Appl. Phys. Lett.* **76**(24):3519-3521.

Cao, J.R., Kuang, W., Choi, S.-J., Lee, P.-T., O'Brien, J.D., Dapkus, P.D., 2003, Threshold dependence on the spectral alignment between the quantum-well gain peak and the cavity resonance in InGaAsP photonic crystal lasers, *Appl. Phys. Lett.* **83**(20):4107-4109.

Chan, S., Horner, S.R., Fauchet, P. M., and Miller, B.L., 2001, Identification of gram negative bacteria using nanoscale silicon microcavities, *J. Am. Chem. Soc.,* **123**(47):11797-11798.

Chang, R.K. and Campillo, A.J. (eds.), 1996, *Optical processes in microcavities*, World Scientific, Singapore.

Chao, C.-Y. and Guo, L.J., 2003, Biochemical sensors based on polymer microrings with sharp asymmetrical resonance, *Appl. Phys. Lett.* **83**(8):1527-1529.

Chern, G.D., Tureci, H.E., Stone, A.D., Chang, R.K., Kneissl M., and Johnson, N.M., 2003, Unidirectional lasing from InGaN multiple-quantum-well spiral-shaped micropillars, *Appl. Phys. Lett.* **83**(9):1710-1712.

Chin M.K. and Ho, S.T., 1998, Design and modeling of waveguide-coupled single-mode microring resonators, *J. Lightwave Technol.* **16**(8):1433-1446.

Chin, M.K., Chu, D.Y., and Ho, S.-T., 1994, Estimation of the spontaneous emission factor for microdisk lasers via the approximation of whispering-gallery modes, *J. Appl. Phys.* **75**(7):3302-3307.

Chin, M.K., Youtsey, C., Zhao, W., Pierson, T., Ren, Z., Wu, S.L., Wang, L., Zhao, Y.G., and Ho, S.T., 1999, GaAs microcavity channel-dropping filter based on a race-track resonator, *IEEE Photon. Technol. Lett.* **11**(12):1620–1622.

Choi, S.J., Peng, Z., Yang, Q., Choi, S.J., and Dapkus, P.D., 2004, An eight-channel demultiplexing switch array using vertically coupled active semiconductor microdisk resonators, *IEEE Photon. Technol. Lett.* **16**(11):2517-2519.

Chou, S.Y., Krauss, P.R., and Renstrom, P.J., 1996, Nanoimprint lithography, *J. Vac. Sci. Technol. B* **14**(6):4129- 4133.

Chow, E., Lin, S.Y., Johnson, S.G., Villeneuve, P.R., Joannopoulos, J.D., Wendt, J.R., Vawter, G.A., Zubrzycki, W., Hou, H., and Alleman, A., 2000, Three-dimensional control of light in a two-dimensional photonic crystal slab, *Nature* **407:**983-986.

Chowdhury, D.Q., Barber, P.W., and Hill, S.C., 1992, Energy-density distribution inside large nonabsorbing spheres by using Mie theory and geometrical-optics, *Appl. Opt.* **31**(18):3518-3523.

Chremmos, I.D., Uzunoglu, N.K., 2004, Transmission and radiation in a slab waveguide coupled to a whispering-gallery resonator: volume-integral-equation analysis, *J. Opt. Soc. Am. A* **21**(5):839-846.

Chu, D.Y., Chin, M.K., Bi, W.G., Hou, H.Q., Tu, C.W., and S.T. Ho, 1994, Double-disk structure for output coupling in microdisk lasers, *Appl. Phys. Lett.* **65**(25):3167-3169.

Chu, S.T. and Chaudhuri, S.K., 1989, A finite-difference time domain method for the design and analysis of guided-wave optical structures, *J. Lightwave Technol.* **7**:2033–2038.

Coccioli, R., Boroditsky, M., Kim, K.W., Rahmat-Samii, Y., and Yablonovitch, E., 1998, Smallest possible electromagnetic mode volume in a dielectric cavity, *IEE Proc.-Optoelectron.* **145**(6):391-397.

Djordjev, K., Choi, S.J., Choi, S.J., and Dapkus P.D., Active semiconductor microdisk devices, *J. Lightwave Technol.* **20**(1):105-113.

Djordjev, K., Choi, S.J., Choi, S.J., and Dapkus, P.D., 2002, High-Q vertically coupled InP microdisk resonators, *IEEE Photon. Technol. Lett.* **14**(3):331-333.

Dumon, P., Bogaerts, W., Wiaux, V., Wouters, J., Beckx, S., Van Campenhout, J., Taillaert, D., Luyssaert, B., Bienstman, P., Van Thourhout, D., and Baets, R., 2004, Low-loss SOI photonic wires and ring resonators fabricated with deep UV lithography, *IEEE Photon. Technol. Lett.* **16**(5):1328-1330.

Einspruch, N.G. and Frensley, W.R. (eds), 1994, *Heterostructure and Quantum Devices*, in series *VLSI Electronics: Microstructure Science* (Academic Press, San Diego).

Eldada, L. and Shacklette, L.W., 2000, Advances in polymer integrated optics, *IEEE J. Select. Topics Quantum Electron.* **6**(1):54-68.

Eldada, L., 2001, Advances in telecom and datacom optical components, *Opt. Eng.* **40**(7):1165–1178.

Elliott, D., 1989, *Integrated circuit fabrication technology*, McGraw-Hill, New York.

Fong, C.Y. and Poon, A.W., 2003, Mode field patterns and preferential mode coupling in planar waveguide-coupled square microcavities, *Opt. Express* **11**(22):2897-2904.

Fong, C.Y. and Poon, A.W., 2004, Planar corner-cut square microcavities: ray optics and FDTD analysis, *Opt. Express* **12**(20):4864-4874.

Foord, J.S., 1997, *Chemical beam epitaxy and related techniques*, John Wiley & Son Ltd.

Foresi, J.S., Villeneuve, P.R., Ferrera, J., Thoen, E.R., Steinmeyer, G., Fan, S., Joannopoulos, J.D., Kimerling, L.C., Smith, H.I., and Ippen, E.P., 1997, Photonic-bandgap microcavities in optical waveguides *Nature* **390**:143–145.

Fujii, M., Freude, W., and Russer, P., 2003, Efficient high-spatial-order FDTD analysis of 3D optical ring resonator filters, *Proc. 19th Annual Rev. Progress in Appl. Comput. Electromagnetics*, 739-744, Monterey, CA.

Fujita, M., Sakai, A., and Baba, T., 1999, Ultrasmall and ultralow threshold GaInAsP-InP microdisk injection lasers: design, fabrication, lasing characteristics, and spontaneous emission factor, *J. Select. Topics Quant. Electron.* **5**(3):673-681.

Fujita, M. and Baba, T., 2001, Proposal and finite-difference time-domain simulation of WG mode microgear cavity, *IEEE J. Quantum Electron.* **37**(10):1253-1258.

Fujita, M. and Baba, T., 2002, Microgear laser, *Appl. Phys. Lett.* **80**(12):2051-2053.

Fukushima T. and Harayama, T., 2004, Stadium and quasi-stadium laser diodes, *IEEE J. Select. Topics Quantum Electron.* **10**(5):1039-1051.

Gardner, D.S., and Brongersma, M.L., 2005, Microring and microdisk optical resonators using silicon nanocrystals and erbium prepared using silicon technology, *Opt. Materials* **27**(5):804-811.

Gastine, M., Courtois, L., and Dormann, J.L., 1967, Electromagnetic resonances of free dielectric spheres, *IEEE Trans. Microwave Theory Tech.* **15**(12):694-700.

Gayral, B., Gerard, J.M., Legrand, B., Costard E., and Thierry-Mieg V., 1998, Optical study of GaAs/AlAs pillar microcavities with elliptical cross section, *Appl. Phys. Lett.* **72**(12):1998, 1421-1423.

Gayral, B., Gerard, J.M., Lemaitre, A., Dupuis, C., Manin, L., and Pelouard, J.L., 1999, High-Q wet-etched GaAs microdisks containing InAs quantum boxes, *Appl. Phys. Lett.* **75**(13):1908-1910.

Gianordoli, S., Hvozdara, L., Strasser, G., Schrenk, W., Faist, J., and Gornik, E., 2000, Long-wavelength (λ= 10 μm) quadrupolar-shaped GaAs–AlGaAs microlasers, *IEEE J. Quantum. Electron.* **36**(4):458-464.

Glisson, A.W., Kajfez, D., and James, J., 1983, Evaluation of modes in dielectric resonators using a surface integral-equation formulation, *IEEE Trans. Microwave Theory Tech.* **31**(12):1023-1029.

Gmachl, C., Cappasso, F., Narimanov, E.E., Nockel, J.U., Stone, A.D., Faist, J., Sivco, D.L., and Cho, A.Y., 1998, High-power directional emission from microlasers with chaotic resonances, *Science* **280:**1556-1564.

Gorodetsky, M.L., Savchenkov, A.A., and Ilchenko, V.S., 1996, Ultimate Q of optical microsphere resonators, *Opt. Lett.* **21**(7):453-455.

Greedy, S.C., Boriskina, S.V., Sewell, P., Benson, T.M., 2001, Design and simulation tools for optical micro-resonators, *Proc. Photonics West Conference*, San Jose, CA.

Grover, R., Absil, P.P., Van, V., Hryniewicz, J.V., Little, B.E., King, O., Calhoun, L.C., Johnson, F.G., and Ho, P.-T., 2001, Vertically coupled GaInAsP–InP microring resonators, *Opt. Lett.* **26**(8):506-508.

Grover, R., Van, V., Ibrahim, T.A., Absil, P.P., Calhoun, L.C., Johnson, F. G., Hryniewicz, J.V., and Ho, P.-T., 2002, Parallel-cascaded semiconductor microring resonators for high-order and wide-FSR filters, *J. Lightwave Technol.* **20**(5):872–877.

Grover, R., Ibrahim, T.A., Kanakaraju, S., Lucas, L., Calhoun, L.C., and Ho, P.-T., 2004, A tunable GaInAsP–InP optical microring notch filter, *IEEE Photon. Technol. Lett.* **16**(2):467-469.

Guo, L.J., 2004, Recent progress in nanoimprint technology and its applications, *J. Phys. D: Appl. Phys.* **37**:R123–R141.

Guo, W.-H., Huang, Y.-Z., Lu, Q.-Y., and Yu, L.-J., 2003, Whispering-gallery-like modes in square resonators, *IEEE J. Quantum Electron.* **39**(9):1563-1566.

Hagness, S.C., Rafizadeh, D., Ho, S.T., and Taflove, A., 1997, FDTD microcavity simulations: design and experimental realization of waveguide-coupled single-mode ring and whispering-gallery-mode disk resonators, *J. Lightwave Technol.* **15**(11):2154-2165.

Hammer M., 2002, Resonant coupling of dielectric optical waveguides via rectangular microcavities: the coupled guided mode perspective, *Opt.Communicat.* **214**(1–6):155–170.

Haroche, S., Kleppner, D., 1989, Cavity quantum electrodynamics, *Phys. Today* **42**(1):24-30.

Harrington, R.F., 1968, *Field computation by Moment Methods* (Krieger, Malabar, Fla.).

Heebner, J.E., Lepeshkin, N.N., Schweinsberg, A., Wicks, G.W., Boyd, R.W., Grover, R., and Ho P.-T., 2004, Enhanced linear and nonlinear optical phase response of AlGaAs microring resonators, *Opt. Lett.* **29**(7):769-771.

Hennessy, K., Reese, C., Badolato, A., Wang, C.F., Imamoglu, A., Petroff, P.M., Hu, E., Jin, G., Shi, S., and Prather, D.W., 2003, Square-lattice photonic crystal microcavities for coupling to single InAs quantum dots, *Appl. Phys. Lett.* **83**(18):3650-3652.

Herman, M.A. and Sitter, H., 1996, *Molecular beam epitaxy*, Springer-Verlag, Berlin.

Hillmer, H. and Germann, R. (eds), 2003, Photonic Materials for Optical Communications, *MRS Bulletin* **5**:340-377.

Hryniewicz, J.V., Absil, P.P., Little, B.E., Wilson, R.A., Ho, P.-T., 2000, Higher order filter response in coupled microring resonators, *IEEE Photon. Technol. Lett.* **12**(3):320-322.

Huang, Y.-Z., Guo, W.-H., and Wang, Q.-M., 2001, Analysis and numerical simulation of eigenmode characteristics for semiconductor lasers with an equilateral triangle micro-resonator, *IEEE J. Quantum Electron.* **37**(1):100-107.

Ilchenko, V.S., Gorodetsky, M.L., Yao, X.S., and Maleki, L., 2001, Microtorus: a high-finesse microcavity with whispering-gallery modes, *Opt. Lett.* **26**(5):256-258.

Ilchenko, V.S., Matsko, A.B., Savchenkov, A.A., Maleki, L., 2002, High-efficiency microwave and millimeter-wave electrooptical modulation with whispering-gallery resonators, *Proc. SPIE* **4629**:158-163.

Jiang, H.X., Lin, J.Y., Zeng, K.C., and Yang, W.D., 1999, Optical resonance modes in GaN pyramid microcavities, *Appl. Phys. Lett.* **75**(6):763-765.

Jones, D.S., 1964, *The Theory of Electromagnetics,* Macmillan, New York.

Kim, S.-H., Ryu, H.-Y., Park, H.-G., Kim, G.-H., Choi, Y.-S., and Lee, Y.-H., 2002, Two-dimensional photonic crystal hexagonal waveguide ring laser, *Appl. Phys. Lett.* **81**(14):2499-2501.

Kim, S.-K., Kim, S.-H., Kim, G.-H., Park, H.-G., Shin, D.-J., and Lee, Y.-H., 2004, Highly directional emission from few-micron-size elliptical microdisks, *Appl. Phys. Lett.* **84**(6): 861-863.

Kippenberg, T.J., Spillane, S.M., Armani, D.K., and Vahala, K. J., 2003, Fabrication and coupling to planar high-Q silica disk microcavities, *Appl. Phys. Lett.* **83**(4):797-799.

Kippenberg, T.J., Spillane, S.M., Armani, D.K., and Vahala K.J., 2004, Ultralow-threshold microcavity Raman laser on a microelectronic chip, *Opt. Lett.* **29**(11):1224-1226.

Klunder, D.J.W., Krioukov, E., Tan, F.S., Van Der Veen, T., Bulthuis, H.F., Sengo, G., Otto, C., Hoekstra, H.J.W.M., Driessen, A., 2001, Vertically and laterally waveguide-coupled cylindrical microresonators in Si_3N_4 on SiO_2 technology, *Appl. Phys. B,* **73**(5-6):603-608.

Kneissl, M., Teepe, M., Miyashita, N., Johnson, N.M., Chern, G.D., and Chang, R.K., 2004, Current-injection spiral-shaped microcavity disk laser diodes with unidirectional emission, *Appl. Phys. Lett.* **84**(14):2485-2487.

Kottmann, J.P. and Martin, O.J.F., 2000, Accurate solution of the volume integral equation for high-permittivity scatterers, *IEEE Trans. Antennas Propagat.* **48**(11):1719-1726.

Krauss, T.F. and De La Rue, R.M., 1999, Photonic crystals in the optical regime - past, present and future, *Progress in Quantum Electronics* **23**:51-96.

Krioukov, E., Klunder, D.J.W., Driessen, A., Greve, J., and Otto, C., 2002, Sensor based on an integrated optical microcavity, *Opt. Lett.* **27**:512–514.

Kucharski, A.A., 2000, Resonances in heterogeneous dielectric bodies with rotational symmetry-volume integral-equation formulation, *IEEE Trans. Microwave Theory Tech.* **48**(5):766-770.

Laine, J.-P., Tapalian, H.C., Little, B.E., and Haus, H.A., 2001, Acceleration sensor based on high-Q optical microsphere resonator and pedestal antiresonant reflecting, *Sensors and Actuators A* **93**:1-7.

Lee, S.Y., Kurdoglyan, M.S., Rim, S., and Kim C.M., 2004, Resonance patterns in a stadium-shaped microcavity, *Phys. Rev. A* **70**(2):id.023809.

Lee, S.-B., Lee J.-H., Chang, J.-S., Moon, H.-J., Kim, S.W., and K. An, 2002, Observation of scarred modes in asymmetrically deformed microcylinder lasers, *Phys. Rev. Lett.*, **88**(3):033903.

Lefevre-Seguin, V., and Haroche, S., 1997, Towards cavity-QED experiments with silica microspheres, *Mater. Sci. Eng. B* **48**:53–58.

Levi, A.F.J., McCall, S.L., Pearton, S.J., and Logan, R.A., 1993, Room temperature operation of submicrometer radius disc laser, *Electron. Lett.* **29**:1666-1667.

Levi, A.F.J., Slusher, R.E., McCall, S.L., Glass, J.L., Pearton, S.J., and Logan, R.A., 1993, Directional light coupling from microdisk lasers, *Appl. Phys. Lett.* **62**:561-563.

Liang, W., Xu, Y., Huang, Y., Yariv, A., Fleming, J. G., and Lin, S.-Y., 2004, Mie scattering analysis of spherical Bragg "onion" resonators, *Opt. Express*, **12**(4):657-669.

Ling, T., Liu, L., Song, Q., Xu, L., and Wang, W., 2003, Intense directional lasing from a deformed square-shaped organic-inorganic hybrid glass microring cavity, *Opt. Lett.* **28**(19):1784-1786.

Little, B.E., Chu, S.T., Haus, H.A., Foresi, J., and Laine, J.-P., 1997, Microring resonator channel dropping filters, *J. Lightwave Technol.* **15**:998–1005.

Little, B.E., Chu, S.T., Pan, W., Ripin, D., Kaneko, T., Kokubun, Y., and Ippen, E., 1999, Vertically coupled glass microring resonator channel dropping filters, *IEEE Photon. Technol. Lett.* **11**(2):215-217.

Little, B.E., Foresi, J.S., Steinmeyer, G., Thoen, E.R., Chu, S.T., Haus, H.A., Ippen, E.P., Kimerling, L.C., Greene, W., 1998, Ultra-compact Si-SiO_2 microring resonator optical channel dropping filters, *Photon. Technol. Lett.* **10**:549-551.

Little, B.E., Haus, H.A., Foresi, J.S., Kimerling, L.C., Ippen, E.P., and Ripin, D.J., 1998, Wavelength switching and routing using absorption and resonance, *IEEE Photon. Technol. Lett.* **10**(6):816-818.

Liu,Y., Safavi-Naeini, S., Chaudhuri, S.K., and Sabry, R., 2004, On the determination of resonant modes of dielectric objects using surface integral equations, *IEEE Trans. Antennas Propagat.* **52**(4): 1062-1069.

Mabuchi, H. and Doherty, A.C., 2002, Cavity quantum electrodynamics: coherence in context, *Science* **298**:1372-1377.

Manolatou, C., Khan, M. J., Fan, S., Villeneuve, P. R., Haus, H. A., and Joannopoulos, J. D., 1999, Coupling of modes analysis of resonant channel add-drop filters, *IEEE J. Quantum Electron.* **35**:1322-1331.

Melloni, A., Costa, R., Monguzzi, P., and Martinelli, M., 2003, Ring-resonator filters in silicon oxynitride technology for dense wavelength-division multiplexing systems, *Opt. Lett.* **28**(17):1567-1569.

Melloni, A., Morichetti, F., and Martinelli, M., 2003, Linear and nonlinear pulse propagation in coupled resonator slow-wave optical structures, *Opt. Quantum Electron.* **35**:365–379.

Mookherjea, S., 2004, Semiconductor coupled-resonator optical waveguide laser, *Appl. Phys. Lett.* **84**(17):3265-3267.

Noda, S., Chutinan, A., and Imada. M., 2000, Trapping and emission of photons by a single defect in a photonic bandgap structure, *Nature* **407:**608–610.

Noeckel, J.U., Stone, A.D., and Chang, R.K., 1994, Q spoiling and directionality in deformed ring cavities, *Opt. Lett.*, **19**:1693-1695.

Noeckel, J.U., 2000, Mode structure and ray dynamics of a parabolic dome microcavity, *Phys. Rev. E* **62**:8677.

Nosich, A.I., 1999, 'The Method of Analytical Regularization in wave-scattering and eigenvalue problems: foundations and review of solutions, *IEEE Antennas Propagat. Mag.* **41**:34-49 (1999).

Olivier, S., Smith, C., Rattier, M., Benisty, H., Weisbuch, C., Krauss, T., Houdre, R., and Oesterl, U., 2001, Miniband transmission in a photonic crystal coupled-resonator optical waveguide, *Opt. Lett.* **26**(13):1019–1021.

Painter O., Srinivasan, K., O'Brien, J.D., Scherer, A., and Dapkus, P.D., 2001, Tailoring of the resonant mode properties of optical nanocavities in two-dimensional photonic crystal slab waveguides, *J. Opt. A: Pure Appl. Opt.* **3**:S161–S170.

Painter, O., Lee, R.K., Scherer, A., Yariv, A., O'Brien, J.D., Dapkus, P.D., and Kim, I., 1999, Two-dimensional photonic band-gap defect mode laser, *Science* **284**:1819-1821.

Painter, O.J., Husain, A., Scherer, A., O'Brien, J.D., Kim, I., and Dapkus, P.D., 1999, Room temperature photonic crystal defect lasers at near-infrared wavelengths in InGaAsP, *J. Lightwave Technol.* **17**(11):2082-2088.

Pan, Y.-L. and Chang, R.K., 2003, Highly efficient prism coupling to whispering gallery modes of a square μ cavity, *Appl. Phys. Lett.* **82**(4):487-489.

Pelton, M., Vuckovic, J., Solomon, G.S., Scherer, A., and Yamamoto, Y., 2002, Three-dimensionally confined modes in micropost microcavities: quality factors and Purcell factors, *IEEE J. Quantum. Electron.* **38**(2):170-177.

Polman, A., Min, B., Kalkman, J., Kippenberg, T.J., Vahala, K.J., 2004, Ultralow-threshold erbium-implanted toroidal microlaser on silicon, *Appl. Phys. Lett.* **84**(7): 1037-1039.

Poon, J.K.S., Scheuer, J., Mookherjea, S., Paloczi, G.T., Huang, Y., and Yariv, A., 2004, Matrix analysis of microring coupled-resonator optical waveguides, *Opt. Express* **12**(1):90-103.

Poon, W., Courvoisier, F., and Chang, R.K., 2001, Multimode resonances in square-shaped optical microcavities, *Opt. Lett.* **26**:632-634.

Poulsen, M.R., Borel, P.I., Fage-Pedersen, J., Hubner, J., Kristensen, M., Povlsen, J.H., Rottwitt, K., Svalgaard, M., and Svendsen, W., 2003, Advances in silica-based integrated optics, *Opt. Eng.* **42**(10):2821–2834.

Purcell, E.M., 1946, Spontaneous emission probabilities at radio frequencies, *Phys. Rev.* **69:**681.

Rabiei, P., and Steier, W.H., 2003, Tunable polymer double micro-ring filters, *IEEE Photon. Technol. Lett.* **15**(9):1968-1975.

Rabiei, P., Steier, W. H., Zhang, C., and Dalton, L. R., 2002, Polymer micro-ring filters and modulators. *J.Lightwave Technol.* **20**:1968–1975.

Rakovich, Y.P. Yang, L., McCabe, E.M., Donegan, J.F., Perova, T., Moore, A., Gaponik, N., and Rogach, A., 2003, Whispering gallery mode emission from a composite system of CdTe nanocrystals and a spherical microcavity, *Semicond. Sci. Technol.* **18**:914-918.

Reithmaier, J.P., Sek, G., Löffler, A., Hofmann, C., Kuhn, S., Reitzenstein, S., Keldysh, L.V., Kulakovskii, V.D., Reinecke, T.L., and Forchel, A., 2004, Strong coupling in a single quantum dot–semiconductor microcavity system, *Nature* **432**:197-200.

Rosenblit, M., Horak, P., Helsby, S., and Folman, R., 2004, Single-atom detection using whispering gallery modes of microdisk resonators, *Phys. Rev. A*, **70**(5):053808.

Rowland, D.R. and Love, J.D., 1993, Evanescent wave coupling of whispering gallery modes of a dielectric cylinder. *IEE Proc.Pt. J* **140**(3):177–188.

Russell, P.S.J., Atkin, D.M., and Birks, T.A., 1996, *Microcavities and Photonic Bandgaps* (Dordrecht: Kluwer) 203–218.

Ryu, H.-Y., Notomi, M., Kim, G.-H. and Lee, Y.-H., 2004, High quality-factor whispering-gallery mode in the photonic crystal hexagonal disk cavity, *Opt. Express* **12**(8):1708-1719.

Sandoghdar V., Treussart, F., Hare, J., Lefèvre-Seguin, V., Raimond, J. -M., and Haroche, S., 1996, Very low threshold whispering-gallery-mode microsphere laser. *Phys. Rev. A* **54**:R1777–R1780.

Santori, C., Fattal, D., Vuckovic, J., Solomon, G.S., and Yamamoto Y., 2004, Single-photon generation with InAs quantum dots, *New J. Physics* **6**:89-105.

Savchenkov, A.A., Ilchenko, V.S., Handley, T., and Maleki, L., 2003, Second-order filter response with series-coupled silica microresonators, *IEEE Photon. Technol. Lett.* **15**(4):543-544.

Savchenkov A.A., Ilchenko V.S., Matsko A.B., and Maleki L., 2004, "KiloHertz optical resonances in dielectric crystal cavities," *Phys. Rev. A*, **70**, 051804.

Savchenkov, A.A., Ilchenko, V.S., Matsko, A.B., and Maleki, L., 2005, High-order tunable filters based on a chain of coupled crystalline whispering gallery-mode resonators, *IEEE Photon. Technol. Lett.* **17**(1):136–138.

Serpenguzel, A., Arnold, S., and Griffel, G., 1995, Excitation of resonances of microspheres on an optical fiber, *Opt. Lett.* **20**:654–656.

Scheuer, J., Green, W.M.J., DeRose, G.A., and Yariv, A., 2005, InGaAsP annular Bragg lasers: theory, applications, and modal properties, *IEEE J. Select. Topics Quantum Electron.*, **11**(2):476-484.

Shima, K., Omori, R., and Suzuki, A., 2001, High- Q concentrated directional emission from egg-shaped asymmetric resonant cavities, *Opt. Lett.* **26**(11):795-797.

Shopova, S.I., Farca, G., Rosenberger, A.T., Wickramanayake, W.M.S., and Kotov, N.A., 2004, Microsphere whispering-gallery-mode laser using HgTe quantum dots, *Appl. Phys. Lett.* **85**(25): 6101-6103.

Smotrova, E.I. and Nosich, A.I., 2004, Mathematical analysis of the lasing eigenvalue problem for the WG modes in a 2-D circular dielectric microcavity, *Opt. Quantum Electron.* **36**(1-3):213-221.

Solomon, G.S., Pelton, M., and Yamamoto, Y., 2001, Single-mode spontaneous emission from a single quantum dot in a three-dimensional microcavity, *Phys. Rev. Lett.* **86**(17):3903-3906.

Spillane, S.M., Kippenberg, T.J., and Vahala, K. J., 2002, Ultralow-threshold Raman laser using a spherical dielectric microcavity, *Nature* **415** (6872): 621-623.

Srinivasan, K., Barclay, P.E., Painter, O., 2004, Fabrication-tolerant high quality factor photonic crystal microcavities, *Opt. Express* **12**(7):1458-1463.

Stratton, J.A., 1997, *Electromagnetic Theory*, Mcgraw Hill, New York.

Tai, C.-Y., Unal, B., Wilkinson, J.S., Ghanem, M.A., and Bartlett, P.N., 2004, Optical coupling between a self-assembled microsphere grating and a rib waveguide, *Appl. Phys. Lett*, **84**(18):3513-3515.

Tishinin, D.V., Dapkus, P.D., Bond, A.E., Kim, I., Lin, C.K., and O'Brien, J., 1999, Vertical resonant couplers with precise coupling efficiency control fabricated by wafer bonding, *IEEE Photon. Technol. Lett.* **11**(8):1003-1005.

Umashankar, K., Taflove, A., and Rao, S.M., 1986, Electromagnetic scattering by arbitrary shaped three-dimensional homogeneous lossy dielectric objects, *IEEE Trans. Antennas. Propagat*. **34**(6):758-766.

Vahala, K.J., 2003, Optical microcavities, *Nature* **424**:839-845.

Vernooy, D. W., Ilchenko, V. S., Mabuchi, H., Steed, E. W., and Kimble, H. J., 1998, High-Q measurements of fused-silica microspheres in the near infrared, *Opt. Lett*. **23**:247–249.

Vernooy, D.W., Furusawa, A., Georgiades, N.P., Ilchenko, V.S., and Kimble, H.J., 1998, Cavity QED with high-Q whispering gallery modes, *Phys. Rev. A* **57**:R2293–R2296.

Vollmer, F., Arnold, S., Braun, D., Teraoka, I., and Libchaber, A., 2003, Multiplexed DNA quantification by spectroscopic shift of 2 microsphere cavities, *Biophys. J.* **85**:1974–1979.

Vuckovic, J., Painter, O., Xu, Y., Yariv, A., and Scherer, A., 1999, Finite-difference time-domain calculation of the spontaneous emission coupling factor in optical microcavities, *IEEE J. Quantum. Electron*. **35**(8):1168- 1175.

Wiersig, J., 2003, Boundary element method for resonances in dielectric microcavities, *J. Opt. A: Pure Appl. Opt*. **5**:53-60.

Xu, Y., Lee, R.K., and Yariv, A., 2000, Finite-difference time-domain analysis of spontaneous emission in a microdisk cavity, *Phys. Rev. A* **61**:033808.

Yamamoto, Y. and Slusher, R.E., 1993, Optical processes in microcavities, *Physics Today*, **46**(6):66-73.

Yanagase, Y., Suzuki, S., Kokubun, Y., and Chu, S. T., 2002, Box-like filter response and expansion of FSR by a vertically triple coupled microring resonator filter, *J. Lightwave Technol.* **20**(8):1525–1529.

Yariv, A., 1973, Coupled-mode theory for guided-wave optics, *IEEE J. Quantum Electron.* **QE–9**(9):919–933.

Yariv, A., Xu, Y., Lee, R.K., and Scherer, A., 1999, Coupled-resonator optical waveguide: a proposal and analysis, *Opt. Lett*. **24**(11):711-713.

Yokoyama, H. and Ujihara, K. (eds), 1995, *Spontaneous emission and laser oscillation in microcavities*, CRC Press, New York.

Yokoyama, H., 1992, Physics and device applications of optical microcavities, *Science* **256**: 66–70.

Yoshie, T., Vuckovic, J., Scherer, A., Chen, H., and Deppe, D., 2001, High quality two-dimensional photonic crystal slab cavities, *Appl. Phys. Lett*. **79**(26):4289-4291.

Zhang, J.P., Chu, D.Y., Wu, S.L., Bi, W.G., Tiberio, R.C., Tu, C.W., and Ho, C.T., 1996, Directional light output from photonic-wire microcavity semiconductor lasers, *IEEE Photon. Technol. Lett*. **8**(8):968-970.

ADVANCED MODELING OF HIGH-CONTRAST PHOTONIC STRUCTURES

Jiří Čtyroký, Ladislav Prkna and Milan Hubálek
Institute of Radio Engineering and Electronics, Academy of Sciences of the Czech Republic, Chaberská 27, 182 51 Praha 8, Czech Republic

Abstract: A choice of methods suitable for modeling high-refractive-index photonic waveguide structures is reviewed. Special attention is paid to methods based on mode matching. Basics of the transfer matrix mode solvers for multilayer structures, the (two-dimensional) bidirectional mode expansion and propagation method, and the film mode matching methods for straight and bent waveguides and circular microresonators with two-dimensional cross-sections are described in some detail.

Key words: eigenmode expansion; microresonators; mode matching; mode solvers; optical waveguide theory, photonic crystals, waveguide bends.

1. INTRODUCTION

With the evolution of guided-wave optics, high refractive-index contrast waveguides play ever more important role in modern photonic devices. "Classical" integrated-optic devices developed mostly for applications in optical communication make use of waveguides with mode field profiles close to those of optical fibers, in order to minimize coupling loss between the integrated-optic chip and the fiber. For example, the change of the refractive index due to titanium diffusion in Ti:$LiNbO_3$ channel waveguides widely used in high-speed electro-optic modulators is typically smaller than 10^{-2}, a value similar to that of standard single-mode step-index fibers. However, it is known that such weakly guiding waveguides are very prone to radiation loss from waveguide bends. To keep the bending loss low, the radii of curvature of waveguide bends have to be of a centimeter scale. This feature makes the high-scale integration of photonic devices difficult. The high-scale integration is the only viable way to cost reduction of integrated

S. Janz et al. (eds.), Frontiers in Planar Lightwave Circuit Technology, 71–100.

photonic components and devices, however, and their high cost is the main obstacle for their mass application in access and metropolitan optical networks. Increased refractive index contrast allows us to dramatically reduce the radius of waveguide bends while keeping the radiation loss low, and thus significantly increase the possible density of photonic integration.

Properties of high-contrast waveguide structures differ from those of low-contrast weakly-guiding waveguides in two main aspects. Mode fields of high-contrast waveguides are usually very strongly localized close to the waveguide 'core', which makes an efficient coupling to standard weakly guiding optical fibers a nontrivial problem. Eigenmodes of high-contrast waveguides have also much more strongly pronounced polarization features. This can be easily understood if we consider the general vectorial equation for the electric field that stems from Maxwell equations:

$$\nabla \times \nabla \times \mathbf{E} = k_0^2 n^2(\mathbf{r})\mathbf{E}, \quad k_0^2 = 2\pi / \lambda\,, \tag{1}$$

where λ is the free-space wavelength of the field, and $n(\mathbf{r})$ is the refractive index distribution. From the Maxwell divergence equation $\nabla \cdot (n^2\mathbf{E}) = 0$ we get $\nabla \cdot \mathbf{E} = -(\nabla \ln n^2) \cdot \mathbf{E}$, so that Eq. (1) can be rewritten in the form

$$\Delta\mathbf{E} + \nabla\left(\nabla \ln n^2 \cdot \mathbf{E}\right) + k_0^2 n^2(\mathbf{r})\mathbf{E} = \mathbf{0}\,. \tag{2}$$

For weakly guiding structures, the second term can be neglected, and we obtain the standard Helmholtz equation in which individual components of the electric field intensity vector $\mathbf{E}$ remain uncoupled. For high contrast waveguides this is clearly not the case. The second term in Eq. (2) in which the transversal electric field components are mutually coupled must be retained.

Before we start to analyze high-contrast structures in more detail, it is worth to clarify the meaning of this term. We will distinguish between *waveguides* alone and more complex *waveguide structures*; waveguides will be considered as structures whose refractive index distribution is independent of *one* spatial coordinate – longitudinal, z for straight waveguides and azimuthal, φ for circularly bent waveguides. In planar waveguides, the refractive index depends on a single transversal coordinate x only. Propagation in waveguides is fully described by the set of their eigenmodes and propagation constants. They are calculated with the help of *mode solvers*. *Waveguide structures* can be considered as structures consisting of a concatenation (or a more general arrangement) of sections of different waveguides. For modeling field distribution in more complex *waveguide structures* excited with a given input field, a number of different versions of *beam propagation methods* are available. For the proper choice of modeling tools it is helpful to distinguish between the refractive-index contrast in the waveguide structures in the *transversal* and *longitudinal* directions. We can then classify the waveguide structures into four groups according to their transversal and longitudinal refractive index contrast. For the calculation of eigenmodes of transversally low-contrast waveguide structures, i.e., weakly guiding structures, scalar or semivectorial mode solvers can be used, while fully vectorial mode solvers are needed for transversally high-contrast waveguides. In structures with small longitudinal refractive index contrast, the back-reflection can be often neglected, and unidirectional tools like standard beam propagation methods are adequate, while for modeling structures with strong and/or rapid variations of the refractive index in the longitudinal direction, bidirectional tools must be used. A typical choice of proper modeling tools for different types of structures is summarized in Table 1.

Table 1. Types of structures and appropriate modeling tools

Type of the waveguide structure	Appropriate modeling tool
low transversal and low longitudinal contrast	scalar (or semivectorial) mode solvers, unidirectional propagation
low transversal contrast, high longitudinal contrast	scalar (or semivectorial) mode solvers, bidirectional propagation
high transversal contrast, low longitudinal contrast	vectorial mode solvers, unidirectional propagation
high transversal contrast, high longitudinal contrast	vectorial mode solvers, bidirectional propagation

This scheme has to be applied deliberately; it is not difficult to find examples that do not fit there, like a waveguide distributed Bragg reflector grating. Its longitudinal refractive index contrast of the grating is typically very low, but a strong total back-reflection can be built up by coherent superposition of back-reflections from its individual grooves. Only bidirectional methods are therefore adequate for its modeling. A good critical overview of mode solvers and beam propagation method can be found in books.[1, 2]. In the following, structures with high transversal contrast corresponding to the last two rows of Table 1 will be preferably considered.

All methods mentioned in Table 1 operate (typically) in the frequency domain; a monochromatic optical wave is usually considered. Two basically different groups of modeling methods are currently used: methods operating in the time domain, and those operating in the spectral domain. The transition between these two domains is generally mediated by the Fourier transform. The time-domain methods became very popular within last years because of their inherent simplicity and generality and due to vast increase in both the processor speed and the memory size of modern computers. The same computer code can be often used to solve many problems with rather different geometries, including the spectral dependence of components and devices and ultrashort pulse propagation even in the presence of optical nonlinearities. The finite-difference time domain method[3] (FDTD) is perhaps the most typical representative of numerical time-domain methods.

In this contribution, we will not consider the time-domain methods, however. We will try to present here a more or less systematic explanation of a modal approach to mode solvers and propagation methods. Modeling software tools are often based on finite difference[4] (FD) or finite-element[5,6] (FE) methods. These methods require discretization in two dimensions for the mode solvers and in three dimensions for the propagation methods. Similarly as in the case of a FDTD method, FD and FE methods are flexible but their requirements on computer resources (memory and speed) are generally high, especially for fully vectorial 3-D methods. The well-known method of lines[7] (MoL) requires discretization in one transversal direction only, but for efficient calculation, its code has to be 'tailored' to the waveguide structure under investigation. We will describe here the principles of methods based on mode matching[8,9] that are known to be very accurate, computationally efficient and have generally low requirements on the computer memory, especially for structures with piecewise constant refractive-index profile. The mode matching method is applicable to lossy and radiating structures whose eigenmodes have complex propagation constants, too, at the expense of longer calculation time. The advantages of the mode matching method for modeling high-contrast waveguides stem mainly from its semi-analytic character. The total vectorial field in the modeled structure is represented by the superposition of eigenmodes of smaller parts (sections, slices) of the structure. These fields can be often expressed analytically, so that they are accurately represented even in very thin layers with strongly different refractive indexes, and the discontinuity of electric field components normal to the interfaces is fully taken into account.

First we briefly present the fundamentals of the well-known transfer matrix method[10] for the calculation of eigenmodes in a 1-D (planar) multilayer waveguide. This method forms the basis for the bidirectional propagation methods as well as for the vectorial mode solvers discussed later. The application of boundary conditions of the type of perfectly matched layers to mode matching methods[11,12] will be briefly discussed, too. Then we explain the basics of the (2-D) bidirectional mode expansion and propagation method[13,14], including modeling of finite periodic structures with the help of Bloch modes. Special attention will then be paid to modeling of (straight and) bent waveguides,[15-19] and ring and disk microresonators with 2-D cross-sections.[20,21]

2. GUIDED WAVES IN A MULTILAYER STRUCTURE

A multilayer structure represents a very general form of a planar (1-D) waveguide. Any graded-index planar waveguide can be approximated by a multilayer waveguide; in this case, the graded-index profile is approximated by a staircase function. It is fair to admit here that the guided modes of a (usually weakly guiding) graded-index waveguide can be more effectively calculated by numerical integration of the pertinent differential equations[22] than by using a staircase approximation. However, high-contrast waveguides that are of interest here are almost exclusively formed by layered structures.

Let us consider a multilayer structure plotted schematically in Fig. 1. It will be taken as a fundamental building block of any more complex guided-wave structure considered here. Let the optical wave propagate along the longitudinal coordinate z, and x is the transversal coordinate.

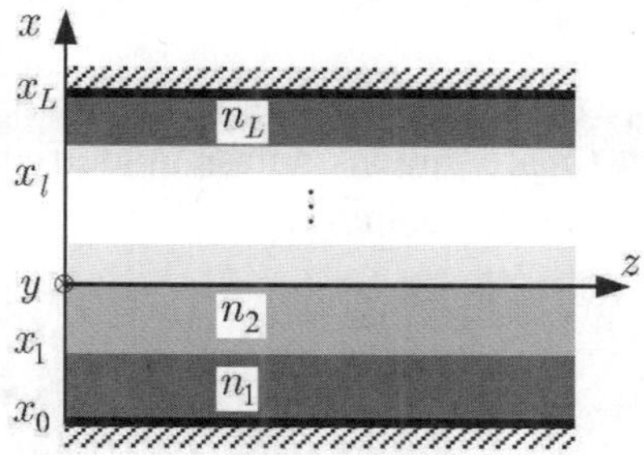

Figure 1. A multilayer structure as an optical waveguide.

We shall consider only multilayers of finite transverse extent. Such structures are formed by L layers of finite thicknesses, with refractive indexes n_l, $l = 1,2,\ldots,L$. The coordinates of the interfaces between adjacent layers are denoted by x_l, $l = 1,2,\ldots,L-1$, and x_0 and x_L are the outer coordinates of the structure. We will suppose for a while that the multilayer is enclosed between a pair of perfect electric or magnetic conductors (electric or magnetic "walls"). More complex boundary conditions will be briefly discussed later.

An efficient formalism for the calculation of eigenmodes of the multilayer is known as the transfer matrix method[10]. We will briefly outline its fundamentals.

For numerical modeling of wave propagation in a structure, it is always advisable to relate the spatial coordinates to the wavelength of the propagating wave. We thus normalize the spatial coordinates as follows:

$$\xi = k_0 x, \quad \xi_l = k_0 x_l, \quad \zeta = k_0 z, \quad k_0 = 2\pi / \lambda, \tag{3}$$

where λ is the free-space wavelength. The time-harmonic dependence of the form $\exp(-i\omega t)$ with $\omega = 2\pi c / \lambda$ is supposed and suppressed, c being the speed of light. Due to the symmetry of the structure, waves of two different polarizations can propagate independently: TE waves with field components E_y, H_x, H_z, and TM waves with components H_y, E_x, E_z. The eigenmode of the structure propagates with its own propagation constant β or its effective refractive index $N = \beta / k_0$. The mode field components can then be normalized using the following relations:

$$\begin{aligned}
&\text{TE polarization} && \text{TM polarization} \\
&E_y = \sqrt{2k_0 Z_0} f(\xi)\exp(iN\zeta), && H_y = \sqrt{2k_0 Y_0} f(\xi)\exp(iN\zeta), \\
&H_x = -\sqrt{2k_0 Y_0} h(\xi)\exp(iN\zeta), && E_x = \sqrt{2k_0 Z_0} h(\xi)\exp(iN\zeta), \\
&H_z = -i\sqrt{2k_0 Y_0} g(\xi)\exp(iN\zeta), && E_z = i\sqrt{2k_0 Z_0} g(\xi)\exp(iN\zeta).
\end{aligned} \tag{4}$$

Here, $Z_0 = Y_0^{-1} = (\mu_0 / \varepsilon_0)^{1/2}$ is the free-space wave impedance, and the functions $f(\xi)$, $g(\xi)$, $h(\xi)$ are dimensionless mode field distributions. From Maxwell equations we easily obtain the following relations among the functions:

$$\frac{df(\xi)}{d\xi} = \varepsilon^{\nu}(\xi) g(\xi), \quad g(\xi) = \frac{1}{\varepsilon^{\nu}(\xi)}\frac{df(\xi)}{d\xi}, \quad h(\xi) = \frac{N}{\varepsilon^{\nu}(\xi)} f(\xi), \tag{5}$$

where $\varepsilon(\xi) = n^2(\xi)$ is the permittivity profile of the structure (a piecewise constant function), and the parameter ν is used to distinguish the polarization: $\nu = 0$ for TE waves, $\nu = 1$ for TM waves.

Within each layer, $\varepsilon(\xi)$ is constant. The pair of first-order differential equations (5) for $f(\xi)$ and $g(\xi)$ can then be solved analytically. We obtain

$$\begin{aligned}
&\begin{pmatrix} f(\xi \pm \Delta\xi) \\ g(\xi \pm \Delta\xi) \end{pmatrix} = \mathbf{M}_l^{\pm}(\xi) \cdot \begin{pmatrix} f(\xi) \\ g(\xi) \end{pmatrix}, \\
&\mathbf{M}_l^{\pm}(\xi) = \begin{pmatrix} \cos \gamma_l \Delta\xi_l & \pm \dfrac{(\varepsilon_l)^{\nu}}{\gamma_l} \sin \gamma_l \Delta\xi_l \\ \mp \dfrac{\gamma_l}{\varepsilon_l^{\nu}} \sin \gamma_l \Delta\xi_l & \cos \gamma_l \Delta\xi_l \end{pmatrix},
\end{aligned} \tag{6}$$

where $\gamma_l = (\varepsilon_l - N^2)^{1/2}$, $\varepsilon_l = n_l^2$. The propagation by the distance $\Delta\xi$ inside the l-th layer can thus be represented by the *transfer matrix* $\mathbf{M}_l^{\pm}(\Delta\xi)$. Both functions $f(\xi)$ and $g(\xi)$ describe *tangential* field components. Because of field continuity conditions at the interfaces, *they are identical at both sides of the interface.* At the outer boundaries, $f(\xi_0) = f(\xi_L) = 0$ for TE polarization at the electric walls and for TM polarization at the magnetic walls, while $g(\xi_0) = g(\xi_L) = 0$ for TE polarization at the magnetic walls and TM polarization at the electric walls.

The eigenmodes of the structure have to satisfy the boundary conditions at both outer boundaries simultaneously. The propagation constant of a mode (or, more practically, its effective refractive index *squared*, N^2) can be found using the following very simple procedure. We start from both outer boundaries where we choose (rather arbitrarily) the values of $f(\xi_0)$ and $f(\xi_L)$, or $g(\xi_0)$ and $g(\xi_L)$, according to the boundary conditions. Then we choose a suitable position $\xi_{l'}$ within the multilayer structure where the field of the mode being calculated is expected to be strong (or at least nonzero). (A good choice might be e.g. one of the interfaces of the layer with the largest refractive index.) The total transfer matrix for the transition from the bottom boundary ξ_0 to this position $\xi_{l'}$ is given by the product $\mathbf{M}^+$ of matrices $\mathbf{M}_l^+$ of each layer with $l \le l'$. Similarly, the matrix representing the transition from the upper boundary ξ_L to $\xi_{l'}$ is given by the product $\mathbf{M}^-$ of matrices $\mathbf{M}_l^-$ of layers with decreasing layer counter, $l = L, L-1, \ldots, l'+1$. The resulting field components $f(\xi_{l'}), g(\xi_{l'})$ at the interface $\xi_{l'}$ must be unique. This condition can be written in the form

$$\mathbf{M}^+ \cdot \begin{pmatrix} f(\xi_0) \\ 0 \end{pmatrix} - \mathbf{M}^- \cdot \begin{pmatrix} f(\xi_L) \\ 0 \end{pmatrix} = \mathbf{0} \tag{7}$$

or a similar one if $g(\xi_0)$ and $g(\xi_L)$ are nonzero. After simple manipulations we obtain from Eq. (7) the dispersion equation in the simple form

$$M_{11}^+ M_{21}^- - M_{21}^+ M_{11}^- = 0 \tag{8}$$

that combines the elements of the matrices $\mathbf{M}^+$ and $\mathbf{M}^-$. Equation (8) has to be numerically solved for the effective refractive index N^2 squared. Having known N^2 and thus N, the mode field distribution can be calculated by the repetitive application of Eq. (6).

It is well known that for lossless media, all squared effective indexes N_m^2 are real, and for any transversally limited structure they form a discrete non-growing sequence. The field distributions $f_m(\xi)$ and $h_m(\xi)$ of the *m*-th mode are mutually orthogonal, have the same phase at each point ξ of the cross-section, and the set of functions corresponding to all modes is complete. The orthogonality relations can be taken in the form

$$\int_{\xi_0}^{\xi_L} f_m(\xi) h_{m'}(\xi) d\xi = \int_{\xi_0}^{\xi_L} f_m(\xi) \frac{N_{m'}}{n^{2\nu}(\xi)} f_{m'}(\xi) d\xi = \delta_{mm'}. \quad (9)$$

The fact that any field in the multilayer structure can be expressed by a superposition of its eigenmodes will be heavily used later as a basic principle of the film mode matching method.

Let us now briefly consider the case of a lossy structure when some of the refractive indexes n_l are complex. The whole algorithm remains unchanged, but the effective refractive indexes of modes N_m as well as the mode field distribution functions $f_m(\xi)$, $h_m(\xi)$ and $g_m(\xi)$ become complex. It means that the roots N_m^2 of the dispersion equation (8) have to be localized in the complex plane. A very simple algorithm of "root tracking" works reasonably well in most cases of planar (1-D) structures. First, the structure is considered as lossless, and all real effective indexes of modes are found. The imaginary parts of n_l^2 are then gradually increased, and the motion of the roots in the complex plane is tracked using a simple Newton (or Newton-Rawson) algorithm that takes into account analytic properties of the dispersion function. Note that the orthogonality condition given by Eq. (9) also remains unchanged, without any complex conjugation sign.

An important problem worth to be briefly discussed here is the problem of boundary conditions. The perfectly conducting walls considered so far behave as perfect reflectors. If an open region has to be simulated, the back-reflected waves must be suppressed. One possibility is to use strongly absorbing layers with complex refractive indexes close to the boundaries. However, a nonzero reflection always takes place at the interface between a lossless and an absorbing layer, due to the difference in the refractive indexes. Thus, a number of layers with gradually increasing absorption has to be used near the boundary, which increases the complexity of the problem (more layers are often required for absorbers than for the structure to be modeled itself). An original and efficient solution was proposed by Bérenger[23]. He proposed to use a layer of a special artificial inhomogeneous medium that absorbs radiation but does not bring about any reflection at the boundary with the neighboring lossless layer. Because of this feature, it is called a perfectly matched layer (PML). From the point of view of classical electrodynamics, it can be considered as medium with nonzero uniaxially anisotropic electric and magnetic conductivities. We will not describe the properties of the PML in details here. Recently it has been shown[24] that from the formal point of view, the PML is fully equivalent to 'stretching' of the transversal spatial coordinate of the isotropic and lossless layer into the complex plane; it means that the thickness of the PML can be considered to be complex[12,25]. Let us take the mode field with the effective index N. In the L-th layer the mode field can be represented by a superposition of two plane waves propagating upwards and downwards, with the spatial dependence of the form $\exp[i(N\zeta \pm \gamma\xi)]$, where $\gamma = \sqrt{n_L^2 - N^2}$. If the upper coordinate of the layer ξ_L is complex, $\xi_L = \xi_L' + i\xi_L''$, the downward propagating wave is attenuated by the factor

$$R_L = \exp(-2\gamma\xi_L'') \qquad (10)$$

with respect to the upward propagating wave. Here, R_L is clearly the resulting amplitude reflectance of the PML. From this very simple consideration it is apparent that the attenuation of any PML is finite but exponentially growths with increasing the imaginary part of the coordinate. It is also the larger the smaller is N^2. It means that the PML attenuation is the smallest for the fundamental mode of the multilayer and rapidly grows with the mode number.

Since PMLs are lossy, the eigenmodes of multilayers with PML are generally complex, so that their effective indexes have to be localized in the complex plane. The great advantage of PML's is that they can be very easily incorporated into the modeling software based on mode matching, and their behavior can be easily understood from their interpretation in terms of complex coordinate stretching.

The transfer matrix method as a mode solver for multilayer structures proved to be efficient and reliable even for structures with extremely large refractive index contrast consisting of the combination of dielectric and metallic layers supporting propagation of surface plasmons[26]. The optical permittivity of such metals is complex, with large negative real part. In combination with high-index dielectric layers they probably represent structures with the largest experimentally attainable refractive index contrast. The possibility to choose the "matching interface" $\xi_{l'}$ properly, i.e., if need be, differently for different modes, makes the method very flexible. Contrary to the application of the method to the longitudinal propagation in the bidirectional mode expansion and propagation method, it is usually very stable. The stability is perhaps the most strongly compromised for the lowest order modes in structures containing thick low-index layers, in which the transversal propagation constants $\gamma_l = \sqrt{n_l^2 - N^2}$ are large imaginary. However, proper formulation of the algorithm can minimize the danger of its failure. Higher-order modes are less prone to instabilities of this kind due to their smaller values of N^2 and thus larger values of γ_l^2.

3. BIDIRECTIONAL MODE EXPANSION AND PROPAGATION

Bidirectional mode expansion and propagation method (BEP) is a rather simple but powerful, general and accurate method for modeling wave propagation in structures in which back-reflections are important[13,14,27,]. For brevity we shall consider here only its two-dimensional version. The waveguide structure to be modeled is schematically depicted in Fig. 2. It consists of a concatenation of longitudinally uniform waveguide sections, each of which is considered as a multilayer waveguide. The lengths of the input and output sections are supposed to be infinite.

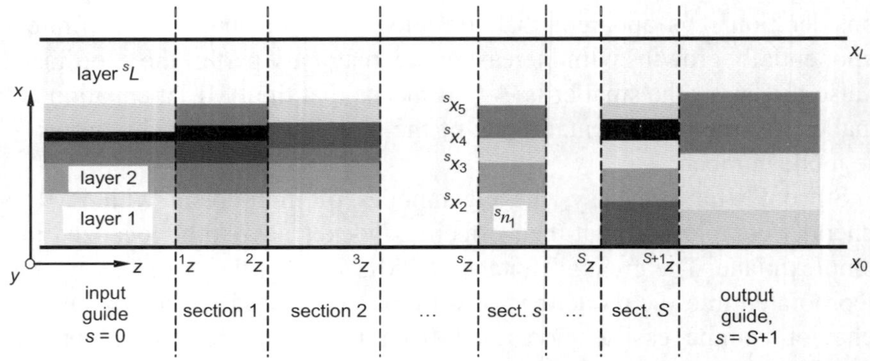

Figure 2. Waveguide structure as a concatenation of several multilayer waveguide sections.

The principle of the BEP method can be easily explained using simple physical arguments. As before, the symmetry of the problem allows us to consider the TE and TM polarizations independently. Suppose that the structure is excited from the left by a forward propagating monochromatic wave in the input section, $s = 0$. Our task is to calculate the field distribution in the whole structure. In each longitudinally uniform section s, $s = 0, 1, \ldots, S+1$, the propagating wave can be decomposed into forward and backward propagating eigenmodes of the section. In the output section, $s = S+1$, only the forward propagating modes can appear. If the same number M of modes is considered in each section, in each 'inward' sections $s = 1, 2, \ldots S$ we have $2M$ unknown amplitudes while in each of the input and output sections there are only M unknown amplitudes (reflected modes in the input section and outgoing modes in the output one). Thus, we have altogether $2(S+1)M$ unknown amplitudes. At each interface between the neighboring sections, two tangential field components (E_y and H_x for the TE polarization and H_y and E_x for the TM polarization) are to be continuous. This gives $2(S+1)$ equations for the field dependences on the transversal coordinate x. Expanding these functions into eigenmodes of the corresponding sections we get altogether $2(S+1)M$ linear equations for unknown mode amplitudes. Having solved this system of linear equations, we have solved the problem.

Different longitudinal sections in the structure may generally have very different refractive index profiles, and thus their eigenmode functions can strongly differ, too. As they have to be well-represented by the superposition of modes of other sections, the number M of considered eigenmodes needs to be sufficiently large (typically from tens to hundreds), and it means that many of these modes are evanescent in the direction of propagation (for their effective indexes, $\mathrm{Re}\{N^2\} < 0$).

The procedure just described, although formally correct, is not suitable for practical implementation. There are several known approaches that do not require to solve the set of a large number of $2(S+1)M$ linear equations simultaneously. The most straightforward approach is based on the transfer matrix method that is formally similar to that explained in the previous section. However, the contributions of both forward and backward propagating modes are combined in the individual transfer matrix elements. For evanescent modes, the contributions from the exponentially decaying modes become soon negligible in comparison with those of the exponentially growing modes, and because of this, the whole procedure becomes numerically unstable. Other approaches are therefore preferred. One of them, known as the impedance method (or more correctly, immittance method) operates with the *ratios* between the field components instead of the components themselves. Even if the components themselves become to be very large or very small, their ratios remain finite and well-behaved. Another approach, the scattering matrix method, will be discussed later in some detail, too.

3.1 Basic considerations

We shall now formulate the problem more rigorously. Let us consider the field in the s-th section. It can be expressed as a superposition of all its modes in the following way:

TE polarization TM polarization

$$\begin{aligned}
E_y &= \sqrt{2k_0Z_0}\sum_{m=0}^{M-1} {}^s f_m(\xi)\,{}^s p(\zeta), & H_y &= \sqrt{2k_0Y_0}\sum_{m=0}^{M-1} {}^s f_m(\xi)\,{}^s p(\zeta),\\
H_x &= -\sqrt{2k_0Y_0}\sum_{m=0}^{M-1} {}^s h_m(\xi)\,{}^s q_m(\zeta), & E_x &= \sqrt{2k_0Z_0}\sum_{m=0}^{M-1} {}^s h_m(\xi)\,{}^s q_m(\zeta),\\
H_z &= -i\sqrt{2k_0Y_0}\sum_{m=0}^{M-1} {}^s g_m(\xi)\,{}^s p(\zeta), & E_z &= i\sqrt{2k_0Z_0}\sum_{m=0}^{M-1} {}^s g_m(\xi)\,{}^s p(\zeta).
\end{aligned} \tag{11}$$

Here, the longitudinal evolution of the modes is described by the functions ${}^s p_m(\zeta)$ and ${}^s q_m(\zeta)$ instead of a simple exponential term as in Eq. (4) since we have to take into account both forward and backward propagation modes. Substituting Eq. (11) into Maxwell equations and taking into account the relations described by Eq. (5) we find that the functions ${}^s p_m(\zeta)$ and ${}^s q_m(\zeta)$ have to satisfy the following very simple set of first-order differential equations:

$$\begin{aligned} \frac{d\,{}^s\mathbf{p}(\zeta)}{d\zeta} &= i\,{}^s\mathbf{N}\cdot{}^s\mathbf{q}(\zeta), \\ \frac{d\,{}^s\mathbf{q}(\zeta)}{d\zeta} &= i\,{}^s\mathbf{N}\cdot{}^s\mathbf{p}(\zeta). \end{aligned} \tag{12}$$

Here, ${}^s\mathbf{N}$ is a diagonal matrix of effective refractive indexes of the modes of the s-th section, and ${}^s\mathbf{p}$ and ${}^s\mathbf{q}$ are the column vectors of mode field amplitudes ${}^s p_m(\zeta)$ and ${}^s q_m(\zeta)$, respectively. From these equations it is easy to find the relations between the mode field amplitudes ${}^s\mathbf{p}(\zeta)$ and ${}^s\mathbf{q}(\zeta)$ at the position ζ and those at the position shifted by $\Delta\zeta$ within the same waveguide section s:

$$\begin{aligned} \begin{pmatrix} {}^s\mathbf{p}(\zeta+\Delta\zeta) \\ {}^s\mathbf{q}(\zeta+\Delta\zeta) \end{pmatrix} &= {}^s\mathbf{A}(\Delta\zeta)\cdot\begin{pmatrix} {}^s\mathbf{p}(\zeta) \\ {}^s\mathbf{q}(\zeta) \end{pmatrix}, \\ {}^s\mathbf{A}(\Delta\zeta) &= \begin{pmatrix} \cos({}^s\mathbf{N}\Delta\zeta) & i\sin({}^s\mathbf{N}\Delta\zeta) \\ i\sin({}^s\mathbf{N}\Delta\zeta) & \cos({}^s\mathbf{N}\Delta\zeta) \end{pmatrix}. \end{aligned} \tag{13}$$

Here, ${}^s\mathbf{A}(\Delta\zeta)$ is the transfer matrix describing the translation by $\Delta\zeta$ within the section s. The mode field amplitudes ${}^s\mathbf{p}(\zeta)$ and ${}^s\mathbf{q}(\zeta)$ can be alternatively expressed in terms of the amplitudes of forward and backward propagating modes ${}^s\mathbf{a}$ and ${}^s\mathbf{b}$,

$$\begin{aligned} {}^s\mathbf{a}(\zeta) &= \exp(i\,{}^s\mathbf{N}\zeta), \quad {}^s\mathbf{p}(\zeta) = {}^s\mathbf{a}(\zeta) + {}^s\mathbf{b}(\zeta), \\ {}^s\mathbf{b}(\zeta) &= \exp(-i\,{}^s\mathbf{N}\zeta) \quad {}^s\mathbf{q}(\zeta) = {}^s\mathbf{a}(\zeta) - {}^s\mathbf{b}(\zeta). \end{aligned} \tag{14}$$

Note that for high-order modes, $\cos({}^s N_m\Delta\zeta) = \cosh(|{}^s N_m|\Delta\zeta)$ and $i\sin({}^s N_m\Delta\zeta) = -\sinh(|{}^s N_m|\Delta\zeta) \approx -\cosh(|{}^s N_m|\Delta\zeta)$, and as a result, the matrix ${}^s\mathbf{A}(\Delta\zeta)$ becomes singular. An alternative approach to solve this problem, the immittance matrix method, will be discussed in the next section, another one, the scattering matrix method, is described in[14].

3.2 Immittance matrix method

In this approach, we introduce the immittance matrix ${}^{s}\mathbf{U}(\zeta)$ by the relation

$$ {}^{s}\mathbf{q}(\zeta) = {}^{s}\mathbf{U}(\zeta) \cdot {}^{s}\mathbf{p}(\zeta). \tag{15} $$

The term 'immittance' is used for the sake of generality; for TE modes, the physical meaning of the matrix ${}^{s}\mathbf{U}$ is namely the transversal admittance (it mutually relates the field components H_x and E_y) while for TM modes it is impedance (the field components E_x and H_y are related). It can be derived from Eq. (13) by algebraic manipulations that upon translation from ζ to $\zeta + \Delta\zeta$, the immittance matrix is transformed as follows:

$$ \begin{aligned} {}^{s}\mathbf{U}(\zeta + \Delta\zeta) &= -i\,{}^{s}\mathbf{t}^{-1} + {}^{s}\mathbf{s}^{-1} \cdot \left[{}^{s}\mathbf{U}(\zeta) - i\,{}^{s}\mathbf{t}^{-1} \right]^{-1} \cdot {}^{s}\mathbf{s}^{-1}, \\ {}^{s}\mathbf{t} &= \tan({}^{s}\mathbf{N}\Delta\zeta), \quad {}^{s}\mathbf{s} = \sin({}^{s}\mathbf{N}\Delta\zeta). \end{aligned} \tag{16} $$

Note that all elements of diagonal matrices ${}^{s}\mathbf{t}^{-1}$ and ${}^{s}\mathbf{s}^{-1}$ remain limited, especially those corresponding to high-order evanescent modes.

As has already been mentioned, at the interface between two neighboring sections, two transversal field components are continuous. Their distributions can be expressed by two expansions corresponding to modes of each section. From Eq. (11) we get for the interface between sections *s* and *t*

$$ \sum_{m=0}^{M-1} {}^{s}f_m(\xi)\,{}^{s}p_m = \sum_{m=0}^{M-1} {}^{t}f_m(\xi)\,{}^{t}p_m, \quad \sum_{m=0}^{M-1} {}^{s}h_m(\xi)\,{}^{s}q_m = \sum_{m=0}^{M-1} {}^{t}h_m(\xi)\,{}^{t}q_m \,. \tag{17} $$

Multiplicating these equations subsequently by functions $f_m(\xi)$ and $h_m(\xi)$ from both sections, integrating over ξ and taking into account the orthogonality conditions of Eq. (9) we finally receive

$$ {}^{t}\mathbf{p} = {}^{\underline{t}s}\mathbf{O} \cdot {}^{s}\mathbf{p}, \quad {}^{t}\mathbf{q} = {}^{t\underline{s}}\mathbf{O} \cdot {}^{s}\mathbf{q}, \tag{18} $$

where ${}^{\underline{t}s}\mathbf{O}$ and ${}^{t\underline{s}}\mathbf{O}$ are matrices of overlap integrals with elements

$$ \begin{aligned} {}^{t\underline{s}}O_{mm'} &= {}^{t}N_{m'} \int_{\xi_0}^{\xi_L} {}^{t}f_m(\xi) \frac{1}{{}^{s}\eta^{2\nu}} {}^{s}f_{m'}(\xi)\, d\xi, \\ {}^{\underline{t}s}O_{mm'} &= {}^{t}N_{m'} \int_{\xi_0}^{\xi_L} {}^{t}f_m(\xi) \frac{1}{{}^{t}\eta^{2\nu}} {}^{s}f_{m'}(\xi)\, d\xi. \end{aligned} \tag{19} $$

From Eq. (18) one easily concludes that the immittance matrix is transformed at the interface in the following way:

$$ {}^{t}\mathbf{U} = {}^{t\underline{s}}\mathbf{O} \cdot {}^{s}\mathbf{U} \cdot {}^{\underline{s}t}\mathbf{O}. \tag{20} $$

The calculation of field distribution in the whole waveguide structure using the immittance matrix starts from the output section. As there are no backward propagating modes in the output section, ${}^{S+1}\mathbf{b} = \mathbf{0}$, it follows from Eq. (14) that ${}^{S+1}\mathbf{p} = {}^{S+1}\mathbf{q}$, from which we get using Eq. (15)

$$ {}^{S+1}\mathbf{U} = \mathbf{I}, \tag{21} $$

where $\mathbf{I}$ is the unity matrix. Having known the immittance matrix at the output section, we can successively calculate its value everywhere in the structure using Eq. (20) for the transformation between the sections and Eq. (16) inside each section. In this way we get the immittance matrix at the input section, $s = 0$. The amplitudes of the incident (forward propagating) modes ${}^{0}\mathbf{a}$ are expected to be known. From Eqs. (14) and (15) one can easily find the amplitudes of the backward propagating modes:

$$ {}^{0}\mathbf{b} = \left(\mathbf{I} - {}^{0}\mathbf{U}\right)^{-1} \cdot \left(\mathbf{I} + {}^{0}\mathbf{U}\right) \cdot {}^{0}\mathbf{a} \tag{22} $$

and thus the amplitude ${}^{0}\mathbf{p} = {}^{0}\mathbf{a} + {}^{0}\mathbf{b}$, too. Then, the field distribution in the whole structure can be calculated using the transformation relations given by Eq. (18) between the sections and the transformation

$$ {}^{s}\mathbf{p}(\zeta + \Delta\zeta) = i\left[{}^{s}\mathbf{U}(\zeta + \Delta\zeta) + i\,{}^{s}\mathbf{t}^{-1}\right]^{-1} \cdot {}^{s}\mathbf{s}^{-1} \cdot {}^{s}\mathbf{p}(\zeta) \tag{23} $$

within the uniform section s. Eq. (23) can be derived from Eqs. (15) and (16) and, in contrast to Eq. (13), it is numerically stable. The value of ${}^{s}\mathbf{q}(\zeta + \Delta\zeta)$ needed for the calculation of the field distribution can be obtained from Eq. (15) since the corresponding value of the immittance matrix ${}^{s}\mathbf{U}(\zeta + \Delta\zeta)$ has already been calculated in the first step.

Although the process just described seems to work well, there are still some details worth mentioning. From Eq. (18) one can easily conclude that

$$ {}^{\underline{t}s}\mathbf{O} = \left({}^{\underline{s}t}\mathbf{O}\right)^{-1}, \quad {}^{t\underline{s}}\mathbf{O} = \left({}^{s\underline{t}}\mathbf{O}\right)^{-1}. \tag{24} $$

But it is true only if the complete system of eigenmodes is considered. In reality, we can use only a finite number $\mathbf{M}$ of eigenmodes in each section, and in this case Eq.(24) is not accurately satisfied. As it has been discussed in detail by Berends[29], if a finite number of modes is considered, there are in fact *four* different possibilities how to transform the mode amplitudes **p** and **q** between the two sections. From these four possibilities, two choices satisfy the reciprocity, the other two not, but even if reciprocity is preserved, it does not mean that that particular algorithm is more accurate than the others. The best way we can go is to apply one of the reciprocal algorithms and use it as consistently as possible.

3.3 Bloch modes in periodic waveguides

Quite often the waveguide structure to be modeled contains one or more parts that exhibit periodicity in the direction of propagation. Typical examples are deeply etched gratings in waveguides[29-31] and waveguides in photonic crystals[14,32]. Such a waveguide structure is schematically plotted in Fig. 3. Besides the input, output and some other sections, it contains also a periodic structure with four periods; each period (incidentally) consists of two waveguide sections.

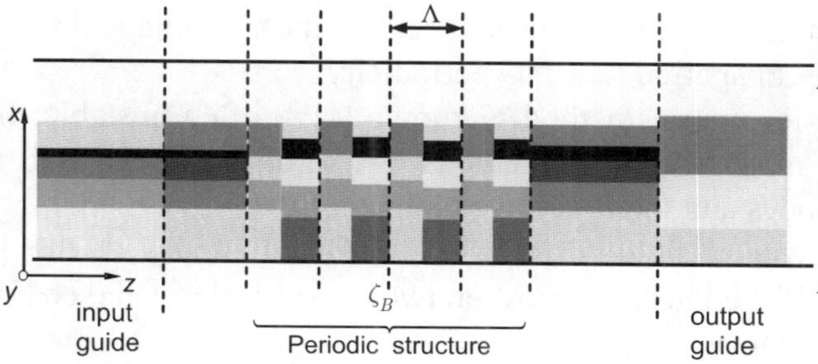

Figure 3. Waveguide with a periodical structure.

Wave propagation in periodic structures can be efficiently modeled using the concept of Bloch (or Floquet-Bloch) modes[33,31]. This approach is also applicable for the calculation of band diagrams of 1-D and 2-D photonic crystals[34,35]. Contrary to classical methods like the plane-wave expansion[36,37], the material dispersion can be fully taken into account without any additional effort. For brevity we present here only the basic principles of the method.

The Bloch mode can be defined as a wave corresponding to the eigenmode of the transfer matrix of one period of the structure. Let **A** is the transfer matrix describing wave transition from the left to the right of one period, calculated by successive applications of Eq. (13) to each section and Eq. (13) to each interface between sections within the period. The Bloch mode then has to satisfy the condition

$$\mathbf{A} \cdot \begin{pmatrix} \mathbf{p}_n \\ \mathbf{q}_n \end{pmatrix} = \exp(i\varphi_n) \begin{pmatrix} \mathbf{p}_n \\ \mathbf{q}_n \end{pmatrix}. \tag{25}$$

The field distribution of the n-th Bloch mode is determined by the field amplitudes $\mathbf{p}_n$ and $\mathbf{q}_n$, its "propagation constant" is $k_n^B = \varphi_B / \Lambda$, where Λ is the period length. It is known that the "propagation constant" k_n^B is defined only in relation to the length of the whole period, Λ; any multiple of the grating wave number $K = 2\pi / \Lambda$ can be added to k_n^B, without any physical effect. The real part of k_n^B is therefore usually selected from the interval $(-K/2, K/2\rangle$ which forms the first Brillouin zone of the 1-D periodicity. In general, even for the lossless structure, k_n^B can be complex, with nonzero imaginary part. It means that the Bloch modes behave like evanescent modes of a standard waveguide, i.e., decay in the direction of propagation. In this case, the frequency of the mode is considered to belong to the forbidden gap (bandgap) of the periodic structure. The dependence of k_n^B on the frequency (or the wavelength) determines the band structure of the 1-D periodicity. The band diagram for many 2-D structures can be calculated using this approach, too, by considering wave propagation along different directions of the 2-D structure and taking the periodicity in the particular direction of propagation as a 1-D periodicity.

The transfer matrix method is known to be often unstable. If this is the case for the matrix **A**, other methods can be alternatively used to calculate the Bloch modes and their "propagation constants"[34,35].

The transversal field distribution of Bloch modes at the longitudinal position ζ_B of the interface between two period is defined (see Fig. 3) can be explicitly written as

$$f_n^B(\xi) = \sum_m P_{mn} f_m(\xi), \quad h_n^B(\xi) = \sum_m Q_{mn} h_m(\xi), \tag{26}$$

where P_{mn} and Q_{mn} are m-th elements of the vectors $\mathbf{p}_n$ and $\mathbf{q}_n$, respectively. The Bloch modes form a new basis that can be used to expand field propagating in a periodic structure:

$$\begin{aligned} \sum_m f_m(\xi)\, p_m(\zeta_B) &= \sum_n f_n^B(\xi)\, p_n^B(\zeta_B), \\ \sum_m h_m(\xi)\, q_m(\zeta_B) &= \sum_n h_n^B(\xi)\, q_n^B(\zeta_B). \end{aligned} \tag{28}$$

From Eqs. (26) it immediately follows that the relations between the "standard" mode amplitudes p_m, $\mathrm{q_m}$ and "Bloch" amplitudes p_n^B, q_n^B are

$$p_m = \sum_n P_{mn} p_n^B, \quad q_m = \sum_n Q_{mn} q_n^B . \tag{29}$$

These amplitudes play in a periodic structure similar roles as "standard" amplitudes in a "standard" waveguide structure. For example, the transformation of the amplitudes $p_n^B(\zeta_B)$, $q_n^B(\zeta_B)$ by propagation through l periods of the structure takes the very simple form

$$\begin{pmatrix} p_n^B(\zeta_B + lK) \\ q_n^B(\zeta_B + lK) \end{pmatrix} = \begin{pmatrix} \cos(l\varphi_n) & i\sin(l\varphi_n) \\ i\sin(l\varphi_n) & \cos(l\varphi_n) \end{pmatrix} \cdot \begin{pmatrix} p_n^B(\zeta_B) \\ q_n^B(\zeta_B) \end{pmatrix}. \tag{30}$$

The approach based on Bloch modes can be easily cast into the stable immittance formalism. We can introduce the "Bloch immittance matrix" similarly as in Eq. (15)

$$\mathbf{q}^B(\zeta_B) = \mathbf{U}^B(\zeta_B) \cdot \mathbf{p}^B(\zeta_B); \tag{31}$$

one can also easily find the transformation from standard to Bloch immittance matrix:

$$\mathbf{U}^B = \mathbf{Q}^{-1} \cdot \mathbf{U} \cdot \mathbf{P}, \tag{32},$$

where $\mathbf{P}$ and $\mathbf{Q}$ are matrices with elements P_{mn} and Q_{mn}, respectively. The immittance matrix is transformed by l periods of the structure in the way analogous to Eq. (16),

$$\begin{gathered} \mathbf{U}^B(\zeta_B + lK) = -i\mathbf{t}^{-1} + \mathbf{s}^{-1} \cdot \left[\mathbf{U}^B(\zeta_B) - i\mathbf{t}^{-1}\right]^{-1} \cdot \mathbf{s}^{-1}, \\ \mathbf{t} = \tan(l\boldsymbol{\varphi}), \quad \mathbf{s} = \sin(l\boldsymbol{\varphi}), \end{gathered} \tag{33}$$

where $\boldsymbol{\varphi}$ is a diagonal matrix with elements φ_n. Thus, the Bloch mode formalism can be easily incorporated into the BEP method for the efficient analysis of periodic waveguide structures with any number of periods. Applications of such an approach can be found in[30,31,34,35].

4. VECTORIAL MODE SOLVERS

In this section we will discuss the mode matching method applied to vectorial modeling of straight and bent channel waveguides with 2-D cross-sections. For straight waveguides, this method is known for more than a decade as the film mode matching method (FMM)[8,9]. Although it is numerically very efficient and accurate, it was generalized for bent waveguides only very recently[20,21,38], mostly because of numerical difficulties with cylindrical functions. In the next two parts we will briefly discuss FMM for both straight and bent waveguides in the same way.

4.1 Straight waveguides

Let us consider a straight channel waveguide with the cross-section schematically shown in Fig. 4. The cross-section can be subdivided by lines parallel with the (vertical) x axis into "slices" $s = 1, 2, \ldots, S$ with interfaces positioned at the lateral coordinates $y_1, y_2, \ldots, y_{S-1}$. Each slice is formed by a vertical stack of layers numbered by l, $l = 1, 2, \ldots, L$. In the vertical direction, the waveguide is bound by perfectly conducting electric or magnetic "walls" positioned at $x = x_0$ and $x = x_L$. If need be, they can be substituted by PMLs as described before.

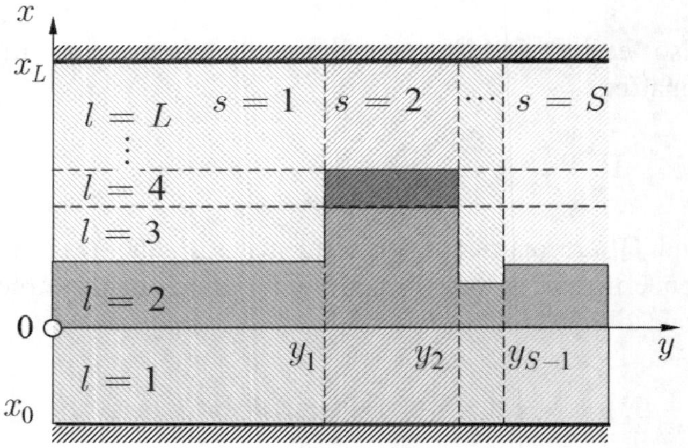

Figure 4. Cross-section of a straight channel waveguide.

The physical principle of the FMM method is quite similar to the BEP method and can be described as follows: the field in each slice can be represented by the superposition of both sets of its TE and TM eigenmodes. Let us suppose that in each slice we take into account M modes of each polarization. Because of field continuity conditions at the interfaces between the slices, the modes within the slices propagate under different angles with respect to the axis of propagation z; these angles are adjusted so that their propagation constants along z direction are all identical. The modes reflect from the interfaces between slices, and as a result, in each of the inner slices, two waves of each mode can propagate, which differ in the sign of their propagation constants in the lateral direction y. The two outermost slices can be left open in the lateral direction, i.e., they are considered as semiinfinite. As a result, each of their modes can propagate as a single (outgoing) wave only. We thus have altogether $4M(S-1)$ unknown mode field amplitudes. From the field continuity conditions at the interfaces we get $4(S-1)$ equations. Expanding them into the corresponding M eigenmodes of the slices we can arrive to the set of $4M(S-1)$ homogeneous linear equations. Zeros of its determinant correspond to the propagation constants of the eigenmodes of the channel waveguide as a whole, and the corresponding field amplitudes determine the mode field distribution.

From this brief description it is apparent that 'slice' modes of both TE and TM polarizations are mutually coupled by the continuity conditions at the interfaces between slices. As a result, the mode fields of a channel waveguide have all components of the electric and magnetic field nonzero. In this way, the vectorial properties of mode fields are taken into account.

The procedure just described is physically easily understandable but not suitable for numerical implementation. We now briefly describe its numerically efficient version based on the immittance matrix approach.

We start again with the normalization of coordinates and fields. Because of two-dimensional cross-section of the waveguide, it is now suitable to introduce the normalized quantities in the following way:

$$\begin{aligned} &x = k_0\xi,\ y = k_0\eta,\ z = k_0\zeta,\quad \bar{\bar{\nabla}} = \frac{1}{k_0}\nabla, \\ &\mathbf{E}(x,y,z) = \sqrt{2Z_0}k_0\mathbf{e}(\xi,\eta,\zeta),\ \mathbf{H}(x,y,z) = \sqrt{2Y_0}k_0\mathbf{h}(\xi,\eta,\zeta). \end{aligned} \tag{34}$$

It can be easily found that Maxwell equations get the simple form

$$\bar{\bar{\nabla}} \times \mathbf{e} = i\,\mathbf{h},\ \bar{\bar{\nabla}} \times \mathbf{h} = -i n^2 \mathbf{e}. \tag{35}$$

It is known[39,19] that the complete electromagnetic field in each slice s can be expressed in terms of Hertz electric and magnetic vectors ${}^s\boldsymbol{\pi}^h$ and ${}^s\boldsymbol{\pi}^e$ that are both parallel with the x coordinate axis. (It is a consequence of the invariance of the refractive index distribution of the slice with respect of the coordinates y and z). The field components in the slice s are given by the expressions

$$\begin{aligned} &{}^s\mathbf{e}^h = i\bar{\bar{\nabla}} \times {}^s\boldsymbol{\pi}^h, && {}^s\mathbf{h}^h = \bar{\bar{\nabla}} \times \bar{\bar{\nabla}} \times {}^s\boldsymbol{\pi}^h, \\ &{}^s\mathbf{e}^e = \overset{\dagger}{\varepsilon}\,\bar{\bar{\nabla}} \times \bar{\bar{\nabla}} \times {}^s\boldsymbol{\pi}^e, && {}^s\mathbf{h}^e = -i\bar{\bar{\nabla}} \times {}^s\boldsymbol{\pi}^e. \end{aligned} \tag{36}$$

It can be shown that Maxwell Eqs. (35) are satisfied if the Hertz vectors have the form

$$\begin{aligned} {}^{s}\boldsymbol{\pi}^{h} &= \mathbf{x}^{0}\sum_{m} {}^{s}f_{m}^{h}(\xi)\, {}^{s}p_{m}^{h}(\eta)\exp(iN_{z}\zeta), \\ {}^{s}\boldsymbol{\pi}^{e} &= \mathbf{x}^{0}\sum_{m} {}^{s}f_{m}^{e}(\xi)\, {}^{s}p_{m}^{e}(\eta)\exp(iN_{z}\zeta), \end{aligned} \tag{37}$$

where $\mathbf{x}^0$ is the unit vector in the direction of the x axis, N_z is the effective refractive index (normalized propagation constant), the functions ${}^{s}f_{m}^{h}(\xi)$ and ${}^{s}f_{m}^{e}(\xi)$ are the field distribution functions of TE and TM modes of the slice s, respectively, and ${}^{s}p_{m}^{h}(\eta)$ and ${}^{s}p_{m}^{e}(\eta)$ are the local amplitudes of the TE and TM modes, respectively. They satisfy the very simple equations

$$\frac{d^2\, {}^{s}p_{m}^{\sigma}}{d\eta^2} + \left({}^{s}N_{ym}^{\sigma}\right)^2 {}^{s}p_{m}^{\sigma} = 0, \quad \left({}^{s}N_{ym}^{\sigma}\right)^2 + N_z^2 = \left({}^{s}N_{xm}^{\sigma}\right)^2, \; \sigma = h,e, \tag{38}$$

where ${}^{s}N_{xm}^{\sigma}$ are effective refractive indexes of the "slice" modes with the field distributions ${}^{s}f_{m}^{\sigma}(\xi)$.

Equations (37) are very similar to Eqs. (11). It is thus not surprising that the evolution of the amplitudes ${}^{s}p_{m}^{\sigma}(\eta)$ can be described in the form of the "lateral transfer matrix", similarly as in Eq. (13):

$$\begin{aligned} &\begin{pmatrix} {}^{s}p_{m}^{\sigma}(\eta+\Delta\eta) \\ {}^{s}q_{m}^{\sigma}(\eta+\Delta\eta) \end{pmatrix} = \\ &= \begin{pmatrix} \cos\left({}^{s}N_{ym}^{\sigma}\Delta\eta\right) & \dfrac{1}{{}^{s}N_{y,m}^{\sigma}}\sin\left({}^{s}N_{ym}^{\sigma}\Delta\eta\right) \\ -{}^{s}N_{y,m}^{\sigma}\sin\left({}^{s}N_{ym}^{\sigma}\Delta\eta\right) & \cos\left({}^{s}N_{ym}^{\sigma}\Delta\eta\right) \end{pmatrix} \cdot \begin{pmatrix} {}^{s}p_{m}^{\sigma}(\eta) \\ {}^{s}q_{m}^{\sigma}(\eta) \end{pmatrix}, \\ &{}^{s}q_{m}^{\sigma}(\eta) = \frac{d\, {}^{s}p_{m}^{\sigma}(\eta)}{d\eta}. \end{aligned} \tag{39}$$

From the field continuity conditions at the interface between the slices s and t we can derive the relations analogous to Eqs. (18) and (19),

$$\begin{aligned} {}^{s}\mathbf{p} &= \left({}^{s}\mathbf{N}_{x}^{2}\right)^{-1} \cdot {}^{st}\mathbf{O} \cdot {}^{t}\mathbf{N}_{x}^{2} \cdot {}^{t}\mathbf{p}, \\ {}^{s}\mathbf{q} &= \left({}^{ts}\mathbf{O}\right)^{T} \cdot {}^{t}\mathbf{q} - {}^{st}\mathbf{X} \cdot {}^{t}\mathbf{p}, \end{aligned} \tag{40}$$

where ${}^{s}\mathbf{p}$ is a column vector

$$ {}^{s}\mathbf{p} = \left({}^{s}p_0^h, {}^{s}p_1^h, \cdots, {}^{s}p_{M-1}^h, {}^{s}p_0^e, {}^{s}p_1^e \cdots, {}^{s}p_{M-1}^e\right)^T, \tag{41} $$

$$ {}^{s}\mathbf{N}_x = diag({}^{s}N_{x,0}^h, {}^{s}N_{x,1}^h, \ldots, {}^{s}N_{x,M-1}^h, {}^{s}N_{x,0}^e, {}^{s}N_{x,1}^e, \ldots, {}^{s}N_{x,M-1}^e), \tag{42} $$

analogous relations hold for ${}^{t}\mathbf{p}$ and ${}^{t}\mathbf{N}_x$, and ${}^{st}\mathbf{O}$ and ${}^{st}\mathbf{X}$ are overlap matrices

$$ {}^{s,t}\mathbf{O} = \begin{pmatrix} {}^{s,t}\mathbf{O}^{hh} & \mathbf{0} \\ \mathbf{0} & {}^{s,t}\mathbf{O}^{ee} \end{pmatrix}, \quad {}^{s,t}\mathbf{X} = \begin{pmatrix} \mathbf{0} & N_z \, {}^{s,t}\mathbf{O}^{he} \\ -N_z \left({}^{t,s}\mathbf{O}^{he}\right)^T & \mathbf{0} \end{pmatrix} \tag{43} $$

with the elements[8,9]

$$ \begin{aligned} {}^{st}O_{mn}^{hh} &= \int_{\xi_0}^{\xi_L} {}^{s}f_m^h \, {}^{t}f_n^h d\xi, \quad {}^{st}O_{mn}^{ee} = \int_{\xi_0}^{\xi_L} \frac{{}^{s}f_m^e \, {}^{t}f_n^e}{{}^{t}\varepsilon} d\xi, \\ {}^{st}O_{mn}^{he} &= \int_{\xi_0}^{\xi_L} \frac{{}^{s}f_m^h}{{}^{t}\varepsilon} \frac{d\,{}^{t}f_n^e}{d\xi} d\xi + \frac{{}^{t}N_{xn}^{e2}}{{}^{s}N_{xm}^{h2}} \int_{\xi_0}^{\xi_L} \frac{d\,{}^{s}f_m^h}{d\xi} \frac{{}^{t}f_n^e}{{}^{t}\varepsilon} d\xi. \end{aligned} \tag{44} $$

Note that the TE and TM modes are now mutually coupled at the interface. The coupling is represented by the matrix ${}^{st}\mathbf{X}$.

For increased numerical stability, we introduce the immittance matrix ${}^{s}\mathbf{U}(\eta)$ by the usual relation

$$ {}^{s}\mathbf{q}(\eta) = {}^{s}\mathbf{U}(\eta) \cdot {}^{s}\mathbf{p}(\eta). \tag{45} $$

The transformation of the immittance matrix within the single slice is given by the relation analogous to Eq. (16),

$$ \begin{aligned} {}^{s}\mathbf{U}(\eta + \Delta\eta) &= {}^{s}\mathbf{t}^{-1} - {}^{s}\mathbf{s}^{-1} \cdot \left[\mathbf{U}(\eta) + \mathbf{t}^{-1}(\eta)\right]^{-1} \cdot {}^{s}\mathbf{s}^{-1}, \\ {}^{s}\mathbf{s} &= {}^{s}\mathbf{N}_y \sin^{-1}\left({}^{s}\mathbf{N}_y \Delta\eta\right), \quad {}^{s}\mathbf{t} = {}^{s}\mathbf{N}_y \tan\left({}^{s}\mathbf{N}_y \Delta\eta\right), \end{aligned} \tag{46} $$

where ${}^{s}\mathbf{N}_y$ is a diagonal matrix with the elements

$$ {}^{s}N_{y,0}^h, {}^{s}N_{y,1}^h, \ldots, {}^{s}N_{y,M-1}^h, {}^{s}N_{y,0}^e, {}^{s}N_{y,1}^e, \ldots, {}^{s}N_{y,M-1}^e, \tag{47} $$

The transformation between the slices s and t reads now

$$^{t}\mathbf{U} = \left(^{st}\mathbf{O}\right)^{T} \cdot \left[^{s}\mathbf{U} \cdot \left(^{s}\mathbf{N}_{x}^{2}\right)^{-1} \cdot {}^{st}\mathbf{O} \cdot {}^{t}\mathbf{N}_{x}^{2} + {}^{st}\mathbf{X}\right]. \tag{48}$$

Since the calculation window is left open in the y direction, all modes in the outermost slices $s=1$ and $s=S$ must be outgoing. A more detailed inspection shows that in this case, the immittance matrices in these slices are diagonal and constant:

$$^{1}\mathbf{U} = -i\,^{1}\mathbf{N}_{y}, \quad ^{S}\mathbf{U}_{y} = i\,^{S}\mathbf{N}_{y}\,. \tag{49}$$

To find the eigenmode, we choose a suitable lateral position η' where the mode field is expected to be nonzero and calculate the values $\mathbf{U}^{+}$ and $\mathbf{U}^{-}$ of the immittance matrix at that position starting from both sides as indicated in Fig. 5. Since the tangential field components at that position must be continuous, we easily arrive to the set of linear equations

$$\left(\mathbf{U}^{+} - \mathbf{U}^{-}\right) \cdot \mathbf{p}\left(\eta'\right) = \mathbf{0}\,. \tag{50}$$

Zeros of its determinant determine the effective refractive indexes N_z of the eigenmodes, and from the corresponding nontrivial solutions $\mathbf{p}\left(\eta'\right)$ we can calculate the complete vectorial field distribution everywhere in the cross-section using Eqs. (36) – (40) or their more stable equivalents.

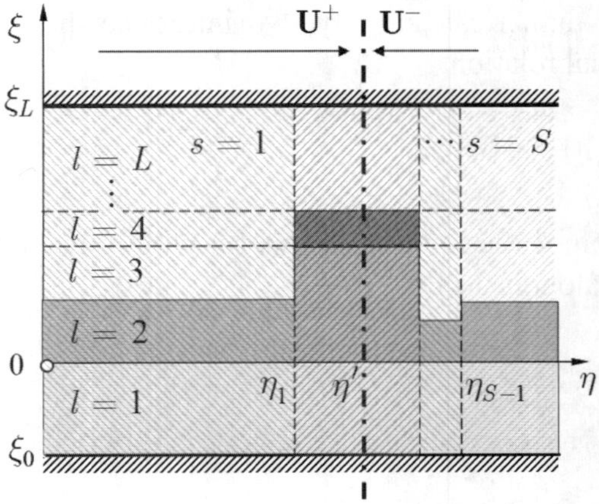

Figure 5. Searching of the eigenmode of a channel waveguide.

It is interesting to note that if the matrix $^{st}\mathbf{X}$ in Eqs. (40) and (48) is neglected, we obtain two independent sets of equations in which TE and TM modes of the slices do not mutually couple. They represent semivectorial approximations to the fully vectorial problem.

An example of the vectorial field distribution of the quasi-TE_{00} mode of a rib waveguide with a comparatively large refractive index profile is shown in Fig. 6.

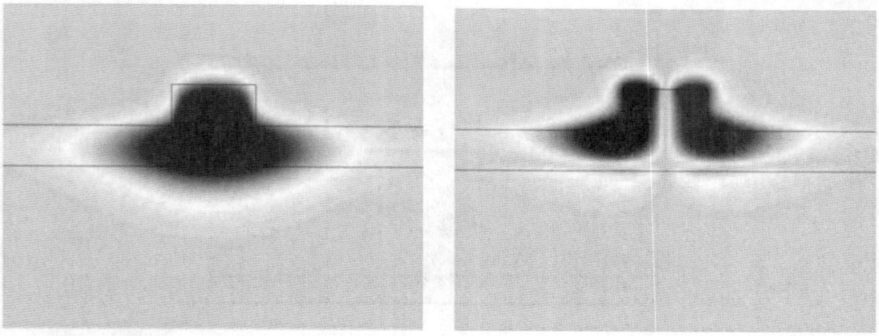

Figure 6. Vertical (left) and horizontal (right) magnetic field components of the quasi-TE_{00} mode of the rib waveguide. Refractive indexes of the substrate, the guiding layer and the superstrate are 1.9, 2.2 and 1.0, respectively, the rib width and height are 1 μm and 0.5 μm, respectively, the wavelength is 1.3 μm.

Fig. 7 presents another example of a very high-contrast waveguide – a rectangular buried waveguide covered by a gold layer. Strong hybridization of the waveguide mode with the antisymmetric surface plasmon is clearly manifested.

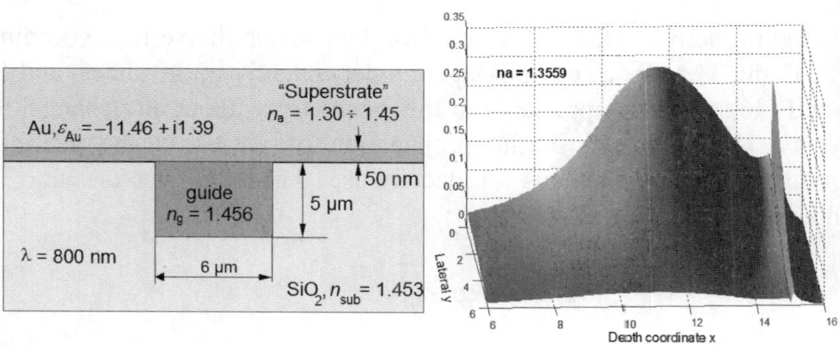

Figure 7. Left: cross-section of a buried waveguide covered by a gold layer. Right: Field distribution of a quasi-TM-polarized mode. (Courtesy of J. Petráček, TU Brno.)

4.2 Circularly bent waveguides and microresonators

The FMM mode solver for circularly bent waveguides can be constructed in a very similar way as for straight waveguides. The cross-section of a bent waveguide is schematically depicted in Fig. 8. We will suppose that the cross-section of an azimuthally invariant waveguide can be subdivided into radially (and azimuthally) uniform slices, each of which consists of a multilayer structure (see Fig. 8). The waveguide is considered to be bound in the axial (vertical) direction by perfect walls or PMLs.

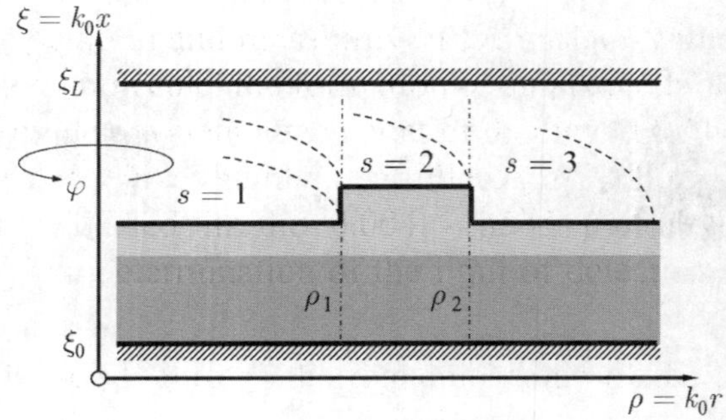

Figure 8. Cross-section of a circularly bent channel waveguide

The cylindrical coordinates and the field vectors can be normalized in the same way as for the straight waveguide:

$$\begin{aligned} &\rho = k_0 r,\ \xi = k_0 x,\ \varphi = \varphi,\quad \bar{\nabla} = \frac{1}{k_0}\nabla, \\ &\mathbf{E}(r,\varphi,x) = \sqrt{2Z_0}k_0\mathbf{e}(r,\varphi,\xi),\ \mathbf{H}(r,\varphi,x) = \sqrt{2Y_0}k_0\mathbf{h}(r,\varphi,\xi). \end{aligned} \tag{51}$$

As the refractive index within the slice depends on the vertical coordinate only, TE and TM modes can propagate independently in the slices, and Eqs (35) and (36) are fully applicable to this geometry without any change, too. Since we are interested in the propagation of waves in the azimuthal direction, it is advantageous to decompose the Hertz vectors into the eigenmodes of the slice as follows:

$$\begin{aligned} {}^s\boldsymbol{\pi}^h &= \mathbf{x}^0 \sum_m {}^s f_m^h(\xi)\, {}^s p_m^h(\rho) \exp(i\nu\varphi), \\ {}^s\boldsymbol{\pi}^e &= \mathbf{x}^0 \sum_m {}^s f_m^e(\xi)\, {}^s p_m^e(\rho) \exp(i\nu\varphi), \end{aligned} \tag{52}$$

where ν is the azimuthal propagation constant of the mode being searched. The mode amplitudes ${}^s p_m^h(\rho)$ and ${}^s p_m^e(\rho)$ are now required to satisfy the Bessel equation

$$\frac{d^2\, {}^s p_m^\sigma}{d\rho^2} + \frac{1}{\rho}\frac{d\, {}^s p_m^\sigma}{d\rho} + \left[\left({}^s N_{xm}^\sigma\right)^2 - \frac{\nu^2}{\rho^2}\right] {}^s p_m^\sigma = 0,\quad \sigma = h, e\,. \tag{53}$$

A general solution of this equation can be written in the form

$$ {}^{s}p_{m}^{\sigma}(\rho) = {}^{s}B_{m}^{\sigma} J_{\nu}({}^{s}N_{xm}^{\sigma}\rho) + {}^{s}C_{m}^{\sigma}\, {}^{s}H_{\nu}^{(1)}({}^{s}N_{xm}^{\sigma}\rho), \tag{54} $$

where J_ν and $H_\nu^{(1)}$ are the Bessel function and the Hankel function of the first kind, respectively. Note that the propagation constant plays the role of the order ν of the cylindrical functions. In the innermost slice $s = 1$ that contains the center of rotation, only the J_ν function is a physically acceptable solution. Similarly, in the outermost slice $s = S$ that is unlimited in the radial direction, only $H_\nu^{(1)}$ can be retained since it describes the radially outgoing wave.

Equation (54) can suffer from numerical instabilities. Therefore we adopt the immittance matrix approach. First we define the column vector of mode amplitudes $\mathbf{p}(\rho)$ in the same way as in Eq. (41). Then the immittance matrix is introduced by the relation

$$ \frac{d\mathbf{p}(\rho)}{d\rho} = \mathbf{U}(\rho)\cdot\mathbf{p}(\rho). \tag{55} $$

From this definition it follows that at the outer interface ρ_1 of the innermost slice $s = 1$, and at the inner interface ρ_{S-1} of the outermost slice $s = S$, the immittance matrix takes the values

$$ \begin{aligned} {}^{1}\mathbf{U}(\rho_1) &= {}^{1}\mathbf{N}_x \cdot J_\nu{}'({}^{1}\mathbf{N}_x\rho_1)\cdot\left[J_\nu({}^{1}\mathbf{N}_x\rho_1)\right]^{-1}, \\ {}^{S}\mathbf{U}(\rho_{S-1}) &= {}^{S}\mathbf{N}_x \cdot H_\nu^{(1)\prime}({}^{S}\mathbf{N}_x\rho_{S-1})\cdot\left[H_\nu^{(1)}({}^{S}\mathbf{N}_x\rho_{S-1})\right]^{-1}, \end{aligned} \tag{56} $$

respectively. From Eqs. (54) and (55) one can derive the transformation of ${}^{s}\mathbf{U}(\rho)$ from ρ_a to ρ_b within any slice. After lengthy but straightforward manipulations one obtains the relation

$$ \mathbf{U}(\rho_b) = \mathbf{G} - \mathbf{H}\cdot(\mathbf{U}(\rho_a) - \mathbf{F})^{-1}\cdot\mathbf{E}, \tag{57} $$

where the matrices $\mathbf{E}, \mathbf{F}, \mathbf{G}, \mathbf{H}$ can be found to be

$$\begin{aligned}
\mathbf{E} &= \mathbf{N}_x \cdot \left[H_\nu^{(1)}(\mathbf{N}_x\rho_b) \cdot J_\nu'(\mathbf{N}_x\rho_b) - H_\nu^{(1)\prime}(\mathbf{N}_x\rho_b) \cdot J_\nu(\mathbf{N}_x\rho_b) \right] \cdot \mathbf{W}^{-1}, \\
\mathbf{F} &= \mathbf{N}_x \cdot \left[H_\nu^{(1)\prime}(\mathbf{N}_x\rho_a) \cdot J_\nu(\mathbf{N}_x\rho_b) - H_\nu^{(1)}(\mathbf{N}_x\rho_a) \cdot J_\nu'(\mathbf{N}_x\rho_b) \right] \cdot \mathbf{W}^{-1}, \\
\mathbf{G} &= \mathbf{N}_x \cdot \left[H_\nu^{(1)}(\mathbf{N}_x\rho_b) \cdot J_\nu'(\mathbf{N}_x\rho_a) - H_\nu^{(1)\prime}(\mathbf{N}_x\rho_a) \cdot J_\nu(\mathbf{N}_x\rho_a) \right] \cdot \mathbf{W}^{-1}, \\
\mathbf{H} &= \mathbf{N}_x \cdot \left[H_\nu^{(1)}(\mathbf{N}_x\rho_a) \cdot J_\nu'(\mathbf{N}_x\rho_a) - H_\nu^{(1)\prime}(\mathbf{N}_x\rho_a) \cdot J_\nu(\mathbf{N}_x\rho_s) \right] \cdot \mathbf{W}^{-1},
\end{aligned} \tag{58}$$

with

$$\mathbf{W} = H_\nu^{(1)}(\mathbf{N}_x\rho_b) \cdot J_\nu(\mathbf{N}_x\rho_a) - H_\nu^{(1)}(\mathbf{N}_x\rho_a) \cdot J_\nu(\mathbf{N}_x\rho_b). \tag{59}$$

Note that for better readability, we omitted the left superscript s that should appear at all matrices in Eqs. (57) to (59).

For the transformation of the immittance matrix between the slices s and t, Eq. (48) is valid without any formal change. Instead of Eq. (43) the matrices of overlap integrals are now given by

$${}^{st}\mathbf{O} = \begin{pmatrix} {}^{st}\mathbf{O}^{hh} & \mathbf{0} \\ \mathbf{0} & {}^{st}\mathbf{O}^{ee} \end{pmatrix}, \quad {}^{st}\mathbf{X} = \begin{pmatrix} \mathbf{0} & \dfrac{\nu}{\rho_s}\,{}^{st}\mathbf{O}^{he} \\ -\dfrac{\nu}{\rho_s}\left({}^{ts}\mathbf{O}^{he}\right)^T & \mathbf{0} \end{pmatrix}. \tag{60}$$

The elements of the matrices are given by Eq. (44) without any change, too.

The remaining part of the mode seeking procedure can formally proceed exactly in the same as for the straight waveguide: starting from both innermost and outermost slices, two values of the immittance matrix in some suitably chosen radial position ρ' are found and are then used to construct the set of homogeneous linear equations for the mode field amplitudes:

$$\left[\mathbf{U}^+(\rho') - \mathbf{U}^-(\rho')\right] \cdot \mathbf{p}(\rho') = \mathbf{0}. \tag{61}$$

Similarly as before, nontrivial solutions to this equation correspond to eigenmodes of the bent waveguide.

This formally simple procedure is very difficult to perform, however. Because of radiation from the bend, the azimuthal propagation constant ν to be found is complex. Since the bend radius of the waveguide is typically larger than the wavelength, the real part of ν can be large, too. Moreover, a number of modes of each slice with very different values of their effective indexes are to be considered simultaneously. It causes very serious numerical problems. Equations (58) and (59) cannot be evaluated on the basis of individual cylindrical functions since it would often lead to products of the type $0 \cdot \infty$. Instead, the cross-products of cylindrical functions in Eqs. (58) have to be calculated using recurrence relations[40,41]. For reliable calculation of cylindrical functions of generally complex argument and order, special procedures based on uniform asymptotic expansions[40] were developed[21,38].

Note that this procedure is suitable also for the calculation of eigenmodes in circular ring and disk microresonators. The cross-sections of disk microresonators are usually simpler, without any mid-slices, which leads to shorter calculation times. However, good estimate of initial values of the propagation constants of the disk modes for their search in the complex plane is more difficult, and their mode spectrum is denser than for ring microresonators with similar radii of curvature.

In Fig. 9, the mode field distributions of the fundamental quasi-TM_{00} modes of ring and disk microresonators are mutually compared. Note that the radiation is directed mainly into the substrate, and it is stronger from the ring since the field in the disk is localized closer to the center.

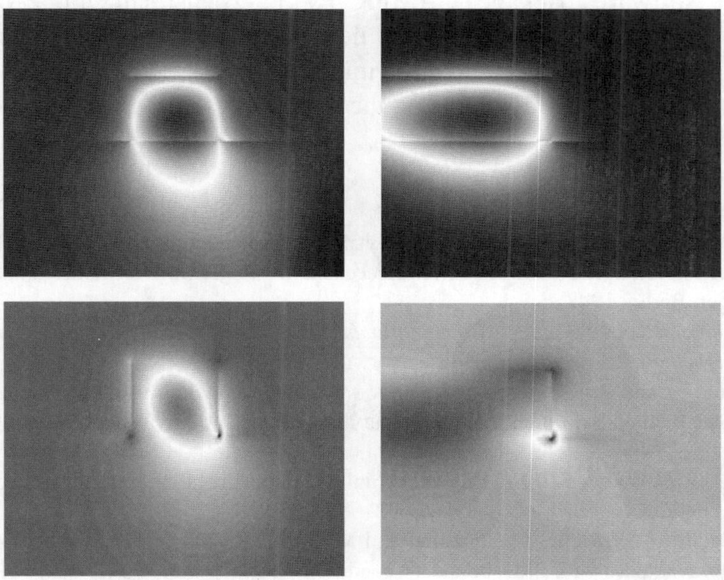

Figure 9. The comparison of mode fields of ring (left) and disk (right) microresonators. Outer radii of both microresonators are R = 50 μm, the height of the guiding layer is 1 μm, the ring width is also 1 μm. The refractive indexes of the substrate, the guiding layer and the superstrate are 1.45, 1.59 and 1.0, respectively, the wavelength is 1.55 μm. Upper figures: vertical electric field component; lower figures: horizontal (radial) electric field component.

5. CONCLUDING REMARKS

In this chapter, a semianalytical approach to the analysis and modeling of high-contrast guided wave photonic structures based on mode matching was briefly reviewed. Due to its principles, it is best suited for modeling structures with piecewise constant refractive index distribution, which is very often the case in modern guided-wave photonics: most photonic wires, disk and ring microresonators and photonic crystal structures belong into this class. Although modeling of truly 3-D structures like devices based on 2-D photonic crystals in planar optical waveguides or membranes remains still rather challenging, for structures with essentially 2-D refractive index distribution the method has proven to be very efficient and accurate. While its application to structures with rectangular geometry has been known for more than a decade, its extension to circular bends and microresonators has appeared only very recently.

Readers may be interested that 2-D and even 3-D modules based on the mode matching method can be found in several commercial software packages for modeling and design of integrated photonic devices.[41,42]. An excellent software package CAMFR by P. Bienstman for 2-D mode expansion and propagation modeling not only for rectangular geometry but also for structures with rotational symmetry like vertical cavity lasers and resonant LEDs is freely available on the web[44].

6. REFERENCES

1. C. Vassallo, *Optical waveguide concept,* (Elsevier, Amsterdam, 1991).
2. G. Guekos, ed., *Photonic Devices for telecommunications: how to model and measure,* (Springer, Berlin, 1998).
3. A. Taflove and S.C. Hagness, *Computational electrodynamics: the finite-difference time-domain method, 2nd ed.*, (ArtechHouse, Norwood, 2000).
4. G.R. Hadley and R.E. Smith, "Full-vector waveguide modeling using an iterative finite-difference method with transparent boundary conditions," *J. Lightwave Technol.* **13**, 465–469 (1995).
5. B.M.A. Rahman and J.B. Davies, "Finite-element solution of integrated optical waveguides," *J. Lightwave Technol.* **2**, 682–688 (1984).
6. 6. S. Selleri, L. Vincetti, A. Cucinotta and M. Zoboli, "Complex FEM modal solver for optical waveguides with PML boundary conditions," *Opt. Quantum Electron.* **33**, 359–371 (2001).
7. U. Rogge and R. Pregla, "Method of lines for the analysis of dielectric waveguides," *J. Lightwave Technol.* **11**, 2015–2020 (1993).
8. A.S. Sudbø, "Film mode matching: a versatile numerical method for vector mode field calculations in dielectric waveguides," *Pure and Appl. Opt.* **2**, 211–233 (1993).
9. A.S. Sudbø, "Improved formulation of the film mode matching method for mode field calculations in dielectric waveguides," *Pure and Appl. Opt.* **3**, 381–388 (1994).

10. J Chilwell and I Hodgkinson, "Thin-films field-transfer matrix theory of planar multilayer waveguides and reflection prism-loaded waveguides," *J. Opt. Soc. Am. A*, **1**, 742–753 (1984).
11. W.C. Chew, J.M. Jin and E. Michelsen, "Complex coordinate stretching as a generalized absorbing boundary condition," *Microwave and Opt. Technol .Lett.* **15**, 363–369 (1997).
12. P. Bienstman, H Derudder, R Baets, F Olyslager and D. De Zutter, "Analysis of cylindrical waveguide discontinuities using vectorial eigenmodes and perfectly matched layers," *IEEE Trans. Microwave Theory Tech.* **49**, 349–354 (2001).
13. G. Sztefka and H-P. Nolting, "Bidirectional eigenmode propagation for large refractive index steps," *IEEE Photonics Technol. Lett.* **5**, 224–557 (1993).
14. P. Bienstman and R. Baets, "Optical modelling of photonic crystals and VCSELs using eigenmode expansion and perfectly matched layers," *Opt. Quantum Electron.* **33**, 327-341 (2001).
15. M. Heiblum and J. H. Harris, "Analysis of curved optical waveguides by conformal transformation," *IEEE. J. Quantum Electron.* **11**, 75–83 (1975).
16. L. Lewin, D. C. Chang and E. F. Kuester, *Electromagnetic waves and curved structures*, (UK: IEE Press, Peter Peregrinus Ltd., Stevenage, 1977).
17. W. Pascher and R. Pregla, "Vectorial analysis of bends in optical strip waveguides by the method of lines," *Radio Science* **28**, 1229–1233 (1993).
18. R. Pregla, "The method of lines for the analysis of dielectric waveguide bends," *J. Lightwave Technol.* **14**, 634–639 (1996).
19. W. Pascher, "Modelling of rib waveguide bends for sensor applications," *Opt. Quantum Electron.* **33**, 433–449 (2001).
20. L. Prkna, M. Hubálek and J. Čtyroký, "Vectorial eigenmode solver for bent waveguides based on mode matching," *IEEE Photonics Technol. Lett.* **16**, 2057-2059 (2004).
21. L. Prkna, M. Hubálek and J. Čtyroký, "Field Modelling of Circular Microresonators by Film Mode Matching," *IEEE J. Sel. Topics on Quantum Electron.* **11**, 217–223 (2005).
22. 22. E. F. Kuester and D. C. Chang, "Propagation, attenuation and dispersion characteristics of inhomogeneous dielectric slab waveguides," *IEEE Trans Microwave Theory Tech.* **23**, 98–106 (1975).
23. J.-P. Bérenger, "A perfectly matched layer for the absorption of electromagnetic waves," *Journal of Computational Physics* **114**, 185–200 (1994).
24. W. C. Chew, J. M. Jin and E. Michielsen, "Complex coordinate stretching as a generalized absorbing boundary condition," *Microwave and Optical Technology Letters* **15**, 383–369 (1997).
25. D. Derudder, D. De Zutter and F Olyslager, "Analysis of waveguide discontinuities using perfectly matched layers," *Electron. Lett.* **34**, 2138-2140 (1998).
26. J Čtyroký, J. Homola, P. V. Lambeck, S. Musa, H.J.W.M. Hoekstra, R.D. Harris, J.S. Wilkinson, B. Usievich, N.M. Lyndin, "Theory and modelling of optical waveguide sensors utilising surface plasmon resonance," *Sensors and Actuators B* **54**, 66-73 (1999).
27. A. Sudbø and P.I. Jensen, "Stable bidirectional eigenmode propagation of optical fields in waveguide devices," Integrated Photonics Research, 27-29 (OSA, Monterey, 1995).
28. J. Čtyroký, J. Homola, M. Skalský, "Modelling of surface plasmon resonance waveguide sensor by complex mode expansion and propagation method," *Opt. Quantum Electron.* **29**, 301–311 (1997).
29. H. Berends, *DBR grating spectral filters for optical waveguide sensors*, PhD thesis, (University of Twente, The Netherlands, 1997).
30. J. Čtyroký, S. Helfert and R. Pregla, "Analysis of a deep waveguide Bragg grating," *Opt. Quantum Electron.* **30,** 343–358 (1998).

31. J. Čtyroký, S. Helfert and R. Pregla, "Bragg waveguide grating as a 1D photonic bandgap structure: COST 268 modelling task," *Opt. Quantum Electron.* **34**, 455–470 (2002).
32. P. Sanchis, P. Bienstman, B. Luyssaert, R. Baetsand J. Marti, "Analysis of butt coupling in photonic crystals," *IEEE J. Quantum Electron.* **40**, 541–550 (2004).
33. A. Yariv and P. Yeh, *Optical Waves in Crystals*. (Wiley, New York, 1984).
34. S.F. Helfert, "Numerical stable determination of Floquet-modes and the application to the computation of band structures," *Opt. Quantum Electron.* **36**, 87–107 (2004).
35. J. Petráček, *Modelling of optical waveguide structures by the mode matching method*, Assoc. Prof. thesis, (Brno Technical University, 2004, in Czech).
36. D. Joannopoulos, R.D. Meade and J.N. Winn, *Photonic Crystals: Molding the Flow of Light* (Princeton University Press, Princeton, 1995).
37. S.G. Johnson and J.D. Joannopoulos, "Block-iterative frequency-domain methods for Maxwell's equations in a planewave basis", *Optics Express*, **8**, 173–190 (2001).
38. L. Prkna, *Rotationally symmetric resonant devices in integrated optics*, PhD thesis, (Charles University in Prague, Czech Republic, 2004).
39. R.E. Collin, *Field theory of guided waves*, 2nd ed., (IEEE Press, New York, 1991).
40. W. Pascher, *Analysis of Waveguide Bends and Circuits by the Method of Lines and the Generalized Multipole Technique*, Assoc. Prof. thesis, (FernUniversität Hagen, Germany, 1998).
41. M. Abramovitz, I.A. Stegun, *Handbook of mathematical functions*, (National Bureau of Standards, Boulder, 1964).
42. http://www.photond.com/
43. http://www.c2v.nl/
44. 44. http://camfr.sourceforge.net/

SURFACE PLASMON RESONANCE (SPR) BIOSENSORS AND THEIR APPLICATIONS IN FOOD SAFETY AND SECURITY

Jiří Homola
Institute of Radio Engineering and Electronics, Academy of Sciences of the Czech Republic, Chaberská 57, Prague, Czech Republic

Abstract: Surface plasmon resonance (SPR) biosensors show vast potential for detection of chemical and biological analytes in numerous important areas. This chapter discusses the underlying principles and configurations of SPR biosensors and reviews applications of SPR biosensor technology in food safety.

Key words: optical sensor; biosensor; surface plasmon; surface plasmon resonance; foodborne pathogen; toxin; food safety; security.

1. INTRODUCTION

In recent years, we have witnessed remarkable progress in the development of affinity biosensors and their applications in areas such as medical diagnostics, drug screening, environmental monitoring, biotechnology, food safety, and security. Illnesses caused by foodborne pathogens and toxins present serious problem with substantial public health impact and economic implications. Therefore, detection of food safety-related substances is of paramount importance to food producers, processors, distributors and regulatory agencies. To address this urgent need, various affinity biosensors such as electrochemical sensors[1], piezoelectric sensors[2], electrical impedance sensors[3], and optical sensors[4] have been developed. Optical affinity biosensors demonstrated for detection of analytes implicated in food safety include in particular fluorescence-based sensors[5] and label-

S. Janz et al. (eds.), Frontiers in Planar Lightwave Circuit Technology, 101–118.

free sensor technologies such as grating coupler sensors[6], resonant mirror sensors[7], and surface plasmon resonance (SPR) sensors[8].

2. FUNDAMENTALS OF SURFACE PLASMON RESONANCE SENSORS

2.1 Surface plasmons and their optical excitation

Surface plasmon-polaritons (SPP), also referred as to surface plasma waves, are special modes of electromagnetic field which can exist at the interface between a dielectric and a metal that behaves like a nearly-free electron plasma. A surface plasmon is a transverse-magnetic mode (magnetic vector is perpendicular to the direction of propagation of the wave and parallel to the plane of interface) and is characterized by its propagation constant and field distribution. The propagation constant, β can be expressed as follows:

$$\beta = \frac{\omega}{c}\sqrt{\frac{\varepsilon_M \varepsilon_D}{\varepsilon_M + \varepsilon_D}}, \qquad (1)$$

where ω is the angular frequency, c is the speed of light in vacuum, and ε_D and ε_M are the dielectric functions of the dielectric and metal[9,10]. This equation describes a surface plasmon that propagates along the interface if the real part of ε_M is negative and its absolute value is smaller than ε_D. At optical wavelengths, this condition is fulfilled for several metals of which gold is most commonly used in SPR biosensors. The real part of the propagation constant is related to the effective refractive index, N:

$$N = \frac{c}{\omega}\mathrm{Re}\{\beta\} = \mathrm{Re}\left\{\sqrt{\frac{\varepsilon_M \varepsilon_D}{\varepsilon_M + \varepsilon_D}}\right\}, \qquad (2)$$

where Re{} denotes the real part of a complex number. The imaginary part of the propagation constant is related to the modal attenuation b:

$$b = \mathrm{Im}\{\beta\}\frac{0.2}{\ln 10} = \mathrm{Im}\left\{\sqrt{\frac{\varepsilon_M \varepsilon_D}{\varepsilon_M + \varepsilon_D}}\right\}\frac{0.2\omega}{c \ln 10}, \qquad (3)$$

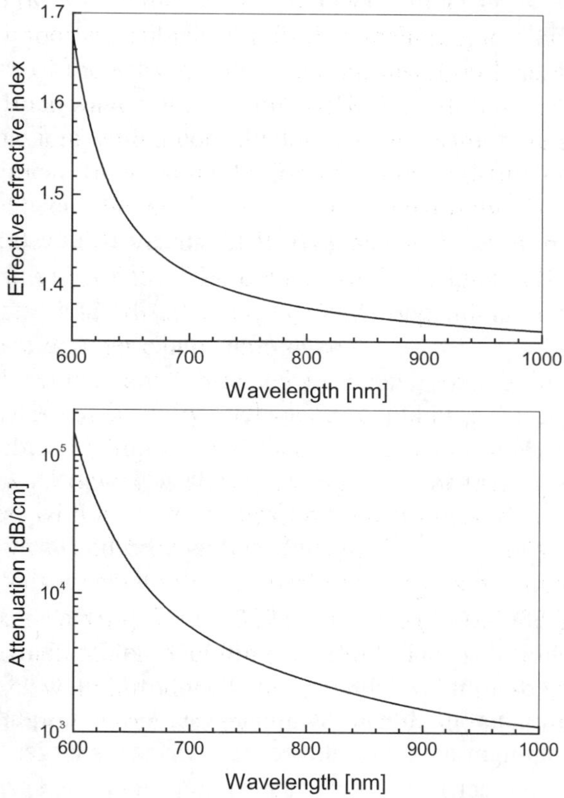

Figure 1. The effective refractive index (upper plot) and attenuation (lower plot) of a surface plasmon as a function of wavelength for a surface plasmon at the interface between gold and a non-dispersive dielectric (dielectric refractive index: n = 1.32).

where Im{} denotes the imaginary part of a complex number. Spectral dependencies of the effective refractive index and mode attenuation for a surface plasmon propagating along the interface between a gold and a non-dispersive dielectric are shown in Fig. 1.Distribution of the electromagnetic field of a surface plasmon is shown in Fig. 2. Apparently, the field of the surface plasmon is concentrated at the metal-dielectric interface and decreases exponentially into both media with an increasing distance from the interface. For a surface plasmon at the gold – aqueous environment interface, the penetration depth (the distance from the interface at which the amplitude of the field falls to 1/e of its value at the surface) is typically 20-30 nm and 100-500 nm in metal and dielectric, respectively, in visible and near infrared regions. The surface plasmon penetration depth in the dielectric is

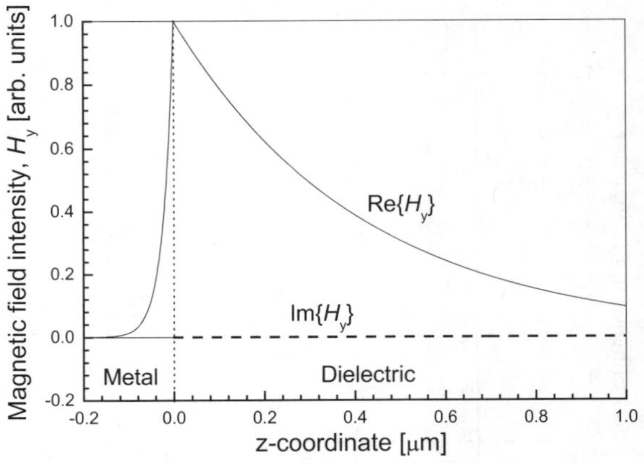

Figure 2. Distribution of the magnetic field intensity of a surface plasmon at the interface between gold and dielectric (refractive index of the dielectric – 1.32) in the direction perpendicular to the interface (x-y plane) for the wavelength of 850 nm.

particularly important for SPR sensing, as it determines the region probed by the SPR sensor.

In optical sensors based on surface plasmon resonance (SPR), surface plasmons are excited by light waves. An optical wave can excite a surface plasmon if the component of the light's wave vector that is parallel to the interface, matches that of the surface plasmon. As the propagation constant of a surface plasmon at the metal-dielectric interface is larger than that of the light wave in the dielectric, the surface plasmon cannot be excited directly by light wave incident on a metal surface. Therefore, the wave vector of light wave needs to be increased to allow excitation of a surface plasmon by the light wave. This can be accomplished, for instance, by passing light wave through an optically denser medium. A light wave passes through a high refractive index prism and is totally reflected at the prism base generating an evanescent wave penetrating a thin metal film (Fig. 3). This evanescent wave propagates along the interface with the propagation constant which can be adjusted to match that of the surface plasmon by controlling the angle of incidence. Thus, the matching condition can be fulfilled allowing light coupling to the surface plasmon. This method is referred to as the attenuated total reflection (ATR) method[10].

Assuming that the prism has only a minor influence on the propagation constant of the surface plasmon at the interface of a metal and a low refractive index dielectric, the coupling condition can be approximately expressed as follows:

$$\sqrt{\varepsilon_P}\sin(\theta) = \mathrm{Re}\left\{\sqrt{\frac{\varepsilon_M \varepsilon_D}{\varepsilon_M + \varepsilon_D}}\right\} \qquad (4)$$

where θ denotes the angle of incidence, ε_P , ε_M and ε_D denote the dielectric functions of the prism, metal film and dielectric medium (sample); $\varepsilon_P > \varepsilon_D$.

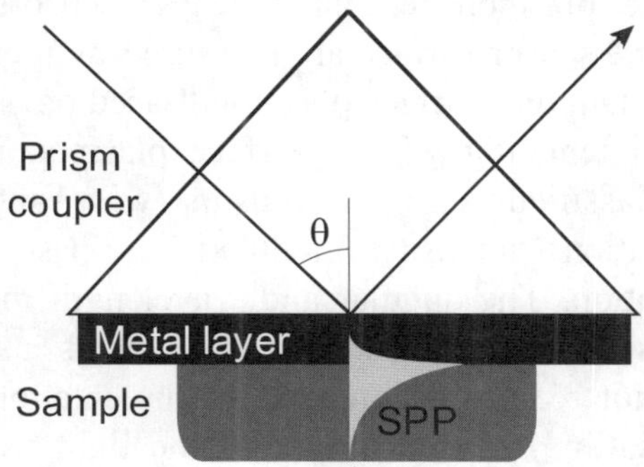

Figure 3. Optical excitation of surface plasmon-polaritons (surface plasmon) by the attenuated total reflection (ATR) method.

The excitation of a surface plasmon is accompanied by the transfer of the light wave energy into the energy of the surface plasmon and its subsequent dissipation in the metal film. This process results in a drop in the intensity of reflected light (Fig. 4).

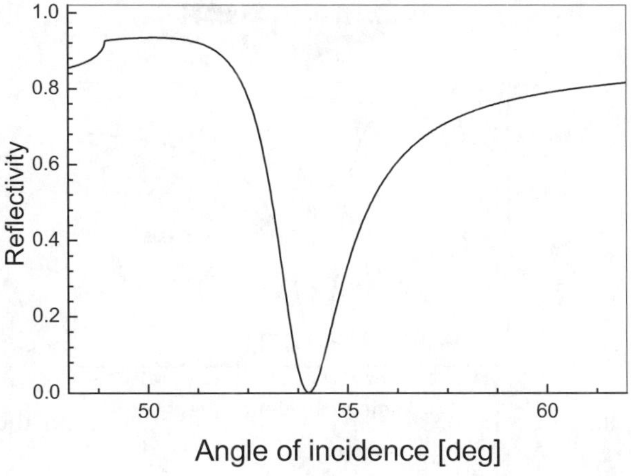

Figure 4. Reflectivity in the Kretschmann geometry of ATR consisting of an SF14 glass prism (refractive index – 1.65), a gold layer (thickness – 50 nm), and a low refractive index dielectric medium (refractive index – 1.32), and wavelength - 800 nm.

2.2 Surface plasmon resonance (SPR) sensors

When a surface plasmon propagates along a metal-dielectric interface, its electromagnetic field probes the dielectric (sample). Any change in the refractive index of the dielectric within the evanescent field of the surface plasmon results in a change in the propagation constant of the surface plasmon. In surface plasmon resonance (SPR) sensors, changes in the refractive index at the sensor surface are measured by measuring changes in the propagation constant of a surface plasmon. Based on which characteristic of the light wave interacting with a surface plasmon is measured, SPR sensors can be classified as SPR sensors with angular modulation, wavelength modulation, intensity modulation, phase modulation, and polarization modulation. The angular and wavelength modulations are the most frequently used modulations in SPR sensors as they offer high resolution and do not require a complex instrumentation. In SPR sensors with angular modulation, the component of the light wave's wave vector parallel to the metal surface matching that of the surface plasmon is determined by measuring the coupling strength at a fixed wavelength and

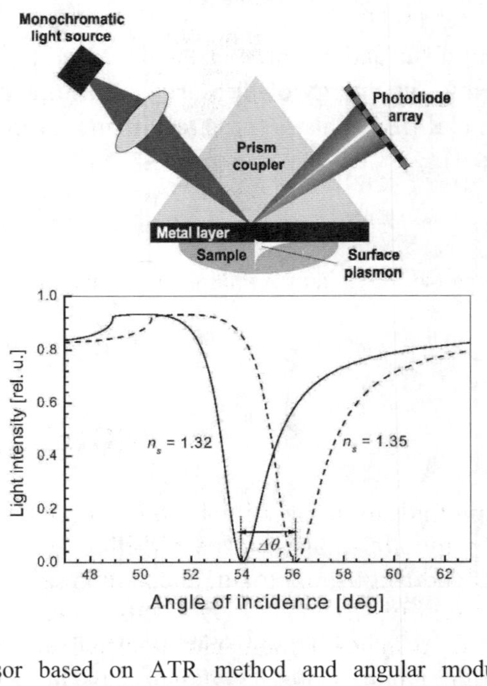

Figure 5. SPR sensor based on ATR method and angular modulations (upper) and corresponding reflectivity calculated for two different refractive indices of sample (lower). Sensor configuration: SF14 glass prism, 50 nm thick gold layer, sample, wavelength – 682 nm.

multiple angles of incidence and determining the angle of incidence yielding the strongest coupling (Fig. 5)[11,12].

In SPR sensors with wavelength modulation, the component of the light wave's wave vector parallel to the metal surface matching that of the surface plasmon is determined by measuring the coupling strength at a fixed angle of incidence and multiple wavelengths and determining the wavelength yielding the strongest coupling (Fig. 6)[13].

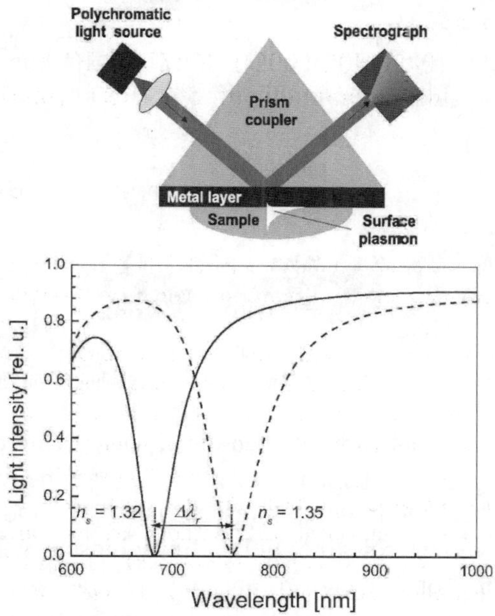

Figure 6. SPR sensor based on ATR method and wavelength modulations (upper) and corresponding reflectivity calculated for two different refractive indices of sample (lower). Sensor configuration: SF14 glass prism, 50 nm thick gold layer, sample, incident angle – 54°.

2.3 Surface plasmon resonance affinity biosensors

In affinity biosensors based on spectroscopy of surface plasmons, biomolecular recognition elements immobilized on the surface of a SPR-active metal layer recognize and capture analyte molecules present in a liquid sample. This analyte binding results in an increase in the refractive index at the sensor surface. The refractive index increase gives rise to an increase in the propagation constant of the surface plasmon which is measured by measuring a change in one of the characteristics of light interacting with the surface plasmon.

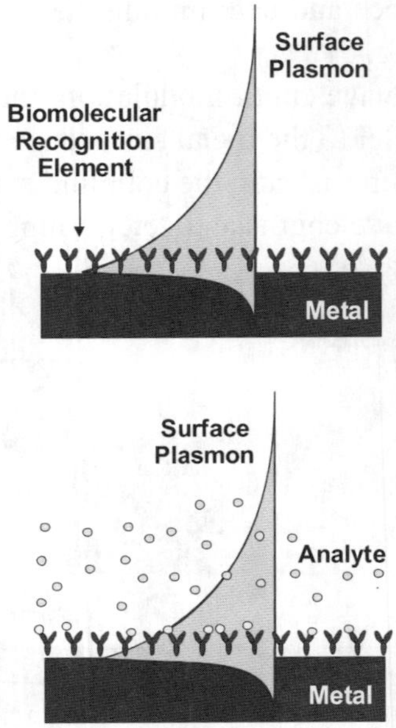

Figure 7. Concept of affinity biosensor based on surface plasmon resonance.

SPR biosensors are label-free detection devices – binding between the biomolecular recognition element and analyte can be observed directly without the use of radioactive or fluorescent labels. In addition, the binding event can be observed in real-time. SPR affinity biosensors can, in principle, detect any analyte for which an appropriate biomolecular recognition element is available. Moreover, analyte molecules do not have to exhibit any special properties such as fluorescence or characteristic absorption or scattering bands.

3. IMPLEMENTATIONS OF SPR BIOSENSORS

An SPR biosensor consists of two key elements - an optical system for excitation and interrogation of surface plasmons and a biospecific coating incorporating biomolecular recognition elements which interact with target molecules in a sample.

3.1 Biomolecular recognition elements

Various types of biomolecular recognition elements have been exploited in affinity biosensors. These include antibodies[14], aptamers[15], peptides[16], and molecularly imprinted polymers[17]. These biomolecular elements have to be immobilized on the sensor surface in a manner ensuring the conservation of biological activity of immobilized biomolecular recognition elements and non-fouling background. Another desirable property is regenerability - the possibility to regenerate the biomolecular recognition elements (*i.e.* break their complex with the analyte molecules and make the biosensor available for another use).

In general, methods for immobilization of biomolecular recognition elements on gold films exploit physicochemical interactions such as chemisorptions[18], covalent binding[19], electrostatic coupling[20], and high-affinity molecular linkers in multilayer systems (*e.g.* streptavidin – biotin[21,22], proteins A or G[23], and complementary oligonucleotides[24]), and photo-immobilization (*e.g.* albumin derivatized with aryldiaziridines as a photo-linker[25]). One of the most remarkable techniques in surface chemistry is the spontaneous self-organization of *n*-alkylthiols or disulfides on gold . Self-

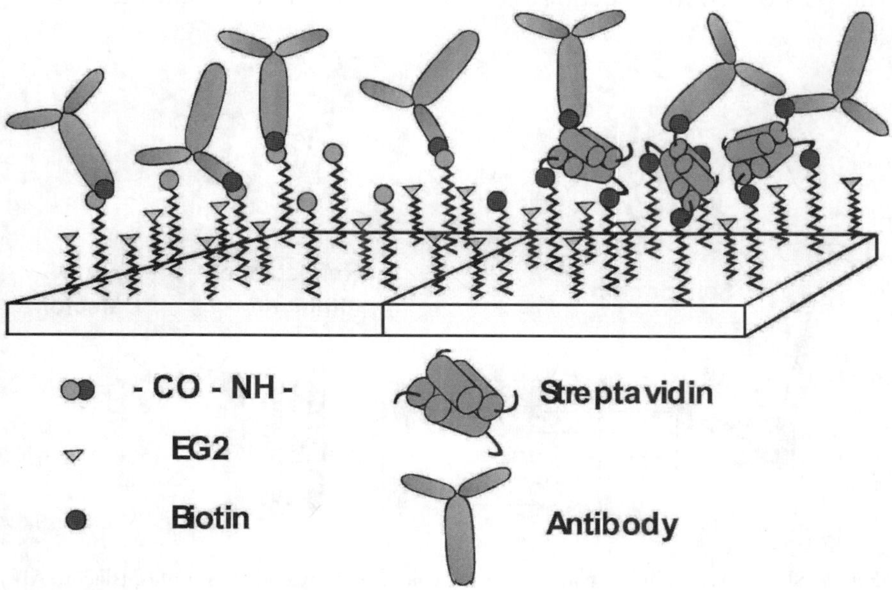

Figure 8. Immobilization of antibodies via self-assembled monolayers of alkylthiols terminated with functional group (left) and self-assembled monolayers of alkylthiols terminated with biotin and subsequent capture of streptavidin (right).

assembled monolayers (SAMs) have been employed in many immobilization methods for spatially-controlled attachment of biomolecular recognition elements to surfaces of sensors[26]. In order to provide a desired surface concentration of biomolecular recognition elements on gold, mixed SAMs of long-chained (n = 12 and higher) n-alkylthiols terminated with functional group for further attachment of biomolecular recognition elements and short-chained alkylthiols for a non-fouling background have been developed[27, 28].

3.2 Multichannel SPR sensor platforms based on ATR

SPR affinity biosensors are based on the measurement of refractive index changes induced by the interaction between the biomolecular recognition element immobilized on sensor surface and analyte in the sample. Therefore, accuracy of SPR measurements can be compromised by interfering effects which produce a change in the refractive index but are not associated with the capture of target analyte (*e.g.* non-specific interaction between the sensor surface and sample, background refractive index variations due to sample temperature and composition fluctuations). The effect of these interferences can be reduced by using multichannel SPR sensors in which certain sensing channels are used for detection and others are used as a reference channels.

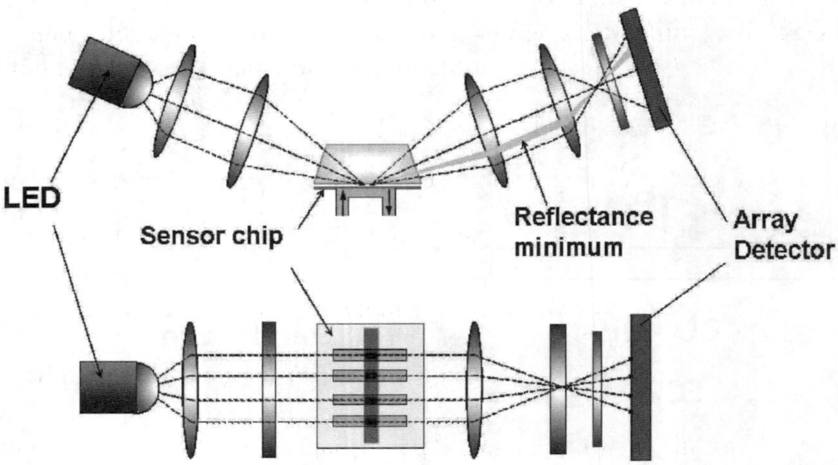

Figure 9. SPR sensor with four parallel sensing channels (provided by S. Löfås, Biacore AB.)

In recent years, numerous configurations of multichannel SPR sensors have been proposed. The most straightforward approach to multichannel sensing is based on parallel architecture of sensing channels. In the prism-based SPR biosensors with angular modulation and parallel architecture, a series of convergent monochromatic beams is focused on a row of sensing channels, where they excite surface plasmons and the reflected light from a row of sensing channels is projected onto a two-dimensional detector array,

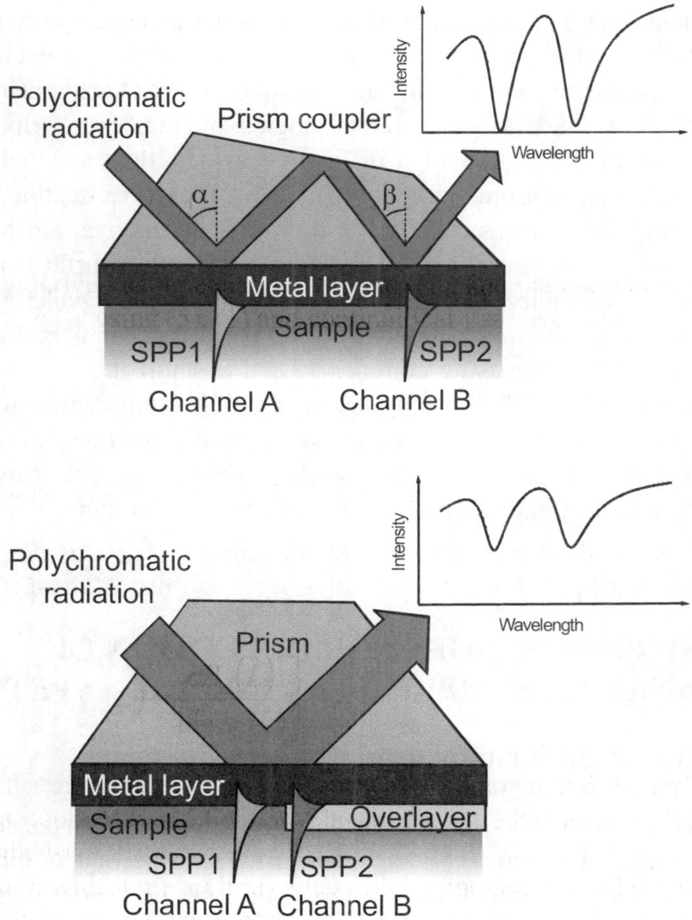

Figure 10. SPR dual-channel sensors based on wavelength division multiplexing (WDM) of sensing channels. WDM of sensing channels by means of altered angles of incidence (upper)[35]. WDM of sensing channels by means of a high refractive index overlayer (lower)[36].

Fig. 9[29-30]. In prism-based SPR sensors with wavelength modulation and parallel architecture, series of collimated polychromatic beams are made incident on a row of sensing channels and SPR spectra encoded into the spectrum of reflected light wave are analyzed by means of multiple spectrographs[31].

Recently, an alternative approach to multichannel sensing based on the wavelength division multiplexing (WDM) technique has been developed. In the WDMSPR sensors, SPR spectra from multiple channels are encoded in different wavelength regions of a single polychromatic light wave. Subsequently, SPR signals for all the sensing channels can be extracted from a single spectrum. This is accomplished by changing the angle of incidence of the incident light beam[32] or by a dielectric overlayer deposited over a part of the SPR-active surface[33], Fig. 10. In the first approach, a polychromatic light wave is made sequentially incident on two different areas of an SPR-active metal film under two different angles of incidence. Upon each incidence, a surface plasmon of a different wavelength is excited. In the latter approach, a polychromatic light beam of a larger cross-section is made incident on an SPR-active metal film, a part of which is overlaid by a thin high refractive index overlayer. This film shifts the surface plasmon occurring at a metal-dielectric overlayer interface to longer wavelengths.

The WDMSPR sensor approach offers the benefit of multichannel performance without increasing complexity and costs of the sensor system. In addition, the WDMSPR sensors make it possible to discriminate effects occurring in the proximity of sensor surface (specific binding, non-specific adsorption) from those occurring in the whole medium (interfering background refractive index fluctuations) which is a prerequisite for robust referencing[34].

4. SPR BIOSENSORS FOR DETECTION OF ANALYTES IMPLICATED IN FOOD SAFETY

4.1 Detection formats used in SPR biosensors

SPR affinity biosensors have been developed to detect an analyte in a variety of formats. The choice of detection format for a particular application depends on the size of target analyte molecules, binding characteristics of available biomolecular recognition element, and range of concentrations of analyte to be measured. The main detection formats used in SPR biosensors include direct detection (Fig. 11), sandwich assay (Fig. 12) and inhibition assay (Fig. 13).

Direct detection is usually preferred in applications, where direct binding of analyte of concentrations of interest produces a sufficient response. If necessary, the lowest detection limits of the direct biosensors can be improved by using a sandwich assay. Smaller analytes (molecular weight $<$ 10,000) are usually measured using inhibition assay.

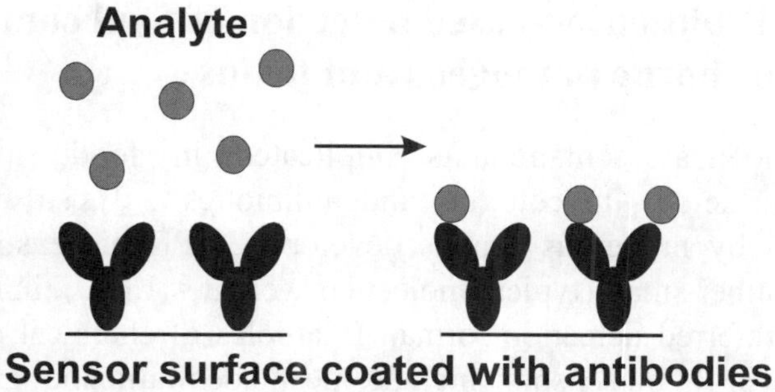

Figure 11. Detection formats used in SPR affinity biosensors: direct detection.

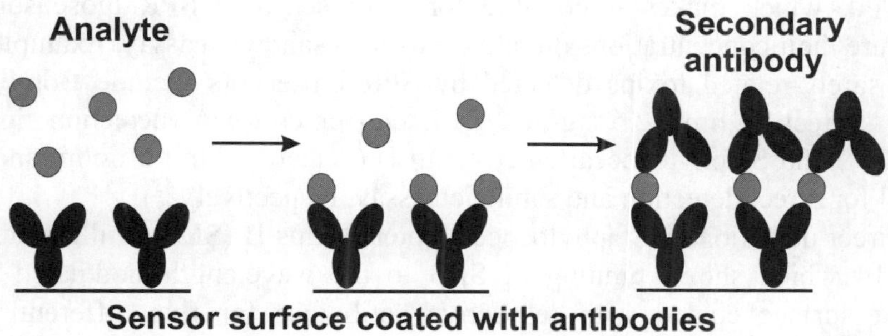

Figure 12. Detection formats used in SPR affinity biosensors: sandwich assay.

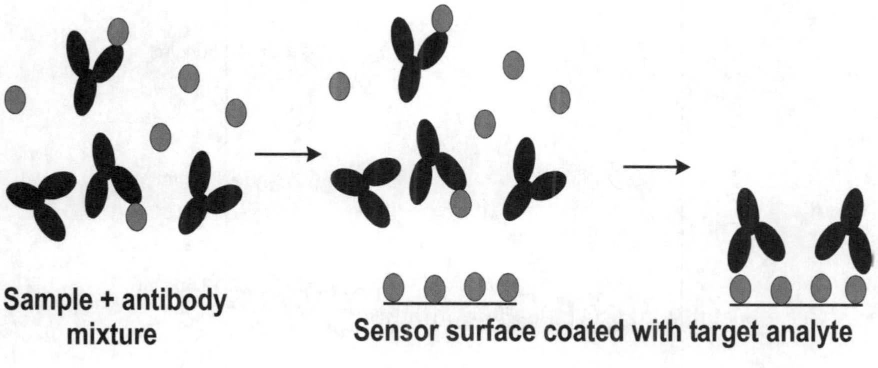

Figure 13. Detection formats used in SPR affinity biosensors: competitive inhibition assay.

4.2 SPR biosensor-based detection of food contaminants, food-borne pathogens and toxins

Major chemical contaminants implicated in food safety include pesticides, herbicides, mycotoxins and antibiotics[35]. These analytes have been targeted by numerous groups developing SPR biosensors. As these analytes are rather small (typical molecular weight < 1,000), inhibition assay has been a preferred detection format. Examples of chemical contaminants detected by SPR biosensors include pesticides atrazine and simazine (detection limits 0.05 ng/ml[36] and 0.1 ng/ml [37], respectively), mycotoxin Fumonisin B1 (detection limit 50 ng/ml [38]), and antibiotics Sulphamethazine, Sulphadiazine (detection limits 1 ng/ml [39] and 20 ng/ml [40], respectively).

Molecular weight of the main bacterial toxins ranges from 28,000 to 150,000, which makes it possible for most sensitive SPR biosensors to measure their concentrations directly or using a sandwich assay. Examples of food safety-related toxins detected by SPR biosensors include Botulinum toxin (detection limit 2.5 μg/ml [41]), *E. coli* enterotoxin (detection limit 6 μg/ml [42]) and Staphylococcal enterotoxin B (detection limit 5 ng/ml and 0.5 ng/ml for direct detection and sandwich assay, respectively [43]).

Direct detection of Staphylococcal enterotoxins B (SEB) is illustrated in Fig. 14 which shows binding of SEB to the wavelength-modulated SPR sensor surface coated with respective antibodies for five different SEB concentrations[43]. Figure 15 shows the sensor response to binding after 30-minute SEB incubation and initial binding rate as a function of SEB

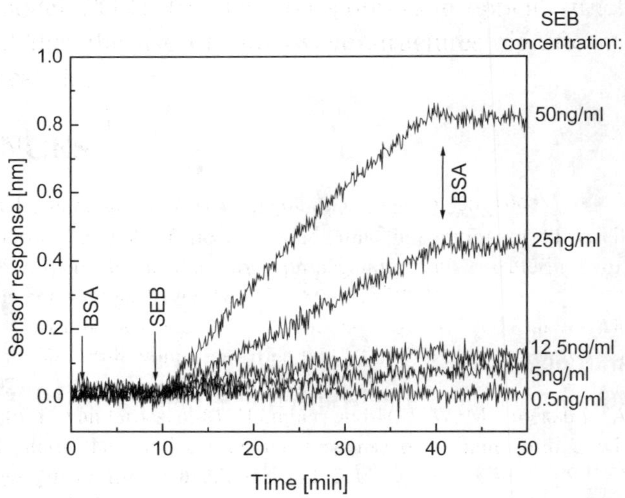

Figure 14. SPR sensor-based detection of Staphylococcal enterotoxin B (SEB)[43].

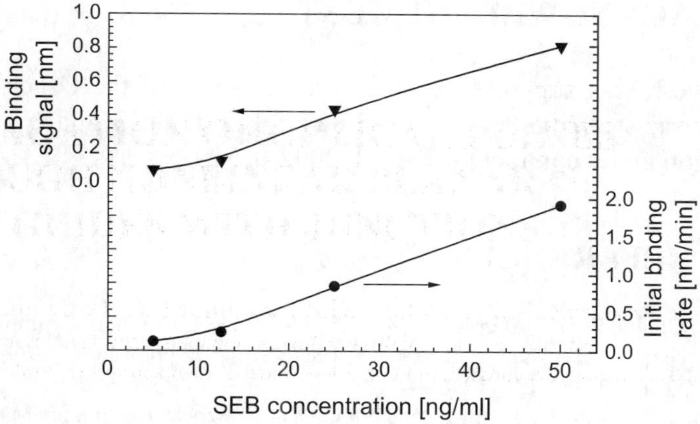

Figure 15. Detection of Staphylococcal enterotoxin B using an SPR biosensor. Sensor response as a function of SEB concentration.

concentration. Both these metrics can be correlated with analyte concentration and used as a "sensor output".

Bacterial pathogens are relatively large targets (> 1μm) and therefore, their presence can be detected directly with an optional amplification by secondary antibodies (sandwich assay). Examples of foodborne bacterial pathogens detected by SPR biosensors include *Escherichia coli* (detection limit 5×10^7 cfu/ml [44]), *Listeria monocytogenes* (detection limit 10^6 cfu/ml [45]) and *Salmonella enteritidis* (detection limit 10^6 cfu/ml [45]).

5. SUMMARY

Early detection and identification of analytes implicated in food safety is of paramount importance in numerous sectors, including agriculture, food industry, regulatory authorities, and security organizations. SPR biosensors emerge as an alternative to time-consuming and laborious laboratory techniques. Major advantages of SPR biosensors are fast response, ability to detect multiple analytes at a time and capacity to perform the measurements in the field. SPR biosensors have been able to detect small and medium size analytes such as food chemical contaminant and bacterial toxins at practically relevant concentrations. Detection limits for large analytes such as bacterial pathogens still need to be improved to meet today needs. The SPR biosensor technology has potential to meet the analytical needs in many other significant areas such environmental monitoring, medical diagnostics, drug development, *etc.*

6. ACKNOWLEDGEMENT

This work was supported by the Grant Agency of the Czech Republic under contracts 202/05/0628, 303/03/0249, 102/03/0633 and by European Commission under contract QLK4-CT-2002-02323.

7. REFERENCES

1. A. L. Ghindilis, P. Atanasov, M. Wilkins, and E. Wilkins, "Immunosensors: Electrochemical Sensing and Other Engineering Approaches," *Biosensors & Bioelectronics* **13**, 113-131 (1998).
2. X. Chu, Z. H. Lin, G. L. Shen, and R. Q. Yu, "Piezoelectric Immunosensor for the Detection of Immunoglobulin M," *The Analyst* **120**, 2829-2832 (1995).
3. P. Pless, K. Futschik, and E. Schopf, "Rapid Detection of Salmonellae by Means of a New Impedance-Splitting Method," *Journal of Food Protection* **57**, 369-376 (1994).
4. G. Gauglitz, *Opto-Chemical and Opto-Immuno Sensors, Sensor Update Vol. 1*, VCH Verlagsgesellschaft, Weinheim (1996).
5. C. A. Rowe-Taitt, J. W. Hazzard, K. E. Hoffman, J. J. Cras, J. P. Golden, and F. S. Ligler, "Simultaneous Detection of Six Biohazardous Agents Using a Planar Waveguide Array Biosensors," *Biosensors & Bioelectronics* **15**, 579-589 (2000).
6. D. Clerc and W. Lukosz, "Integrated Optical Output Grating Coupler as Biochemical Sensor," *Sensors and Actuators* **B19**, 581-586 (1994).
7. R. Cush, J. M. Cronin, W. J. Stewart, C. H. Maule, J. Molloy, and N. J. Goddard, "The Resonant Mirror: A Novel Optical Biosensor for Direct Sensing of Biomolecular Interactions Part I: Principle of Operation and Associated Instrumentation," *Biosensors & Bioelectronics* **8**, 347-354 (1993).
8. J. Homola, S. Yee, and G. Gauglitz, "Surface Plasmon Resonance Sensors: Review," *Sensors and Actuators* **B54**, 3-15 (1999).
9. A. D. Boardman (editor), *Electromagnetic Surface Modes* (John Wiley & Sons, Chichester, 1982).
10. H. Reather, *Surface Plasmons on Smooth and Rough Surfaces and on Gratings*, Springer Tracks in Modern Physics, (Springer Verlag, Berlin, 1983).
11. K. Matsubara, S. Kawata, and S. Minami, "Optical Chemical Sensor based on Surface Plasmon Measurement," *Applied Optics* **27**, 1160-1163 (1988).
12. B. Liedberg, I. Lundstrom, and E. Stenberg, "Principles of Biosensing with an Extended Coupling Matrix and Surface Plasmon Resonance," *Sensors and Actuators* **B11**, 63-72 (1993).
13. L. M. Zhang and D. Uttamchandani, "Optical Chemical Sensing Employing Surface Plasmon Resonance," *Electronics Letters* **23**, 1469-1470 (1988).
14. B. Hock, "Antibodies for Immunosensors. A Review," *Analytica Chimica Acta* **347**, 177 - 186 (1997).
15. S. D. Jaysena, "Aptamers: An Emerging Class of Molecules that Rival Antibodies in Diagnostics,"*Clinical Chemistry* **45**, 1628-1650 (1999).

16. G. J. Wegner, H. J. Lee, and R. M. Corn, "Characterization and Optimization of Peptide Arrays for the Study of Epitopeantibody Interactions Using Surface Plasmon Resonance imaging," *Analytical Chemistry* **74**, 5161-5168 (2002).
17. R. J. Ansell, O. Ramström, and K. Mosbach, "Towards Artificial Antibodies Prepared by Molecular Imprinting," *Clinical Chemistry* **42**, 1506-1512 (1996).
18. R. C. Nuzzo and D. L. Allara, Adsorption of Bifunctional Organic Disulfides on Gold Surfaces, *Journal of American Chemical Society* **105**, 4481 – 4483 (1983).
19. B. Johnsson, S. Löfas, G. Lindquirst, A. Edström, R. M. Müller Hillgren, and A. Hansson, "Comparison of Methods for Immobilization to Carboxymethyl Dextran Sensor Surfaces by Analysis of the Specific Activity of Monoclonal Antibodies," *Journal of Molecular Recognition: JMR* **8**, 125-131 (1995).
20. M. Houska, E. Brynda, and K. Bohatá, "The Effect of Polyelectrolyte Chain Length on Layer-by-layer Protein/Polyelectrolyte Assembly – an Experimental Study," *Journal Of Colloid And Interface Science* **273**, 140-147 (2004).
21. S. Tombelli, M. Mascini, and A. P. F. Turner, "Improved Procedures for Immobilisation of Oligonucleotides on Gold-coated Piezoelectric Quartz Crystals," *Biosensors & Bioelectronics* **17**, 929-936 (2002).
22. S. Busse, E. Scheumann, B. Menges, and S. Mittler, "Sensitivity Studies for Specific Binding Reactions using the Biotin/Streptavidin System by Evanescent Optical Methods " *Biosensors & Bioelectronics* **17**, 704 – 710 (2002).
23. G. P. Anderson, M. A. Jacoby, F. S. Ligler, and K. D. King, "Effectiveness of Protein A for Antibody Immobilization for a Fiber Optic Biosensor,' *Biosensors & Bioelectronics* **12**, 329–336 (1997).
24. J. Ladd, C. Boozer, Q. Yu, S. Chen, J. Homola, and S. Jiang, Langmuir, "DNA-Directed Protein Immobilization on Mixed Self-Assembled Monolayers via a Streptavidin Bridge," *The ACS Journal of Surfaces and Colloids* **20**, 8090 – 8095 (2004).
25. H. Gao, M. Sänger, R. Luginbühl, and H. Sigrist, "Immunosensing with Photo-Immobilized Immunoreagents on Planar Optical Wave Guides," *Biosensors & Bioelectronics* **10**, 317–328 (1995).
26. W. Knoll, M. Liley, D. Piscevic, J. Spinke, and M. J. Tarlov, "Supramolecular Architectures for the Functionalization of Solid Surfaces," *Advances In Biophysics* **34**, 231 – 251 (1997).
27. L. S. Jung, K. E. Nelson, P. S. Stayton, and C. T. Campbell, "Binding and Dissociation Kinetics of Wild-Type and Mutant Streptavidins on Mixed Biotin-Containing Alkylthiolate Monolayers," *Langmuir* **16**, 9421 – 9432 (2000).
28. K. E. Nelson, L. Gamble, L. S. Jung, M. S. Boeckl, E. Neemi, S. L. Golledge, T. Sasaki, D. G. Castner, C. T. Campbell, and P. S. Stayton, "Surface Characterization of Mixed Self-Assembled Monolayers Designed for Streptavidin Immobilization," *Langmuir* **17**, 2807 – 2816 (2001).
29. S. Löfas, M. Malmqvist, I. Rönnberg, E. Stenberg, B. Liedberg, and I. Kundström, "Bioanalysis with Surface Plasmon Resonance," *Sensors and Actuators* **B5**, 79 – 84 (1991).
30. M. J. O'Brien, V. H. Pérez-Luna, S.R.J. Brueck, and G.P. López, "A Surface plasmon Resonance Array Biosensor Based on Spectroscopic Imaging," *Biosensors & Bioelectronics* **16**, 97-108 (2001).
31. J. Homola, J. Dostálek, S. F. Chen, A. Rasooly, S. Jiang, and S. S. Yee, "Spectral Surface Plasmon Resonance Biosensor for Detection of Staphylococcal Enterotoxin B(SEB) in Milk," *International Journal of Food Microbiology* **75**, 61-69 (2002).

32. J. Homola, J. Dostálek, and J. Čtyroký, "Novel Approach to Surface Plasmon Resonance Multichannel Sensing," *SPIE Proceedings* **4416**, 86-89 (2001).
33. J. Homola, H. B. Lu, and S. S. Yee, "Dual-Channel Surface Plasmon Resonance Sensor with Spectral Discrimination of Sensing Channels Using a Dielectric Overlayer," *Electronics Letters* **35**, 1105-1106 (1999).
34. J. Homola, H. B. Lu, G. G. Nenninger, J. Dostálek, and S. S. Yee, "A Novel Multichannel Surface Plasmon Resonance Biosensor," *Sensors and Actuators* **B76,** 403-410 (2001).
35. P. D. Patel, "(Bio)sensors for Measurement of Analytes Implicated in Food Safety: a Review," *TRAC Trends in Analytical Chemistry* **21**, 96-115 (2002).
36. M. Minunni and M. Mascini, "Detection of Pesticide in Drinking Water Using Real-Time Biospecific Interaction Analysis (BIA)," *Analytical Letters* **26**, 1441-1460 (1993).
37. R. D. Harris, B. J. Luff, J. S. Wilkinson, J. Piehler, A. Brecht, G. Gauglitz, and R. A. Abuknesha, "Integrated Optical Surface Resonance Immunoprobe for Simazine Detection," *Biosensors & Bioelectronics* **14**, 3635-3641 (1999).
38. W. Mullett, P. C. Edward, and M. J. Yeung, "Immunoassay of Fumonisins by a Surface Plasmon Resonance Biosensor," *Analytical Biochemistry* **258**, 161-167 (1998).
39. A. Sternesjo, C. Mellgren, and L. Bjorck, "In Immunoassays for Residue Analysis: Food Safety," ed. R. C. Beier and L. H. Stanker, *ACS Symposium Series,* No. 621, American Chemical Society, Washington DC, 463-470 (1996).
40. C.T. Elliot, G. A. Baxter, S. R. H. Crooks, and W. J. McCaughey, "The Development of a Rapid Immunobiosensor Screening Method for the Detection of Residues of Sulphadiazine," *Food and Agricultural Immunology* **11**, 19-27 (1999).
41. K. Choi, W. Seo, S. Cha, and J. Choi, "Evaluation of Two Types of Biosensors for Immunoassay of Botulinum Toxin," *Journal of Biochemistry and Molecular Biology* **31**, 101-105 (1998).
42. B. D. Spangler,E. A. Wilkinson, J. T. Murphy, and B. J. Tyler, "Comparison of the Spreeta Surface plasmon Resonance Sensor and a Quartz Crystal Microbalance for Detection of Escherichia Coli heat-labile Enterotoxin," *Analytica Chimica Acta* **444**, 149-161 (2001).
43. J. Homola, J. Dostálek, S. Chen, A. Rasooly, S. Jiang, and S. Yee, "Spectral Surface Plasmon Biosensor for Detection of Staphylococcal Enterotoxin B in Milk," *International Journal Of Food Microbiology* **75**, 61-69 (2002).
44. P. M. Fratamico, T. P. Strobaugh, M. B. Medina, and A. G. Gehring, "Detection of *Escherichia coli* O157:H7 Using a Surface Plasmon Resonance Biosensor," *Biotechnology Techniques* **12**, 571-576 (1998).
45. V. Koubová, E. Brynda, L. Karasová, J. Škvor, J. Homola, J. Dostálek, P. Tobiška, and J. Rošický, "Detection of Foodborne Pathogens Using Surface Plasmon Resonance Biosensors," *Sensors and Actuators* **B74**, 100-105 (2001).

OPTICAL BIOSENSOR DEVICES AS EARLY DETECTORS OF BIOLOGICAL AND CHEMICAL WARFARE AGENTS

L.M. Lechuga, E. Mauriz, B. Sepúlveda, J. Sánchez del Río, A. Calle, G. Armelles and C. Domínguez
Microelectronics National Center, CSIC, Spain

Abstract: The present article reviews three different optical biosensing mechanisms that can be implemented in a multibiosensor technology platform: surface plasmon resonance, magnetooptical surface plasmon resonance and optoelectronic biosensing (evanescent wave detection) using integrated Mach-Zehnder interferometer devices. In the last case, the use of standard silicon microelectronics technology opens the possibility for integration of optical, fluidics based and electrical functions within a single optical sensing circuit leading to a complete lab-on-a-chip design solution. All three mechanisms can be efficiently used as an early warning system for biological and/or chemical warfare.

Key words: optical biosensor; surface plasmon resonance; magnetooptical surface plasmon resonance; multibiosensor technological platform

1. INTRODUCTION

There exists a critical need for field deployable biosensors that are able to detect in a fast and highly sensitive way biological and chemical warfare agents in air and water samples. There is also agreement that the best defence against these type of agents is early detection and/or identification. Therefore, biological and chemical warfare (BCW) are fields where new type of analyzers (faster, direct, smaller and cheaper than conventional ones) are demanded. In order to have reliable diagnostic tools for the rapid detection and identification of bio and chemical warfare agents, new methods allowing label-free and real time measurement of simultaneous

S. Janz et al. (eds.), Frontiers in Planar Lightwave Circuit Technology, 119–140.

interactions (such as harmful agent/receptor or DNA recognition) must be developed[1,2]. Biosensing devices, mainly those fabricated with micro/nanotechnologies, are powerful devices which can fulfill these requirements and have the additional value of being portable to perform "point-of-care" analysis[3]. These devices can also have the multiplexing capability of a single biodetection platform for performing pathogen identification much faster than in routine laboratory analysis. The biological receptor for the detection of BCW agents could be either DNA strands that can bind to a specific pathogen present in the environment, either antibodies that can recognize specific sites on bacteria or bind to surface proteins[4]. The optical biosensors to be developed today and in the near future should be able to work with both types of receptors as well as to trigger a signal after their specific biomolecular interaction.

In order to conceptually analyze the main components and functionality of a multibiosensor technological platform that could be used as an early warning system for biological and/or chemical warfare, we will review three different optical biosensor technologies:

- a platform based on a portable surface plasmon resonance sensor (SPR) (actually in commercialization);
- a platform based on a newly developed concept of biosensing using a novel magnetooptical surface plasmon resonance (MOSPR);
- a platform based on optoelectronic biosensors (evanescent wave detection) using integrated Mach-Zehnder interferometer devices.

In the last case, the use of standard silicon microelectronics technology allow the possibility for integration of optical, fluidics and electrical functions on a single optical sensing circuit leading to a complete lab-on-a-chip technological solution. With this sensor a detection limit in the femtomole range is achievable in a direct format.

2. OPTICAL BIOSENSORS

Optical biosensors are providing increasingly important analytical technologies for the detection of biological and chemical species[5,6]. Most of the optical biosensors make use of optical waveguides as the basic element of their structure for light propagation and are based on the same operation principle - the evanescent field sensing. In an optical waveguide, light travels inside the waveguide confined within the structure by total internal reflection (TIR). Light travels inside the waveguide in the form of "guided modes". Although most of the light power is confined within the guided modes, there is a part of it (evanescent field, EW) that travels into the outer medium surrounding the waveguide core, which is usually some hundred nanometers thick. This EW field can be used for sensing purposes in the following way:

when a receptor layer is immobilized onto the waveguide, the exposure of such a surface to the partner analyte molecules produces a biochemical interaction, which induces a change in its optical properties. This change is detected by the evanescent wave. The extent of the optical change will depend on the concentration of the analyte and on the affinity constant of the interaction, obtaining, in this way, a quantitative sensor of the interaction. The evanescent wave decays exponentially as it penetrates the outer medium and, therefore, only detects changes taking place on the surface of the waveguide since the intensity of the evanescent field is much higher in this particular region. For that reason it is not necessary to carry out a prior separation of non-specific components (as in conventional analysis) because any change in the bulk solution will hardly affect the sensor response. In this way, evanescent wave sensors are selective and sensitive devices for the detection of very low levels of chemicals and biological substances and for the measurement of molecular interactions in-situ and in real time[7].

The advantages of the optical sensing are significantly improved when the above approach is used within an integrated optics context. Integrated optics technology allows the integration of passive and active optical components (including fibres, emitters, detectors, waveguides and related devices) onto the same substrate, allowing the flexible development of miniaturised compact sensing devices, with the additional possibility to fabricate multiple sensors on a single chip. The integration offers some additional advantages to the optical sensing systems such as miniaturization, robustness, reliability, potential for mass production with consequent reduction of production costs, low energy consumption and simplicity in the alignment of the individual optical elements[8].

3. BIOCHEMICAL ASPECTS OF OPTICAL BIOSENSORS

Biosensor devices must operate in liquids as they measure effects at a liquid-solid interface. Then, the immobilization of the receptor molecule on the sensor surface is a key step for the efficient performance of the sensor[8]. When the complementary analytes are flowing over the surface, they can be directly recognized by the receptor through a change in the physico-chemical properties of the sensor. In this way, the interacting components do not need to be labeled and complex samples can be analyzed without purification.

The chosen immobilization method must retain the stability and activity of the bound biological receptor. Generally, direct adsorption is not adequate and leads to significant losses in biological activity and random orientation of the receptors. Despite these difficulties, direct adsorption is widely employed since it is simple, fast and does not required special reagents.

Immobilization methods based on covalent coupling are preferable, as covalent coupling gives a stable immobilization and the receptor do not dissociate from the surface or exchange with other proteins in the solution.

Various surface chemistries can be employed for covalent bonding[9] but one of the most used one is to form, by chemical modification of the surface[10], a self-assembled monolayer (SAM) and couple the receptor using the end of the SAM via a functional group (-NH_2, -COOH, or other). The functionalization of the transducer by SAMs also depends on the surface type. Transducers with gold are usually modified with thiols while oxidized surfaces (as SiO_2 or Si_3N_4) can easily be modified with silane chemistry. For example, a widespread method is the functionalization of single-strand DNA (ss-DNA) with an alkane chain terminated by a thiol (-S-H) or disulfide group (-S-S). Sulphurs form a strong bond with gold, thus thiol-derivatized ss-DNA spontaneously forms a single self-assembled monolayer upon immersion on clean gold surfaces. Several aspects must be taken into account in the development of the immobilization procedures such as the non-specific interactions, the optimization of the surface density of the receptor in order to prevent steric hindrance phenomenon, or the regeneration of the receptor[10].

We have developed immobilization procedures at the nanometer-scale attempting to fulfill all the requirements described above, based on thiol-chemistry and silanization depending on the type of sensor surface type and the specific application[11]. Different examples will be shown in the descriptions of each of the biosensor platforms which are given below.

4. SURFACE PLASMON RESONANCE BIOSENSOR

One of the most employed and comprehensively developed optical biosensors is the SPR sensor, because of its sensitivity and simplicity[12,13]. The SPR is an optical phenomenon due to the charge density oscillation at the interface of two media with dielectric constants of opposite signs (as for example a metal and a dielectric). Optical excitation of a surface plasmon can be achieved when a TM polarized light beam propagates at a given angle (the angle of resonance) across the interface between a thin metal layer and a dielectric medium (see Figure 1).

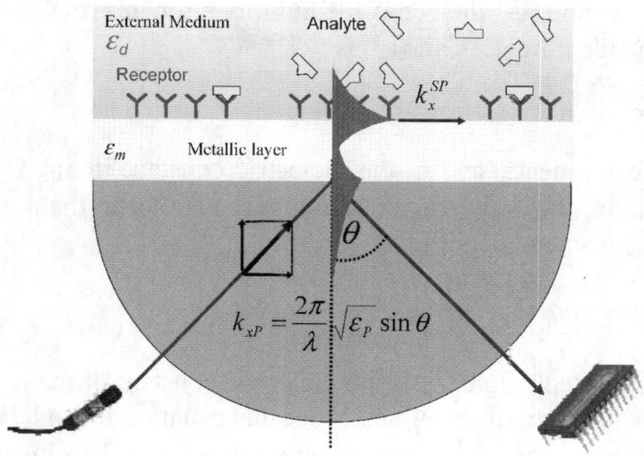

Figure 1. Schematic diagram showing the working principle of a surface plasmon resonance biosensor.

When resonance occurs, a sharp minimum in the intensity of the reflected light at the angle of resonance is observed: a plot of reflectivity vs. incident angle in Fig. 2 shows the dip at that angle. To excite a surface plasmon wave with an electromagnetic wave incident at the interface, the resonant condition has to be fulfilled: the propagating vectors of the surface plasmon (κ_{sp}) and the electromagnetic wave ($\kappa_{x,d}$) must be equal (see Fig. 1). It should be noted that surface plasmons are TM waves and, therefore, can only be excited by p-polarized light. The notation used in the following will be:

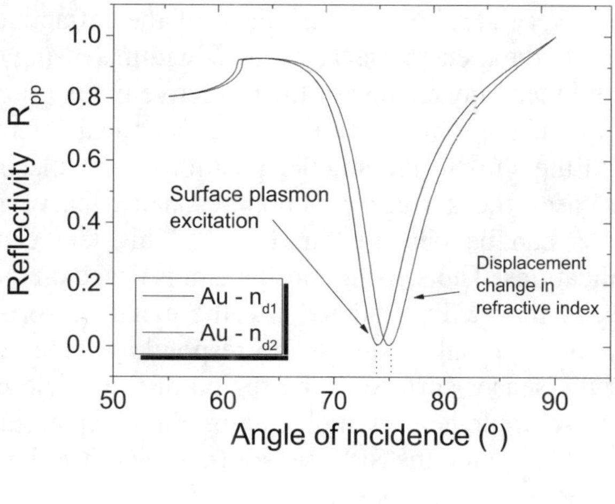

Figure 2. Resonant angle shift due to a change in the outer refractive index.

sp = surface plasmon, d= dielectric medium, o = air, m= metal. The wave vector of the incident light in air (κ_o) is :

$$k_o = \frac{\omega}{c}\sqrt{\varepsilon_o}$$

where ω is the frequency and ε_o the dielectric constant in air. Once in the dielectric medium, the wave vector component parallel to the surface ($\kappa_{x,d}$) must satisfy the resonant condition:

$$k_{xd} = \frac{\omega}{c}\sqrt{\varepsilon_d} \cdot sin\theta = \frac{2\pi}{\lambda}\sqrt{\frac{\varepsilon_m \cdot \varepsilon_d}{\varepsilon_m + \varepsilon_d}}$$

where θ is the incident angle, ε_m is the dielectric constant of the metal and ε_d is the dielectric constant of the prism. From this equation it can be concluded that the SPR propagation can supported only if $\varepsilon_{mr} < -\varepsilon_d$. This means that the surface plasmon can only exist if the dielectric permeability of the metal and dielectric medium are of opposite sign. This condition is only achieved at frequencies in the infrared to visible part of the spectrum by several metals of which gold and silver are the most employed. The use of silver gives more sensitive devices but with less stability than for gold. In general, surface plasmons are generated within the visible light frequency range with rapidly increasing loss in the IR. The thickness of the metal film is critical for the minimum reflectance value. The optimal thickness depends on the optical constants of the boundary media and on the wavelength of light. For gold, the optimal thickness is 45 nm at λ=790 nm .

The most common method of optical excitation to achieve the resonant condition is to use total internal reflection in prism-coupler structures, as already shown in Fig. 1. This is called the Kretschmann configuration. The resonant angle is very sensitive to variations of the refractive index of the medium adjacent to the metal surface, which is within sensing distance of the plasmon field and then, any change of the refractive index (due, for example, to a homogeneous change of material or to a chemical interaction) can be detected through the shift in the angular position of the plasmon resonance angle. In both cases the surface plasmon resonance curve shifts towards higher angles, as can be observed in Fig. 2. This fact can be used for biosensing applications. The sensing mechanism is based on variations of the refractive index of the medium adjacent to the metal sensor surface during the interaction of an analyte to its corresponding receptor, previously immobilized at the sensor surface in the region of the evanescent field. The recognition of the complementary molecule by the receptor causes a change in the refractive Index and the SPR sensor monitors that change. After the

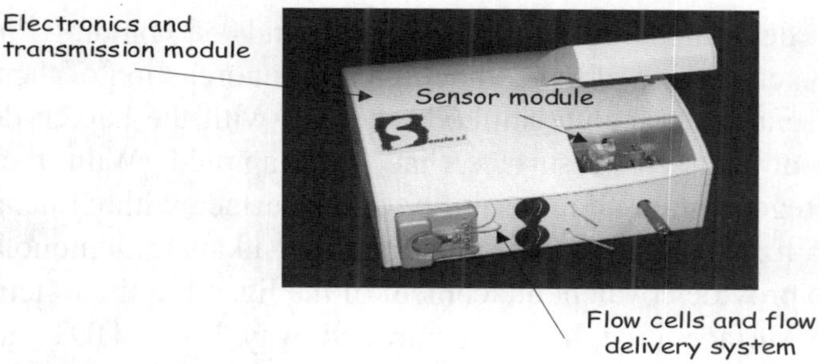

Figure 3. Portable SPR sensor prototype system including sensor, optics, electronics and flow delivery system.

molecular interaction, the surface can be regenerated using a suitable reagent to remove the bound analyte without denaturing the immobilized receptor. SPR biosensors have been used to measure a large variety of small size compounds, as for example environmental pollutants[14,15] and the same principle can be applied to the detection of harmful pathogens. These assays require the use of antibodies (monoclonal or polyclonal), which are the key components of all immunoassays, since they are responsible for the sensitive and specific recognition of the analyte. The application of immunoassays to environmental monitoring also involves the design of hapten derivatives of low molecular weight molecules, such as pesticides or harmful chemical compounds, to determine the antibody recognition properties[16]. Once hapten synthesis and monoclonal antibody production have been accomplished, the use of SPR biosensing technique provides a real-time monitoring of binding interactions without the need of labeling biomolecules.

We have developed a portable SPR sensor prototype (see Fig.3) as a highly sensitive field analytical method for environmental monitoring[17]. As a proof of its utility for the detection of pathogens, we have detected several pesticides including the chlorinated compound DDT, the neurotoxins of carbamate type (carbaryl), and organophosphorus compounds (chlorpyrifos). Organophosphate compounds are well-known irreversible inhibitors of acetylcholinesterase at cholinergic synapses that cause the disruption of nerve signal transduction. They have been considered as powerful neurotoxins with a threatening potential as chemical warfare agents. The chemical structure of these compounds is shown in Fig. 4.

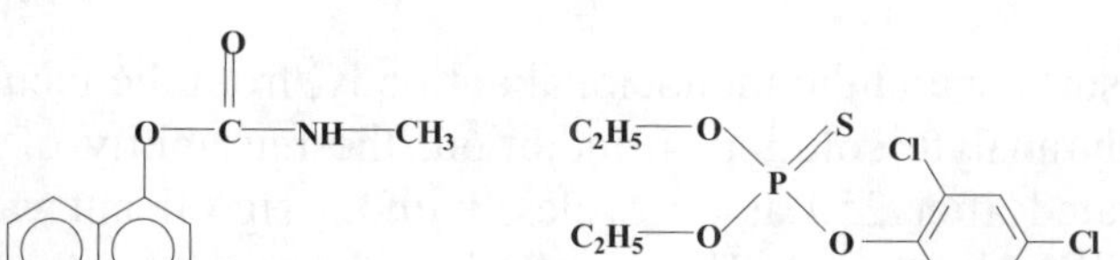

Carbaryl **Chlorpyrifos**

Figure 4. Chemical structures of the neurotoxins pesticides evaluated.

For the determination of these compounds a binding inhibition immunoassay, consisting of the competitive immunoreaction of the unbound antibody present in an analyte-antibody mixture with the hapten derivative immobilized at the sensor surface, has been applied. With the aim of assuring the regeneration and reusability of the surface without denaturation of the immobilized molecule, the formation of an alkanethiol monolayer was carried out to provide covalent attachment of the ligand to the functionalized carbodiimide surface in a highly controlled way. For DDT, the assay sensitivity was evaluated in the 0.004 - 3545 µg/l range of pesticide concentration by the determination of the limit of detection 0.3 µg/l and the I_{50} value 4.2 µg/l.

For carbaryl, (see Figure 5) the dynamic range of the sensor is 0.12-2 µg/l, with an I_{50} value for standards in buffer of 0.38 µg/l and a detection limit of 0.06 µg/l. Likewise the immunoassay for chlorpyrifos determination, afforded a high sensitivity (I_{50}= 0.11 µg/l) working in the 0.02-1.3 µg/l range.

The performance of the inhibition immunoassay enables the SPR biosensor to monitor the immunoreaction between the hapten immobilized

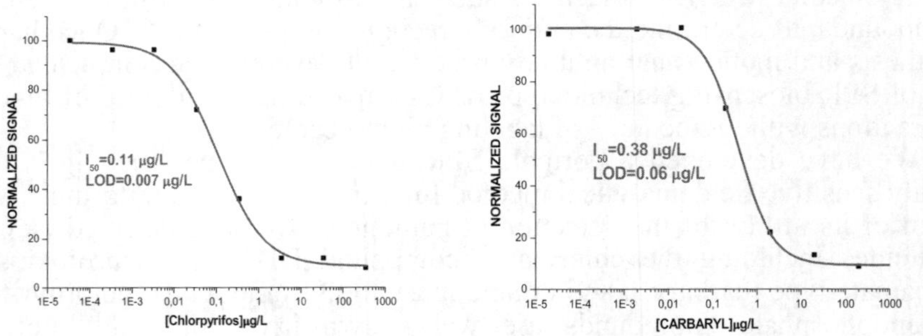

Figure 5. Calibration curve for the immunoassays determination by using SPR technology of (left) Chlorpyrifos (right) Carbaryl.

on the sensor surface and the monoclonal antibody, from the incubation of a mixed antibody-analyte solution. In addition, the reusability of the sensor was demonstrated after 250 assay cycles, without significant variations of the average maximum signal. The reusability of the sensor combined with the small time of response (approximately 15 min), makes the SPR immunosensing a valuable method for real-time and label-free analysis of environmental samples. This immunnosensing technique together with the portable surface plasmon resonance sensor developed can be applied as a fast and cost-effective field-analytical method for the monitoring of chemical and biological warfare agents if the corresponding receptor is available.

5. MAGNETOOPTICAL SURFACE PLASMON RESONANCE SENSOR

In order to achieve a higher sensitivity than the one provided by the SPR technology, we have developed a new biosensor device called Magneto-Optic Surface Plasmon Resonance (MOSPR) sensor[18]. The novel MOSPR sensor is based on the combination of the magneto-optic (MO) effects of the magnetic materials and the surface plasmon resonance. This combination can produce a great enhancement of the magneto-optic effects in the p-polarized light when the resonant condition is satisfied. Such enhancement is very localized at the surface plasmon resonance and depends strongly on the refractive index of the dielectric medium, allowing its use for optical biosensing.

The MOSPR sensor can be designed by means of the introduction of a magnetic layer and measuring the changes of the MO effects using the typical designs of the SPR sensors. In a Kretschmann arrangement, the enhancement of the MO Kerr effects in the reflected light can be achieved through the combination of Co/Au multilayers. Such MO Kerr effects will depend on the direction of the magnetization of the magnetic layer with respect to direction of propagation of the incident light. This behavior is due to the anisotropy introduced by the magnetization in the dielectric tensor of the magnetic layer. For example, if the magnetization is in the plane of the magnetic layer and perpendicular to the propagation plane of the light (transversal configuration), the MO effect produces a relative change of the reflectivity of the p-polarized light (R_{pp}):

$$\frac{\Delta R_{pp}}{R_{pp}} = \frac{R_{pp}(M) - R_{pp}(-M)}{R_{pp}(0)}$$

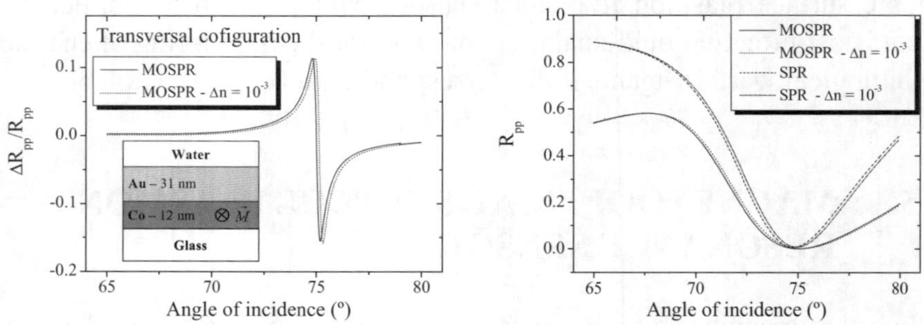

Figure 6. Left: MO effects of a Co/Au multilayer with a 12 nm thick Co layer and a gold layer of 31 nm, in the transversal configuration of the magnetization. Right: reflectivity of the multilayer compared with the case of a gold layer of 50 nm.

Therefore, when the SPR is excited, the $R_{pp}(0)$ is largely reduced while the magneto-optic components (ΔR_{pp}) are maintained, producing the sharp enhancement of the MO effect in the reflected light.

Figure 6 shows the MO effects of the p-polarized light in transversal configurations for a system formed by a Co/Au multilayer in a Kretschmann arrangement similar to the inset of Fig. 6, left, in which the incident medium is glass ($n_g = 1.5200$), and the outer medium is water ($n_w = 1.3323$). These calculations assume that the thickness of cobalt and gold are 12 and 31 nm, respectively, while their refractive index are $n_{Co} = 2.24+ i\ 4.06$ and $n_{Au} = 0.20+ i\ 3.08$ at a wavelength of the light of 632 nm.

Figure 6 also compares the reflectivity of this multilayer structure with the reflectivity of a gold layer of 50 nm (Fig 6, right). It can be observed that the introduction of cobalt widens the reflectivity curve due to its higher absorption leading to the reduction of the sensitivity of the sensor in the reflectivity measurements. However, the MO effects are very localized and show a very sharp curve. As a consequence, small variations of the refractive index will induce large changes in the MO response, as Fig. 6, left, illustrates with the representation of the changes of the MO effect under a variation of 10^{-3} of the refractive index.

The experimental set-up of the MOSPR is very similar to that of standard SPR sensors, the only difference being the introduction of a magnetic layer and a magnetic actuator to control the magnetization state of the magnetic layer. The measurements of the relative variations of the reflectivity require a change in the state of the magnetization as can be deduced from equation given earlier. Such change can be performed easily with a coil or with magnets since the saturation magnetic field of the Co layers is low. In this configuration, the incident monochromatic light is *p*-polarized and prism coupled to produce the surface plasmon, and the reflected light is collected with a photodiode. A flow cell and a peristaltic pump are used to control the different solutions required in the experiments. And, finally, the sensor is mounted on a rotation stage for the angle interrogation.

The evaluation of the sensitivity and the limit of detection of the device have been performed passing solutions of water and ethanol of different refractive index (from 1.3323 to 1.3343), and measuring in real time the variations of $\Delta R_{pp}/R_{pp}$ at a fixed angle of incidence. In Figure 7 shows the MO response of the MOSPR normalized by the minimum detectable signal, which was $1x10^{-3}$ in these experiments. These results are compared with the evaluation of the sensitivity of a standard SPR, using the same experimental set-up. In this situation the minimum detectable reflectivity was $5x10^{-4}$. The comparison of both measurements shows that the sensitivity in term of changes of refractive index of the MOSPR sensor is approximately three times the sensitivity of the SPR (see Fig. 7). These sensitivities represent a limit of detection of the MOSPR and SPR sensors of $5x10^{-6}$ and $1.5x10^{-5}$, respectively. Once the sensitivity in changes of refractive index is evaluated, we test the biosensing response of the MOSPR device with the detection of the physical adsorption of bovine serum albumin (BSA)

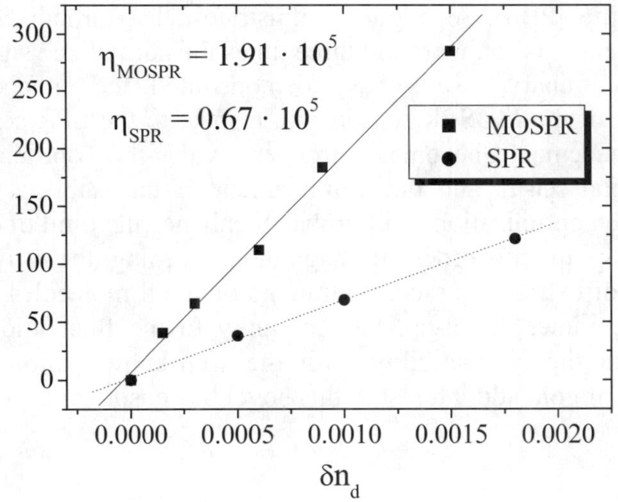

Figure 7. Comparison of the normalized signal of the MOSPR and the SPR sensors due to refractive index changes and evaluation their sensitivity.

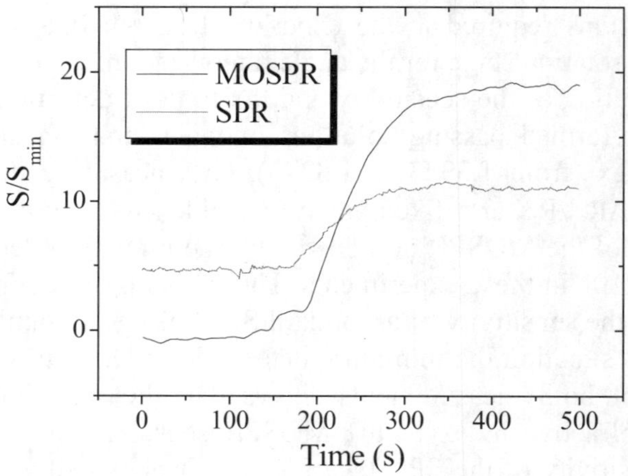

Figure 8. Comparison of the response of the SPR and MOSPR sensor to the adsorption of BSA proteins (10 μg/ml in PBS).

molecules, flowing a solution of 10 μg/ml of biomolecules in phosphate buffer saline (PBS). As Figure 8 illustrates, the normalized signal of the MOSPR sensor is again around three times the normalized signal of the SPR. These preliminary results have demonstrated the enhancement of the sensitivity of the MOSPR sensor with respect to the SPR sensor. However, this enhancement becomes stronger with the optimization of the experimental set-up and the improvement of the stability of the metallic layers. Such optimizations will probably enhance the limit of detection of the MOSPR up to one order of magnitude, opening the way to the direct detection of extreme low concentrations of small molecules. The simplicity of the experimental set-up, the immunity to the fluctuations of the light source and the compatibility with the well-known thiol immobilization chemistry of gold add interest to this novel biosensor concept.

6. MACH-ZEHNDER INTERFEROMETRIC BIOSENSOR

In a Mach-Zehnder interferometer (MZI) device[19] the light from a laser beam is divided into two identical beams that travel the MZI arms (sensor and reference areas) and are recombined again into a monomode channel waveguide giving a signal which is dependent on the phase difference

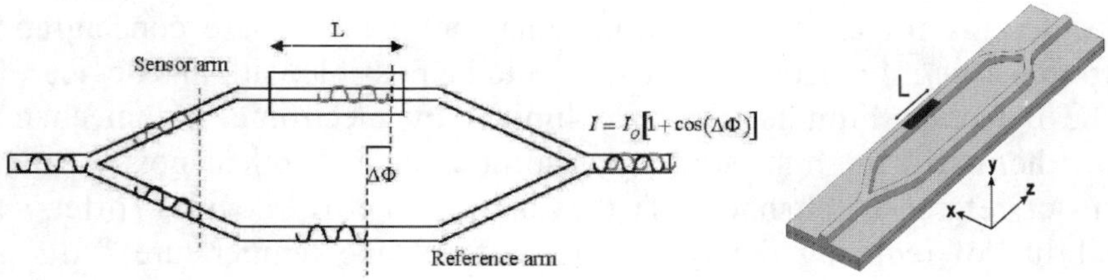

Figure 9. Mach-Zehnder interferometer configuration scheme and its working principle (left); 3-D view of the MZI structure (right).

between the sensing and the reference branches. Any change in the sensor area (in the region of the evanescent field) produces a phase difference (and therein a change of the effective refractive index of the waveguide) between the reference and the sensor beam and then, in the intensity of the outcoupled light. A schematic of this sensor is shown in Figure 9.

When a chemical or biochemical reaction takes place in the sensor area, only the light that travels through this arm will experience a change in its effective refractive index. At the sensor output, the intensity (I) of the light coming from both arms will interfere, showing a sinusoidal variation that depends on the difference of the effective refractive indexes of the sensor ($N_{eff,S}$) and reference arms ($N_{eff,R}$) and on the interaction length (L) :

$$I = \frac{1}{2} I_o \; 1 + \cos \; \frac{2\pi}{\lambda}\left(N_{eff,S} - N_{eff,R}\right)L$$

Here λ is the wavelength. This sinusoidal variation can be directly related to the concentration of the analyte to be measured.

For evaluation of specific biosensing interactions, the receptor is covalently attached to the sensor arm surface, while the complementary molecule binds to the receptor from free solution. The recognition of the complementary molecule by the receptor causes a change in the refractive index and the sensor monitors that change. After the molecular interaction, the surface can be regenerated using a suitable reagent in order to remove the bound analyte without denaturing the immobilized receptor as in the case of the surface plasmon resonance sensor.

The interferometric sensor platform is highly sensitive and is the only one that provides an internal reference for compensation of refractive-index fluctuations and unspecific adsorption. Interferometric sensors have a broader dynamic range than most other types of sensors and show higher sensitivity as compared to other integrated optical biosensors[19,20]. Due to the high sensitivity of the interferometer sensor the direct detection of small molecules (as for example environmental pollutants where concentrations down to 0.1 ng/ml must be detected) would be possible with this device.

The detection limit is generally limited by electronic and mechanical noise, thermal drift, light source instabilities and chemical noise. But the intrinsic reference channel of the interferometric devices offers the possibility of reducing common mode effects like temperature drifts and non-specific adsorptions. Detection limit of 10^{-7} in refractive index (or better) can be achieved with these devices[21] which opens the possibility of development of highly sensitive devices, for example, for in-situ chemical and biologically harmful agent detection.

For biosensing applications the waveguides of the MZI device must be designed to work in the monomode regime and to have a very high surface sensitivity of the sensor arm towards the biochemical interactions. If several modes were propagated through the structure, each of them would detect the variations in the characteristics of the outer medium and the information carried by all the modes would interfere between them. The design of the optical waveguide satisfying the above requirements and the dimensions of the Mach-Zehnder structure, is performed by using home-made modeling programs such as the finite differences methods in non-uniform mesh, effective index method and the beam propagation method. Parameters such as propagation constants, attenuation and radiation losses, evanescent field profile, modal properties and field evolution have been calculated[21]. In order to accomplish the above requirements two MZI technologies have been developed.

6.1 MZI nanodevice based on TIR waveguide

If we want to use TIR waveguides for the sensors, we must come to an agreement between single-mode behaviour, low attenuation losses for the fundamental mode and high surface sensitivity. For those reasons, the structure that has been finally chosen[22], for an operating wavelength of 0.633 μm, has the following geometry: (i) a conducting Si wafer of 500 μm thickness, (ii) a 2 μm thick thermal silicon-oxide layer on top with a refractive index of 1.46, (iii) a LPCVD silicon nitride layer of 250 nm thickness and a refractive index of 2.00, which is used as a guiding layer. To achieve monomode behaviour is needed to define a rib structure, with a depth of only 4 nm, on the silicon nitride layer by a lithographic step. This rib structure is performed by RIE and is the most critical step in the microfabrication of the device. Finally, a silicon-oxide protective layer is deposited by LPCVD over the structure with a 2 μm thickness and a refractive index of 1.46, which is patterning and etching by RIE to define the sensing arm of the interferometer. The final devices (within all its fabrication processes) are CMOS compatible. In Figure 10 the cross-section of the MZI TIR waveguide is shown.

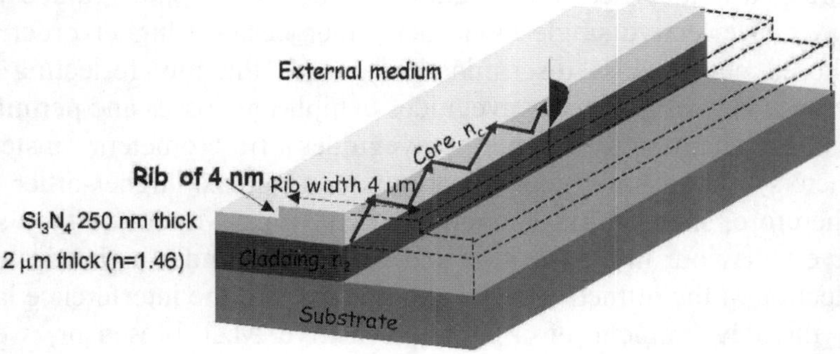

Figure 10. Cross-section of the optical waveguides used in the Mach-Zehnder interferometer. Note that a rib of only 4 nm in needed for monomode and high biomolecular sensitivity characteristics.

6.2 MZI Microdevice based on ARROW waveguide

In the ARROW (Antiresonant Reflecting Optical Waveguides) configuration, light confinement is based on Anti-Resonant Reflections rather than total internal reflection (TIR). The optical confinement of light in these waveguides is based on the total internal reflection at the air-core interface and a very high reflectivity, of 99.96%, at the two interference cladding layers underneath the core[23] (see Fig. 11).

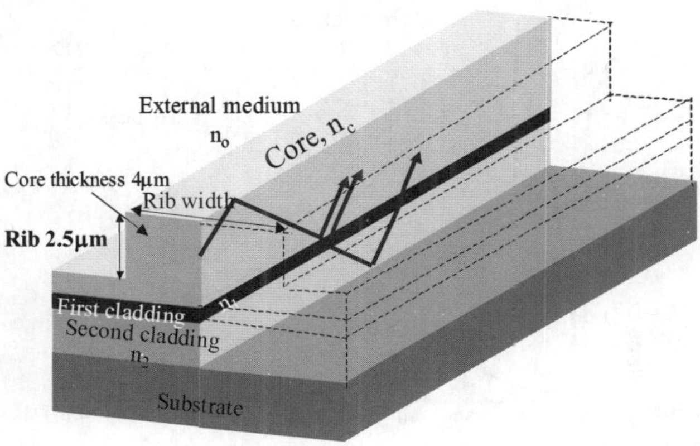

Figure 11. Cross-section of the optimized ARROW structure for MZI biosensing.

Each layer behaves as a quarter wave plate and, for a given wavelength, their refractive indexes and thickness have to be accurately selected. This structure is a kind of leaky waveguide (it does not support guided modes) that have an effective single-mode behaviour because higher order modes are filtered out by loss discrimination due to the low reflecting of the interference cladding. These waveguides exhibit low losses and permit larger dimensions than conventional waveguides (micrometers instead of nanometers). They have good discrimination against higher-order modes and, therefore, show virtual monomode behaviour. Moreover, they show a selective behaviour in polarization and wavelength and a high tolerance for the selection of the refractive index and thickness of the interference layers.

For the development of a highly sensitive MZI biosensor, we have designed an ARROW configuration which verifies the two conditions of monomode behaviour and high surface sensitivity[24]. This optimized waveguide consist on a rib-ARROW structure with a silicon oxide core layer (n_{core}=1.485) and thickness higher than 2 μm; a silicon oxide second cladding layer with a refractive index of 1.46 and a fixed thickness of 2 μm and a silicon nitride first cladding layer, 0.12 μm thick, with a refractive index of 2.00. The waveguide is overcoated with a thin silicon nitride layer (n_{ov} =2.00) and with a silicon oxide layer (n=1.46) with a thickness of 2 μm. The rib depth is 60% of the core thickness and the rib width should be lower than 8 μm to obtain single-mode behaviour. The structure of the ARROW device can be seen in Fig. 11.

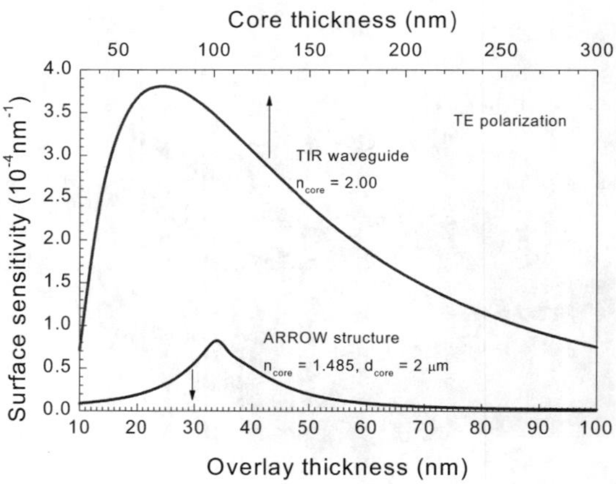

Figure 12. Comparison of sensitivities between TIR and ARROW optical waveguides for biosensing.

This ARROW waveguiding concept solves the two main limitations of the conventional TIR waveguides: the reduced dimensions for monomode behaviour (an important subject for further technological development and mass-production of the sensors) and the high insertion-losses in the optical-interconnects fiber-waveguide. With ARROW structures we can have monomode behaviour with a core thickness of the same size as the core of a single-mode optical fiber, suitable for efficient end-fire coupling. But a comparison between the evaluated surface sensitivity for both types of optical waveguides (TIR versus ARROW), as can be seen in Fig. 12, shows the higher sensitivity of the TIR device. Both designs, TIR and ARROW have advantages and disadvantages, and depending on the specific application one or another can be employed.

All the MZI devices are fabricated in clean room facilities. The wafer with the sensors is cut in individual pieces containing 14 devices each and polished for light coupling by end-face. A schematic and a photograph of one integrated MZI device is shown in Fig. 13. For measuring, the devices are implemented with a microfluidics unit, electronics, data acquisition and software for optical and biochemical testing. Chemical characterization for evaluating the sensor sensitivity was performed by using sugar solution with refractive indexes varying from 1.3325 to1.4004 (Δn 0.0002), as determined by an Abbe refractometer operating at 25ºC. The solutions with varying refractive indexes were introduced alternatively. With these measurements, a sensitivity calibration curve was evaluated, where the phase response of the

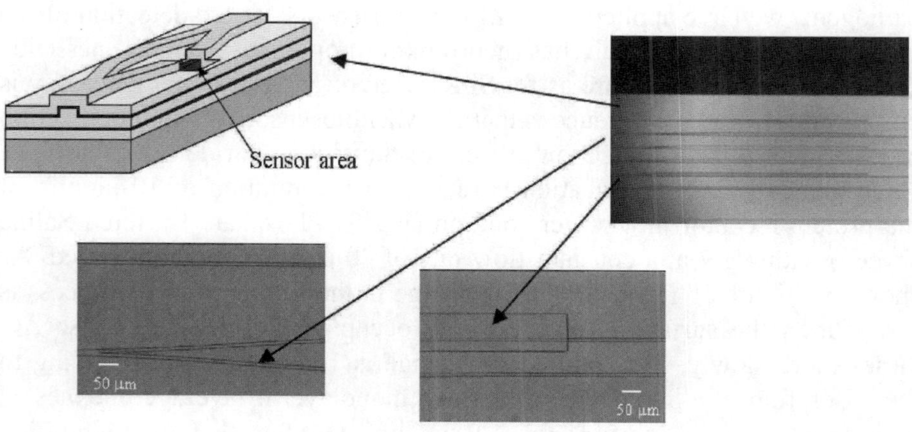

Figure 13. Photographs of the integrated Mach-Zehnder interferometer: details of the MZI Y-junction and sensor area.

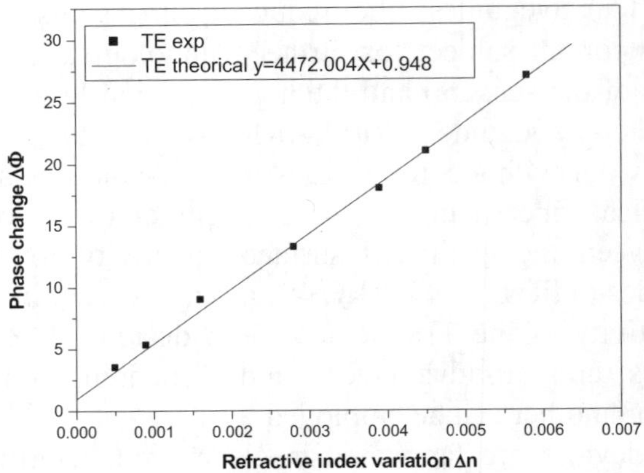

Figure 14. Sensitivity evaluation of a MZI nanodevice by using solutions of varying refractive indices.

sensor is plotted versus the variation in the refractive index, as is depicted in Fig. 14.For this device, the lower detection limit measured was $\Delta n_{o,min} = 2.5x10^{-6}$ corresponding to an effective refractive index change of $\Delta N = 1.4x10^{-7}$. We estimate that the lowest phase shift measurable would be around 0.03x2π. The detection limit value corresponds to a very high surface sensitivity around 2.10^{-4} nm^{-1}.

As a proof of the utility of MZI technology towards detection of pathogens, we have applied the MZI nanobiosensors for the detection of the insecticide carbaryl which has neurotoxin properties. For the pesticide analysis the same procedure as for SPR based on inhibition immunoassay is employed. The only difference is that for MZI biosensors the immobilization procedure used is a silanization process of the silicon nitride sensor surface. Once this chemical modification is ready, a concentration of 10 $\mu g \cdot ml^{-1}$ of the protein receptor in a buffer solution (PBST, Phosphate Buffered Saline Tween) with pH 7 at a constant flow rate of 20 $\mu l \cdot min^{-1}$ was introduced. As shown in Figure 15 (left), the phase change in the first part of the process is fast, but as the surface is progressively occupied the phase response ΔΦ varies more slowly. The total phase change is $16.2 \times 2\pi$, corresponding to the adsorption of a homogeneous antigen monolayer of average thickness d ≈ 3.2 nm (surface covering of 1.9 $ng \cdot mm^{-2}$). After that, the inhibition immunoanalysis is performed. Fig. 15 (right) shows the calibration curve obtained for the analysis of carbaryl pollutant by using the MZI nanobiosensor.

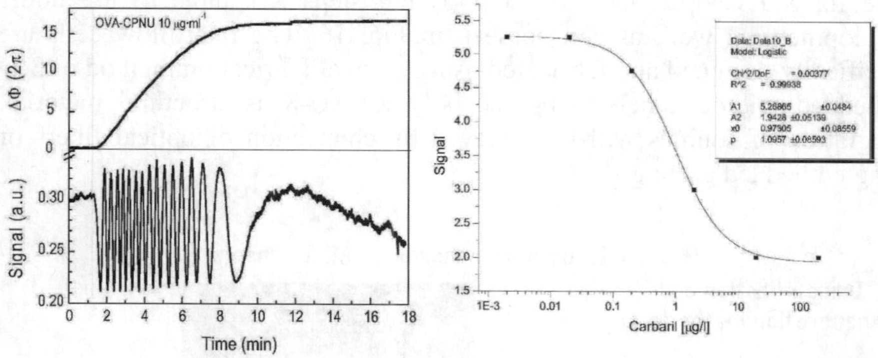

Figure 15. Covalent binding of the protein receptor (left); Calibration curve of the immunoassay of the pollutant carbaryl by using the MZI nanodevice (right).

7. TOWARDS A CMOS COMPATIBLE LAB-ON-A-CHIP MICROSYSTEM

The main advantage of the development of Mach-Zehnder devices fabricated with standard microelectronics technology comes from the possibility to develop a complete "lab-on-a-chip" by optoelectronic integration of the light source, photodetectors and sensor waveguides on a single semiconductor package together with the flow system and the CMOS electronics. The reagent receptor deposition can be performed by ink-jet, screen-printing or other technologies. A complete system fabricated with integrated optics will offer low complexity, robustness, a standardized device and, what is more important, portability. Devices for on-site analysis or point-of-care operations for biological and chemical warfare detection are geared for portability, ease of use and low cost. In this sense, integrated optical devices have compact structure and could allow for fabricating optical sensor arrays on a single substrate for simultaneous detection of multiple analytes. Mass-production of sensors will be also possible with the fabrication of miniaturised devices by using standard microelectronics technology[25].

For the development of a complete MZI microsystem several units must be incorporated on the same platform: (i) the micro/nanodevices, (ii) the flow cells and the flow delivery system, (iii) a modulation or compensation system for translating the interferometric signals in direct ones, (iv) integration of the light sources and the photodetectors (v) CMOS processing electronics. Several steps are undergoing for achieving that. Until now, we have fully developed the first step (i) but steps (ii) and (iii) are under development as well, as can be seen in Fig. 16. The microflow cells are specifically designed and fabricated using a novel fabrication method of 3-D embedded microchannels using the polymer SU-8 as structural material. Integration of sources will be achieved by connection of optical fibers or using embedded gratings.

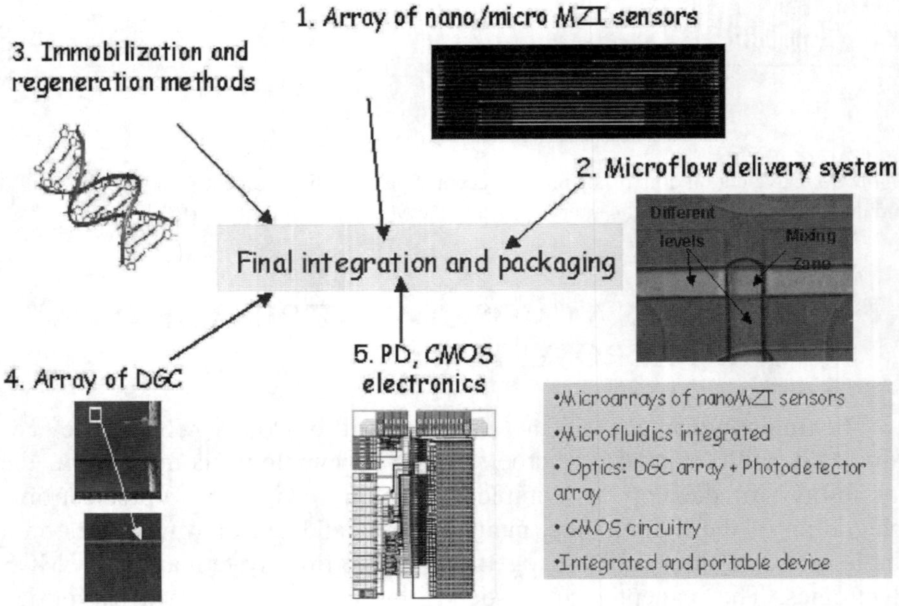

Figure 16. Steps towards the optoelectronics integration of MZI biosensors in a "lab-on-a-chip" microsystem.

8. CONCLUSIONS

For the rapid detection and identification of biological and chemical warfare agents, reliable multi-biosensor systems allowing label-free and real time measurement of simultaneous interactions must be developed. We have discussed the advances in the development of different optical biosensor platforms: a portable surface plasmon resonance Sensor (actually in commercialization) and an integrated Mach-Zehnder interferometer device made on Si technology. The feasibility of the different biosensor platforms have been proved by the immunological recognition of several pesticides including the chlorinated compound DDT, and the neurotoxins of carbamate type (carbaryl) and organophosphorus type (chlorpyrifos which resembles SOMAN chemical warfare agent). These results open the way for further development of portable and multianalyte platform for the detection of several biological and chemical warfare agents in-situ and in real-time.

ACKNOWLEDGEMENTS

This work has been supported by the national projects BIO2000-0351-P4. MAT2002-04484 and DESREMOL ((2004 20FO280). The authors want to thank to Dr. A. Montoya (UPV) for the immunoreagents.

REFERENCES

1. M. D. Wheeler, Photonics on the Battlefield, *Photonics Spectra*, 124-132 (1999).
2. R. A. Lewis, Microarrays take on bioterrorism, *Biophotonics International*, Jan/Feb, 40-41 (2002)
3. E. Croddy, C. Perez-Armendariz and J. Hart, (Eds.), *Chemical and Biological Warfare*, (Copernicus Books, Springer-Verlag, New York, 2002).
4. S. S. Iqbal, M. W. Mayo, J. G. Bruno, B. V. Bronk, C.A. Batt and J. P. Chambers, A review of molecular recognition technologies for detection of biological threat agents, *Biosens & Bioelec* **15**, 459-578 (2000).
5. C. A. Rowe-Taitt, J.W. Hazzard, K.E. Hoffman, J.L. Cras, J.P. Golden and F.S. Ligler, Simultaneous detection of six biohazardous agents using a planar waveguide array biosensor, *Biosens. & Bioelec.* **15,** 579-589 (2000).
6. F. S. Liegler and C.R. Taitt, (Eds), *Optical Biosensors: Present and future* (Elsevier, Amsterdam, Netherlands, 2002).
7. L. M. Lechuga, F. Prieto and B. Sepúlveda, Interferometric Biosensors for environmental pollution detection, in *Optical Sensors for Industrial, Environmental and Clinical Applications*, Springer Series on Chemical Sensors and Biosensors, Vol. 1, (Springer Verlag, New York, 2003).
8. L. M. Lechuga, Optical Biosensors, in *Biosensors and Modern Biospecific Analytical Techniques*, Comprehensive Analytical Chemistry Series, L. Gorton, Ed., (Elsevier Science BV, Amsterdam, Netherlands, 2005).
9. R. L. Rich and D.G. Myszka, Spying on HIV with SPR, *Trends in Microbiology* **11**, 124-133 (2003).
10. G. C. Nenninger, M. Piliarik and J. Homola, Data analysis for optical sensors based on spectroscopy of surface plasmons, *Meas. Sci. Technol.* **13** 2038-2046 (2002).
11. R. P. H. Kooyman and L.M. Lechuga, in *Handbook of Biosensors: Medicine, Food and the Environment*, E. Kress-Rogers (Ed.) (CRC Press, Florida, USA, 1997) 169-196.

12. J. Homola, S. Yee and D. Myszka, Surface Plasmon Biosensors, in F. S. Liegler and C. R. Taitt (Eds), *Optical Biosensor : Present and future* (Elsevier, Amsterdam, Netherlands, 2002).
13. J. Homola, Present and future of surface plasmon resonance biosensors, *Anal. Bioanal. Chem.* **377**, 528-539 (2003)
14. R. L. Rich and D. G. Myszka, Advance in surface plasmon resonance biosensor analysis, *Current Opinion in Biotechnology* **11** 54-61 (2000).
15. R. L. Rich and D. G. Myszka, Survey of the year 2003 commercial optical biosensor literature, *J. Mol. Recognit.* **18,** 1-39; 273-294 (2005).
16. A. E. Botchkareva, F. Fini, S. Eremin, J. V. Mercader, A. Montoya, S. Girotti, Development of a heterogeneous chemiluminescent flow immunoassay for DDT and related compounds, *Anal. Chim. Acta*, 43-52 (2002).
17 www.sensia.es 18 Patent application P200401437.
19. D.P. Campbell and J. McCloskey, Interferometric Biosensor, In "Optical Biosensors: present and future". Ed. F. Liegler and C.Rowe, (Elsevier, Amsterdam, Netherlands, 2002), 277-304.
20. R. G. Heideman and P. V. Lambeck, Remote opto-chemical sensing with extreme sensitivity: design, fabrication and performance of a pigtailed integrated optical phase-modulated Mach–Zehnder interferometer system, Sensors and Actuators B **61,** 100-127 (1999).
21. F. Prieto, A. Llobera, A. Calle, C Domínguez and L.M. Lechuga, Design and analysis of silicon antiresonant reflecting optical waveguides for highly sensitive sensors, J. Lightwave Tech. **18** (7), 966-972 (2000).
22. F. Prieto, B. Sepúlveda, A. Calle, A. Llobera, C. Domínguez, A. Abad, A. Montoya and L.M. Lechuga, An integrated Optical interferometric Nanodevice based on silicon technology for biosensor applications, *Nanotechnology* **14**, 907 (2003).
23. F. Prieto, L. M. Lechuga, A. Llobera, A. Calle and C Domínguez, Optimised Silicon Antiresonant Reflecting Optical Waveguides for Sensing Applications. *J. of Lightwave Tech.* **19** (1), 75-83 (2001).
24. F. Prieto, B. Sepúlveda, A. Calle, A. Llobera, C. Domínguez and L.M. Lechuga, Integrated Mach-Zehnder Interferometer based on ARROW structures for biosensor applications, *Sensors and Actuators B* **92,** 151-158 (2003).
25. C. Domínguez, L.M. Lechuga and J.A. Rodríguez, Integrated optical chemo- and Biosensors, in *Integrated Analytical Systems*, S. Alegret (Ed.) (Elsevier Science BV, Amsterdam, Netherlands, 2003).

MODELING OF ONE-DIMENSIONAL NONLINEAR PERIODIC STRUCTURES BY DIRECT INTEGRATION OF MAXWELL'S EQUATIONS IN THE FREQUENCY DOMAIN

Jiří Petráček
Brno University of Technology, Technická 2, 616 69 Brno, Czech Republic

Abstract: A simple numerical method for direct solving Maxwell's equations in the frequency domain is described. The method is applied to the modelling of one-dimensional nonlinear periodic structures.

Key words: Kerr nonlinearity, nonlinear optics, nonlinear periodic structures, optical bistability

1. INTRODUCTION

Many interesting phenomena can arise in nonlinear periodic structures that possess the Kerr nonlinearity. For analytic description of such effects, the slowly varying amplitude (or envelope) approximation[1] is usually applied. Alternatively, in order to avoid any approximation, we can use various numerical methods that solve Maxwell's equations or the wave equation directly. Examples of these rigorous methods that were applied to the modelling of nonlinear periodical structures are the finite-difference time-domain method,[2] transmission-line modelling[3] and the finite-element frequency-domain method.[4]

Here I describe a simple numerical method for solving Maxwell's equation in the frequency domain. As the structure to be analysed is one-dimensional, Maxwell's equations turn into a system of two coupled ordinary differential equations that can be solved with standard numerical routines.

S. Janz et al. (eds.), Frontiers in Planar Lightwave Circuit Technology, 141–146.

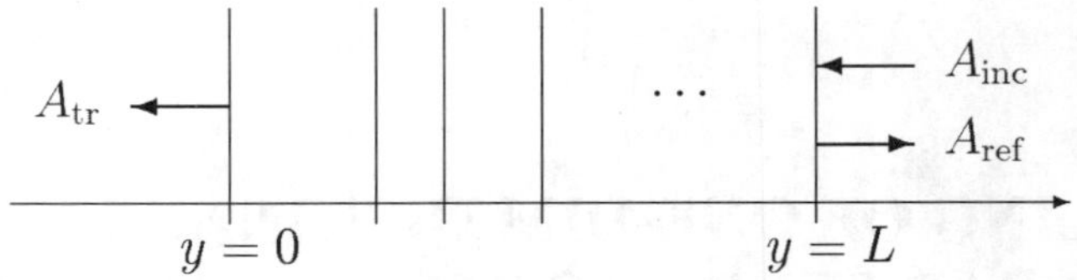

Figure 1. Schematic view of the one-dimensional layered structure. The vertical lines denote discontinuities of refractive index distribution. The monochromatic plane wave is incident in the y direction with the electric field amplitude $A_{\rm inc}$. The amplitudes of reflected and transmitted waves are $A_{\rm ref}$ and $A_{\rm tr}$, respectively.

2. FORMULATION

Consider a one-dimensional optical structure that consists of nonlinear layers. The structure is illustrated in Fig. 1. Assuming no field variation in the x and z directions and the convention $\exp(i\omega t)$ for the time dependence of the fields, Maxwell's equations take the form

$$\frac{\partial}{\partial y}E_x(y) = ikcB_z(y) \tag{1}$$

$$\frac{\partial}{\partial y}\left[cB_z(y)\right] = ikn^2\left(y, E_x\right)E_x(y) \tag{2}$$

Here $k \equiv \omega/c$ is the vacuum wavenumber, the refractive index

$$n\left(y, E_x\right) = n_0(y) + n_2(y)E_x(y)E_x^*(y) \tag{3}$$

changes in y direction only and functions $n_0(y)$ and $n_2(y)$ are the linear and Kerr nonlinear indices, respectively.

The total length of the structure is L. For sake of simplicity the semi-infinite outer layers are supposed to be linear and homogeneous so that electric field inside of the outer layers can be expressed using the amplitudes of incident, reflected and transmitted waves (denoted as $A_{\rm inc}$, $A_{\rm ref}$ and $A_{\rm tr}$, respectively)

$$E_x(y) = A_{\rm tr}\exp\left(inky\right), \quad y < 0 \tag{4}$$

$$E_x(y) = A_{\rm inc}\exp\left[ink\left(y-L\right)\right] + A_{\rm ref}\exp\left[-ink\left(y-L\right)\right], \quad y > L \tag{5}$$

Now we can find boundary conditions for the coupled system (1) and (2). $E_x(y)$ and $cB_z(y)$ are continuous for any y. The conditions at $y = 0$

$$E_x(0) = A_{\text{tr}} \tag{6}$$

$$cB_z(0) = n(y < 0)A_{\text{tr}} \tag{7}$$

are easily derived using equations (4) and (1), and the continuity of fields at $y = 0$. Similarly using (5), (1) and continuity at $y = L$ we obtain

$$E_x(L) = A_{\text{inc}} + A_{\text{ref}} \tag{8}$$

$$cB_z(L) = n(y > L)A_{\text{inc}} - n(y > L)A_{\text{ref}} \tag{9}$$

Here $n(y<0)$ and $n(y>L)$ denote the refractive indices in the outer layers. The last two equations can be solved with respect to the amplitude A_{inc} with the result

$$A_{\text{inc}} = \frac{1}{2}\left[E_x(L) + \frac{cB_z(L)}{n(y > L)}\right] \tag{10}$$

Usually the amplitude of the incident wave is known and the task is to find the other amplitudes and field profiles. In this case the numerical method can be described as follows.

We start with some arbitrary guess for A_{tr} and use (6) and (7) as the initial conditions for numerical integration of Eqs. (1) and (2). The integration can be easily carried out with common numerical routines provided that $n^2(y,E_x)$ is continuous i.e. inside of the layers. Therefore at the boundary between adjacent layers the continuity condition for the field components should be used. At the end of the integration $E_x(L)$ and $cB_z(L)$ are known and (10) is used to calculate $A_{\text{inc}}(A_{\text{tr}})$. This amplitude depends on the initial guess A_{tr} and is generally different from the known amplitude of the incident wave that we denote $A_{\text{inc},0}$. In order to find the correct value of A_{tr} we solve the algebraic equation

$$\left|A_{\text{inc}}(A_{\text{tr}})\right|^2 - \left|A_{\text{inc},0}\right|^2 = 0 \tag{11}$$

The solution is simple provided that the structure does not exhibit multistable behavior: the equation has just one real solution in the interval

$(0, |A_{inc,0}|)$ and we can directly use a root finding routine (e.g. Brent method). The opposite case is more complicated as it is necessary to carefully scan the interval $(0, |A_{inc,0}|)$ in order to get multiple solutions.

3. EXAMPLE: SLOW-WAVE OPTICAL STRUCTURE

To demonstrate the method an example of a slow-wave optical structure is modelled. Such structures consist of a cascade of directly coupled optical resonators in order to enhance the nonlinear effects.[5] The structure used here was recently defined within Working Group 2 of the European Action COST P11 (http://w3.uniroma1.it/energetica/slow_waves.doc). One period of the structure consists of one-dimensional Fabry-Perot cavity placed between two distributed Bragg reflectors (DBR) and can be described by the sequence

$$H^{1/2}(LH)^{26}LH^4L(HL)^{26}H^{1/2} \qquad (12)$$

where H and L are quarter-wave layers at wavelength $\lambda_0 = 1.55$ μm with $n_0 = 2.60$ and $n_0 = 2.36$, respectively. Both layers are nonlinear with the same value of n_2. There are $H^{1/2}$ layers at the ends in order to form an uninterrupted DBR when joining two or more periods.

One period of such a structure can be considered as a finite periodic grating with a defect layer in its middle and that is why a very narrow resonance (defect mode) appears inside of the band gap.

The task is to calculate the wavelength dependence of intensity transmission coefficient $T \equiv |A_{tr}/A_{inc}|^2$ near resonance λ_0 for various levels of nonlinearity and various numbers of periods. Here I present the results for structures of finite length (1, 2 or 3 periods) that are surrounded with semi-infinite linear medium with the same linear refractive index as H.

The results are shown in Figure 2. Spectra are plotted for various values of normalised input intensity $\gamma_{inc} \equiv n_2|A_{inc}|^2$ which can be considered as the strength of the nonlinearity. A positive nonlinearity, $\gamma_{inc} > 0$, that shifts the spectra to the right was used. As expected the nonlinearity (spectral shift) is enhanced with increasing number of the periods. Moreover the structures with 2 and 3 periods exhibit bistable behavior for $\lambda > \lambda_0$. In this case bistable switching is controlled by frequency tuning while input intensity is fixed.

Bistability controlled by input intensity (wavelength fixed) can also be observed. Figure 3 shows input-output intensity characteristics ($\gamma_t \equiv n_2|A_{tr}|^2$ being the normalised intensity of transmitted wave) when $\lambda > \lambda_0$ and demonstrates that the use of slow-wave structures significantly affects appearance of bistability and its threshold.

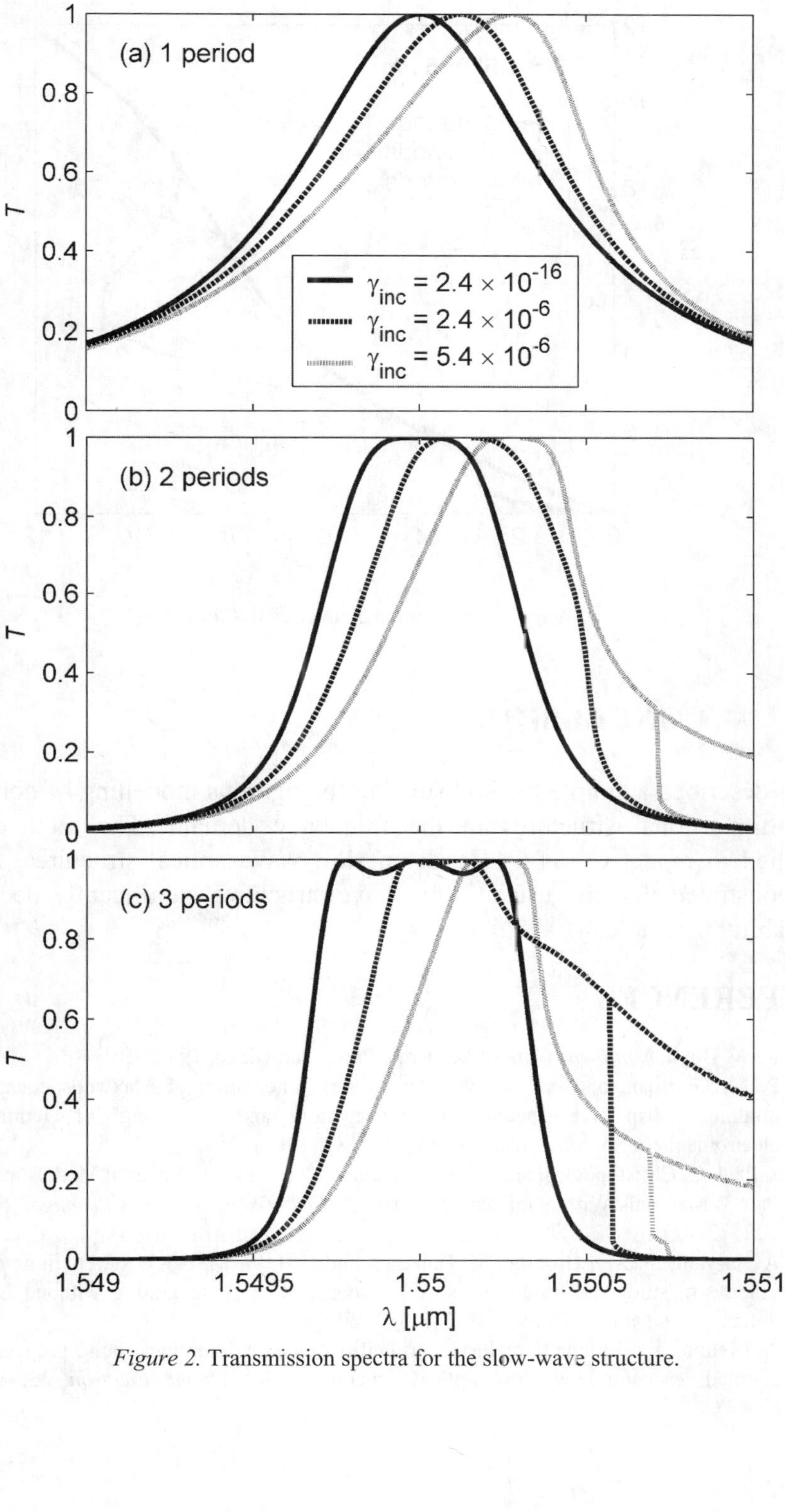

Figure 2. Transmission spectra for the slow-wave structure.

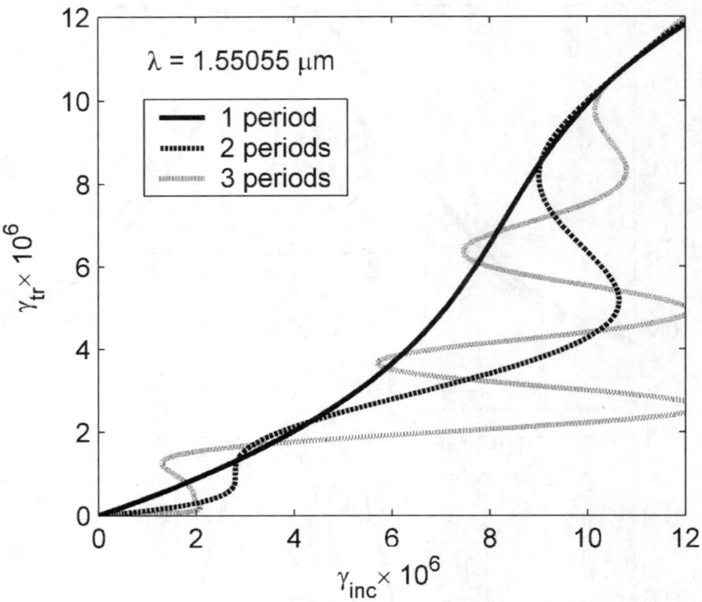

Figure 3. Input-output characteristics for $\lambda > \lambda_0$.

4. CONCLUSIONS

I described a simple method suitable for rigorous modelling of nonlinear one-dimensional structures in the frequency domain. The method was applied to model COST P11 task on slow-wave optical structures. It was demonstrated that the use of slow-wave structures significantly decreases bistablity threshold.

REFERENCES

1. R. W. Boyd, *Nonlinear Optics* (Academic Press, San Diego, 1992).
2. P. M. Goorjian, and A. Taflove, "Direct time integration of Maxwells equations in nonlinear dispersive media for propagation and scattering of femtosecond electromagnetic solitons," *Opt. Lett* **17**, 180-182 (1992).
3. J. Paul, C. Christopoulos, and D. W. P. Thomas, "Generalized Material Models in TLM – Part 3: Materials With Nonlinear Properties," *IEEE Trans. Antennas Propagat.* **50**, 997-1004 (2002).
4. A. Suryanto, E. van Groesen, M. Hammer, and H. J. W. M. Hoekstra, "A finite element scheme to study the nonlinear optical response of a finite grating without and with defect," *Opt. Quant. Electron.* **35,** 313-332 (2003).
5. A. Melloni, F. Morichetti, and M. Martinelli, "Linear and nonlinear pulse propagation in coupled resonator slow-wave optical structures," *Opt. Quant. Electron.* **35,** 365-379 (2003).

PROPAGATION OF OPTICAL PULSES THROUGH NONLINEAR PLANAR WAVEGUIDES WITH JUNCTIONS

E.A. Romanova[1], L.A. Melnikov[1], S.B. Bodrov[2], A.M. Sergeev[2]
[1]*Saratov State University, Astrakhanskaya 83, 410026, Saratov, Russia;*
[2]*Institute of Applied Physics RAS, Ulyanova 46, 603950 , Nizhny Novgorod, Russia*

Abstract: Investigation of spatiotemporal dynamics of optical pulse in step-index planar waveguide with Kerr nonlinearity is presented which is based on numerical solution of a (2D+T) wave equation for slowly varying amplitude of the total field. Effect of emission of radiation field that is specific for open dielectric waveguides is taken into account. This emission can be observed, first, as a result of light beam propagation through waveguide junctions, and second, due to some effects that vary the temporal distribution of an ultra-short optical pulse propagating in a regular nonlinear waveguide. Two types of junctions in weakly-guiding planar waveguides are under consideration: both waveguides have the same width and refractive index profile but possess different non-linear properties, and both waveguides have the same refractive index profile and nonlinearity but their widths are different. In the quasi-static approximation, the problem of optical pulse propagation through the junctions is reduced to the solution of a 2D equation for the pulse envelope with time coordinate given as a parameter. Spatial transformations of the stationary components of the pulse behind the junctions are studied in detail depending on their power and waveguide width. The approach is based on general methods of the theory of Hamiltonian dynamical systems and consists of the following steps: (i) finding the set of stationary nonlinear modes, (ii) investigation of power-dispersion diagrams, and (iii) investigation of global dynamics. Transmittance versus power dependencies demonstrate the applicability of the junctions for pulse shaping and power controlling. In the case of ultrashort optical pulses, self-steepening effect and second-order group velocity dispersion effect are shown to prevent formation of stable spatiotemporal pulse distribution owing to the emission of radiation field outside the guiding region.

Key words: Kerr nonlinearity, planar waveguides, optical pulses, numerical modeling, nonlinear modes, waveguide junctions.

S. Janz et al. (eds.), Frontiers in Planar Lightwave Circuit Technology, 147–187.

1. INTRODUCTION

Dielectric waveguides are now widely used in optical networks and optoelectronic devices. Contrary to closed metallic waveguides, the dielectric waveguides are open structures, so that a part of light beam power can be radiated out the guiding region into the outer space. Investigation of electromagnetic field transformation on macroscopic discontinuities of open waveguides is still important, in spite of the fact that many relevant theoretical problems were under consideration during the last decades.

Radiation propagating through a dielectric waveguide is concentrated in a micrometer-size spatial domain. The intensity of the electrical field can be very high over a long length, so that some nonlinear phenomena become well observable: second harmonic generation, Raman scattering and supercontinuum generation, formation and propagation of optical solitons.

Recent interest in the study and elaboration of novel guiding structures and materials is conditioned mostly by the availability of laser sources generating ultrashort optical pulses. Until recently, the studies of laser pulse self-effects in dielectric waveguides have been mainly concerned with the time-domain interactions or "time focusing", while the effect of spatial self-focusing was considered negligible because the Kerr constant in dielectrics is, as a rule, small (for fused silica, the Kerr constant $n_K \sim 10^{-16}$ cm^2/W), so that the transverse structure of the mode field was thought to be stable for intensities ~ MW/cm^2 of picosecond pulses[1,2]. This was a reason to consider self-phase modulation and diffraction of optical pulses separately. By this way, propagation of picosecond optical pulses in one-dimensional nonlinear medium with dispersion of dielectric permittivity has been well studied. In the case of a positive Kerr nonlinearity and anomalous group-velocity dispersion (GVD) the formation of temporal solitons has been proved.

However, as was shown in recent experiments on propagation of femtosecond optical pulses, the self-focusing effect is observed in bulk fused silica, and, for pulse peak intensity about 100 GW/cm^2, complicated dynamics of the spatiotemporal distribution gives rise to pulse splitting and to breakup of Gaussian beam[3]. In addition, for new materials with strong, in comparison with silica waveguides, Kerr nonlinearity (in chalcogenide glasses, $n_K \sim 10^{-14}$ cm^2/W, and in organic materials, $n_K \sim 10^{-12}$ cm^2/W), non-linear contribution to the refractive index in the core of a waveguide made of such a material may be of the same order as the contrast of the linear refractive index. Thus, it is reasonable to consider not only the temporal but also the spatial effects arising upon propagation of femtosecond pulses in waveguides made of conventional and novel materials.

Theoretical treatment of the femtosecond pulse propagation in a bulk Kerr medium with the dispersion of dielectric permittivity was based on the solution to the generalised nonlinear Schrodinger equation (GNLSE) with account of transverse spatial derivatives. In a general case, the main tool of the analysis is numerical simulation, mostly based on grid methods with the use of split-step technique[3,4,5].

Concerning to optical waveguides, such numerical modelling has been performed in the papers[6,7]. The spatiotemporal effects observed at modelling of bandwidth-limited femtosecond pulse propagation in nonlinear step-index optical waveguide with account of material dispersion (MD), finite time of a nonlinear response of waveguide material (FTNR) and the dependence of the group velocity on the field intensity (self-steepening effect (SS)) have been analysed in detail. A stable spatiotemporal distribution behind a region of unsteady-state regime in nonlinear waveguide has been observed just in the quasi-static approximation, when the MD, FTNR and SS effects have not been taken into consideration. With account of these effects, the pulse envelope continuously varies while the pulse propagates in nonlinear waveguide. The variation leads to the emission of some power fraction from the guiding region and prevents formation of a stable spatiotemporal distribution.

The main purpose of this paper is to consider a two-dimensional non-stationary (2D+T) problem of a nonlinear waveguide excitation by a non-stationary light beam and to study spatiotemporal phenomena arising upon propagation of the beam in a step-index waveguide, first, in the quasi-static approximation and, second, with account of MD and SS effects.

The theoretical approach is based on the solution to the mixed type linear/nonlinear generalized Schrodinger equation for spatiotemporal envelope of electrical field with account of transverse spatial derivatives and the transverse profile of refractive index. In the quasi-static approximation, this equation is reduced to the linear/nonlinear Schrodinger equation for spatiotemporal pulse envelope with temporal coordinate given as a parameter. Then the excitation problem can be formulated for a set of stationary light beams with initial amplitude distribution corresponding to temporal envelope of the initial pulse.

We consider two examples of waveguide junctions (Fig.1): structure A, when both waveguides have the same width and refractive index profile but possess different nonlinear properties (for simplicity, we assume that input waveguide is linear), and structure B, when both waveguides have the same refractive index profile and nonlinearity but their widths are different.

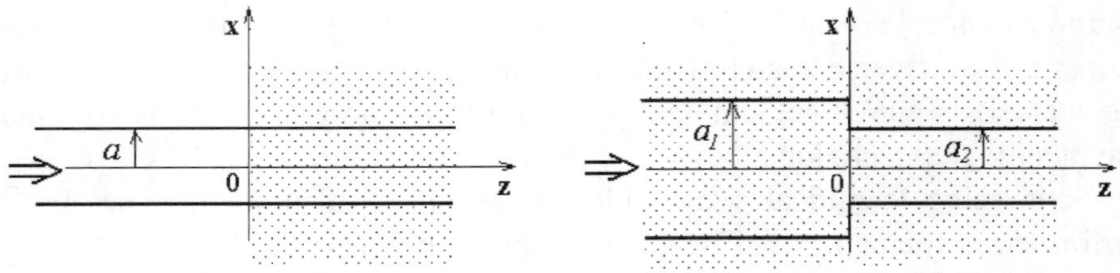

Figure 1. Structures under consideration.

For the problem of a nonlinear waveguide excitation by a stationary light beam we apply a convenient mathematical formalism based on the solution to the mixed-type linear/nonlinear Schrodinger equation for the slowly varying amplitude of electrical field. Stationary localised solutions of the equation are usually referred to as nonlinear modes and have been analysed for different kinds of planar waveguides[8] (predominantly for thin linear films bounded by nonlinear media). The global dynamics case, when the deviation from stationary solutions is large, is still under investigation, however. Global dynamics are directly related to the problem of light beam diffraction at nonlinear waveguide discontinuities, but until recently investigations of diffraction at discontinuities in linear optical waveguides, on one hand, and of nonlinear phenomena in optical waveguides, on the other, existed as two separate areas of research.

Concerning to the stationary light beams, our approach is based on general methods of the theory of Hamiltonian dynamical systems and consists of the following steps: finding the set of stationary nonlinear modes, investigation of their stability and investigation of global dynamics. Results of transmittance evaluation for each of the junctions can be generalized in the quasi-static approximation to the case of non-stationary light beam.

Propagation of optical pulses through the junctions is simulated numerically. Modelling technique based on the finite-difference beam propagation method (FD-BPM)[9] is used. Variation of the pulse duration behind the junctions is evaluated, first, in the quasi-static approximation. Next, the second-order GVD and SS effects are accounted for in the numerical algorithm via insertion of higher-order terms responsible for nonlinearity and group velocity dispersion into the wave equation. In this case, the electrical field amplitude is assumed to be slowly varying both in longitudinal direction and time.

2. METHODOLOGY

A common way to treat the problem of a picosecond pulse propagation in regular dielectric waveguide with Kerr nonlinearity was to solve the non-linear Schrodinger equation (NLSE) for the slowly varying temporal amplitude of electrical field[1,2].

For femtosecond optical pulses, the GNLSE with higher order nonlinear and dispersion terms was used. In some cases, both the NLSE and the GNLSE are integrable, and the dynamics of their solutions have been studied quite well[1]. When the self-focusing effect is taken into account, some additional terms appear in the GNLSE so that the equation may become non-integrable. Then the dynamics of its solutions cannot be studied by analytical methods, for example, by the inverse scattering technique[8]. In a general case, the main instrument of analysis is numerical modelling based predominantly on the grid methods with the split-step technique applied[3,4,5].

Propagation of non-stationary light beam in a nonlinear medium with material dispersion is described by the scalar wave equation for the linearly-polarized *y*-component of electrical field *E*(*x*,*z*,*t*):

$$\frac{\partial^2 E}{\partial z^2} + \frac{\partial^2 E}{\partial x^2} - \frac{1}{c^2} \cdot \frac{\partial^2 D}{\partial t^2} = 0 \tag{2.1}$$

with the electrical displacement

$$D(x, y, t) = D_l(x, y, t) + D_{nl}(x, y, t) \tag{2.2}$$

where the linear part is

$$D_l(x,z,t) = \int_0^\infty n_0^2(x,z,t')E(x,z,t-t')dt' \tag{2.3}$$

and the nonlinear part is

$$D_{nl}(x,z,t) = 2n_0 \Delta\tilde{n} E(x,z,t)\,. \tag{2.4}$$

Here n_0 is the linear refractive index and $\Delta\tilde{n}$ is the nonlinear part of the refractive index. Linear dispersion of the refractive index is accounted for in (2.2), (2.3). In general, nonlinear dispersion has to be taken into consideration, too. However, due to non-resonant character of the electronic nonlinearity in fused silica it is possible, as a first approximation, to neglect the nonlinear dispersion.

Non-stationary self-effects of the light beam depend on the pulse duration with respect to the time τ_R of nonlinear response of the medium. When the response is instantaneous, the refractive index at the time t is defined by the value of electrical field at the same moment. If the time of a nonlinear response is finite, the nonlinear part of the refractive index satisfies the following reduced matter equation[1,2]:

$$\tau_R \frac{\partial \Delta \tilde{n}}{\partial t} + \Delta \tilde{n} = n_K I \tag{2.5}$$

where $I = I(x,z,t) = (\varepsilon_0 / \mu_0)^{1/2} \beta / 2k \mid E(x,z,t) \mid^2$ is the intensity of the light beam.

Rigorous treatment of the self-action problem needs the transformation of Eq.(2.1), (2.5) into a system of integro-differential equations. However, if just some orders of group velocity dispersion and nonlinearity are taken into account, an approximate approach can be used based on differential equations solution[1]. When dealing with the 1D+T problem of optical pulse propagation in a dielectric waveguide, one comes to the wave equation with up to the third order GVD terms taken into account:

$$\begin{aligned} &2ik\left(\frac{\partial \tilde{E}}{\partial z} + \frac{1}{u}\frac{\partial \tilde{E}}{\partial t}\right) + 2i\varepsilon_1 \frac{\partial (\mid \tilde{E} \mid^2 \tilde{E})}{\partial t} + \varepsilon_0 \left(\mid \tilde{E} \mid^2 \tilde{E} - \tau_R \frac{\partial \mid \tilde{E} \mid^2}{\partial t} \tilde{E}\right) - \\ &-\varepsilon_2 \frac{\partial^2 \tilde{E}}{\partial t^2} + i\varepsilon_3 \frac{\partial^3 \tilde{E}}{\partial t^3} + \left(\frac{\partial^2 \tilde{E}}{\partial z^2} - \frac{n^2}{c^2}\frac{\partial^2 \tilde{E}}{\partial t^2}\right) - \frac{\varepsilon_0}{k^2 c^2}\frac{\partial^2 (\mid \tilde{E} \mid^2 \tilde{E})}{\partial t^2} = 0 \end{aligned} \tag{2.6}$$

where $\varepsilon_0 = 2n_0 n_K k^2$, $\varepsilon_1 = n_K k / (c\tau_0)$, $\varepsilon_2 = k_2 k / \tau_0^2$, $\varepsilon_3 = k_3 k / (3\tau_0^3)$, k is the mean wavenumber of the pulse, c is light velocity in vacuum, u is the group velocity, $k_2 = (\partial^2 k / \partial \omega^2)_{\omega_0}$ is the second-order dispersion coefficient, $k_3 = (\partial^3 k / \partial \omega^3)_{\omega_0}$ is the third-order dispersion coefficient, τ_0 is the initial pulse duration (the time coordinate t and τ_R are normalized to τ_0).

In the moving coordinate system ($z = z$, $t = t_0 - z / u$), under the assumption that $E(z,t) = \tilde{E}(z,t)\exp(i\omega_0 t - i\beta z)$ where $\tilde{E}(z,t)$ is a slowly varying function of both z and t so that the second derivatives in Eq.(2.6) can be neglected, $\tilde{E}(z,t)$ satisfies the so-called reduced wave equation :

$$2ik\frac{\partial \tilde{E}}{\partial z} + 2i\varepsilon_1 \frac{\partial(|\tilde{E}|^2 \tilde{E})}{\partial t} + \varepsilon_0 \left(|\tilde{E}|^2 \tilde{E} - \tau_R \frac{\partial |\tilde{E}|^2}{\partial t} \tilde{E} \right) - \\ -\varepsilon_2 \frac{\partial^2 \tilde{E}}{\partial t^2} + i\varepsilon_3 \frac{\partial^3 \tilde{E}}{\partial t^3} = 0 \tag{2.7}$$

Dealing with the (2D+T) problem of TE wave propagation, one has to take into account some additional terms responsible for light beam diffraction in the waveguide with linear refractive index profile $n_l(x)$:

$$2i\beta\frac{\partial \tilde{E}}{\partial z} + \frac{\partial^2 \tilde{E}}{\partial x^2} + (k^2 n_l^2(x) - \beta^2)\tilde{E} + 2i\varepsilon_1 \frac{\partial(|\tilde{E}|^2 \tilde{E})}{\partial t} + \\ +\varepsilon_0 \left(|\tilde{E}|^2 \tilde{E} - \tau_R \frac{\partial |\tilde{E}|^2}{\partial t} \tilde{E} \right) - \varepsilon_2 \frac{\partial^2 \tilde{E}}{\partial t^2} + i\varepsilon_3 \frac{\partial^3 \tilde{E}}{\partial t^3} = 0 \tag{2.8}$$

Here complex amplitude of the pulse envelope $\tilde{E}(x, z, t)$ is also a slowly varying function of z and t. Spatiotemporal distribution of the electric field is described by $E(x, z, t) = \tilde{E}(x, z, t)\exp(i\omega_0 t - i\beta z)$, β being the longitudinal wavenumber of the waveguide mode at the pulse peak.

In the quasi-static approximation ($\varepsilon_1 = \varepsilon_2 = \varepsilon_3 = \tau_R = 0$)[1], Eq.(2.8) can be written as follows:

$$2i\beta\frac{\partial \tilde{E}}{\partial z} + \frac{\partial^2 \tilde{E}}{\partial x^2} + (k^2 n_l^2(x) - \beta^2)\tilde{E} + \varepsilon_0 |\tilde{E}|^2 \tilde{E} = 0 \tag{2.9}$$

where the temporal dependence of $\tilde{E}(x, z, t)$ gives the pulse envelope with time coordinate t as a parameter. Formally, Eq.(2.9) is the same as the wave equation for stationary light beam propagating in a nonlinear planar waveguide (see Eq. 3.1 below).

In the problem of pulse diffraction on waveguide junctions, the quasi-static approximation is feasible if the diffraction length of the light beam is much shorter than the characteristic length of the pulse variation owing to the mentioned above MD, FTNR and SS effects which influence the pulse envelope. Then the results obtained for stationary light beam can be used in the analysis of the non-stationary beam self-focusing.

If the pulse width is of the order of picoseconds or shorter, the terms with $\varepsilon_1 \neq 0$ have to be taken into account. In the case of plane waves, this effect (usually referred to as the SS effect) leads to a change in the shape of the temporal envelope and to the shift of the pulse peak with respect to the center of the moving coordinate system. Another effect significant for ultrashort pulses is the time of finite nonlinear response of the medium τ_R (for electronic nonlinearity in fused silica it can be about 10^{-14} seconds).

As Eq.(2.8) is not integrable, numerical methods are to be used. In an open dielectric waveguide, some part of power can be lost via emitting radiation field on waveguide discontinuities. As it is well known, propagation of radiation field can be simulated by different methods based on expansion in terms of normal modes or orthogonal functions.[10] If, however, dielectric permittivity of the medium depends on the wave intensity, such methods are not efficient. In our opinion, for simulation of the spatial and temporal transformations in a nonlinear waveguide, it is most expedient to use the representation of the total field without any expansion of the sought solution. One of the most efficient and widely used numerical techniques based on total field representation is the finite-difference beam propagation method (FD-BPM)[9].

In our numerical model, Eq.(2.8) was transformed into a six-point finite-difference equation using the alternative direction implicit method (ADIM)[9]. At the edges of the computational grid $(-X, X)$ radiation conditions were applied in combination with complex scaling over a region $|x|>X_2$, where $(-X_2, X_2)$ denotes the transverse computational window. For numerical solution of the obtained tridiagonal system of linear equations, the sweep method[11] was used.

The nonlinear part of the susceptibility was introduced into the quasi-linear finite-difference scheme via iterations, so that at any longitudinal point, the magnitude of $|\tilde{E}|^2$ calculated at the previous longitudinal point was used as a zero approximation. This approach is better than the split-step method[1,2] since it allows one to jointly simulate both the mode field diffraction on irregular sections of the waveguide and the self-action effect by introducing the nonlinear permittivity into the implicit finite-difference scheme which describes the propagation of the total field.

In the solution of the Cauchy problem for Eq.(2.8) and Eq.(2.9), the initial field distribution was taken Gaussian in time:

$$\tilde{E}(x,0,t) = \psi(x)\exp(-(t-t_c)^2/2) \tag{2.10}$$

with the transverse profile corresponding to the fundamental linear TE mode of a planar waveguide:

$$\psi(x) = \begin{cases} \cos(ux/a)/\cos(u), & x \le a \\ \exp(-wx/a)/\exp(-w), & x > a \end{cases} \tag{2.11}$$

where u and w are the core and cladding parameters of a linear mode, respectively. In the time envelope (2.10), t_c corresponds to the central frequency of the pulse and is located in the middle of the computational grid $(t_c = 0)$. Note that the waveguide dispersion is not taken into account in the initial distribution because the Gaussian envelope of the bandwidth-limited pulse corresponds to the absence of group-velocity dispersion[1] (the longitudinal wavenumber β is assumed to be frequency-independent).

In order to analyze pulse shaping, the following parameters have been evaluated:

the pulse duration: $\tau_n(z) = \sqrt{2M_2(X_n, z) / M_0(X_n, z)}$ (2.12)

the pulse displacement: $D_n(z) = M_1(X_n, z) / M_0(X_n, z)$ (2.13)

the pulse asymmetry:

$$A_n(z) = (M_3(X_n, z) - M_1^3(X_n, z))\sqrt{2\left[M_2(X_n, z) - M_1^2(X_n, z)\right]^3} \quad (2.14)$$

where

$$M_i(X_n, z) = \int_{-X_n}^{X_n} dx \int_{-t_b}^{t_b} t^i dt \mid \tilde{E}(x, z, t) \mid^2, \ i = 0,1,2,3\,, \quad (2.15)$$

with integration within the limits of the computational window over the time coordinate $(-t_b, t_b)$ and over the transverse coordinate $(-X_n, X_n)$, $n = 0,1,2$. If $n = 0$, then $X_n = 0$; the above parameters are calculated at the waveguide axis. If $n = 1$ then $X_1 = a$, if $n = 2$ then the spatial integral in (2.15) is calculated over the computational window $(-X_2, X_2)$.
As output parameters we have also used the relative variation of the pulse duration,

$$\Delta\tau_n(z) = \left[\tau_n(z) - \tau_n(0)\right] / \tau_n(0) \quad (2.16)$$

and the normalized energy of the pulse:

$$T_n(z) = \int_{-t_b}^{t_b} \int_{-X_n}^{X_n} dtdx \mid \tilde{E}(x, z, t) \mid^2 \Bigg/ \int_{-t_b}^{t_b} \int_{-X_n}^{X_n} dtdx \mid \tilde{E}(x, 0, t) \mid^2 \quad (2.17)$$

$(n = 1, 2)$, which is transferred in the positive z-direction within the waveguide core $(X_1 = a)$, and also within the transverse computational window $(X_2 = 20a)$.

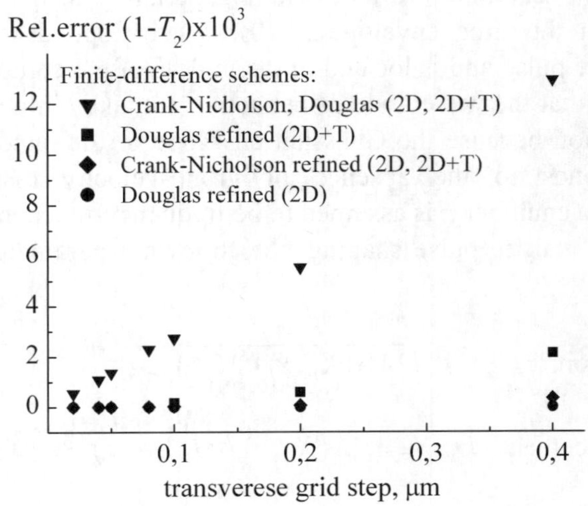

Figure 2. Relative computational error upon solving Eq.(2.8), (2.9); *z*=2mm, *a*=2 m.

As initial distribution corresponds to the linear mode (2.11) of the given waveguide, the deviation of $T_n(z)$ with respect to unity may be considered as a measure of the error in this method. The results presented in Fig.2 allow one to analyze the accuracy of the method depending on the type of finite-difference scheme (Crank-Nicholson[11] or Douglas[12] schemes have been applied) and on the method of simulation of conditions at the interface between the core and the cladding for both (2D+T) and 2D problems.

Refined finite-difference schemes were obtained by introducing corrections at the interface between the core and the cladding according to the method proposed in[13] (the value of the transverse grid step Δx was chosen to meet the condition $d = N\Delta x$ where N is an integer).

The following values of parameters have been used in the calculations: the refractive index of the waveguide core n_{co} = 1.491, the refractive index of the waveguide cladding n_{cl} = 1.487, the wavelength at the maximum of the initial temporal distribution λ = 1.53μm, $X_2 / X = 2/3$, $X = 80$ μm, $t_b = 8$, $\Delta z = 2$ μm, and $\Delta t = 0.04$.

We do not consider here the dependence of the group velocity on the beam divergence and the related spatiotemporal effects in the nonlinear medium[1] leading to additional changes in the pulse shape. In the region of the core, these effects are small, and the radiation field power for the levels of the input pulse power considered here is low.

3. STATIONARY LIGHT BEAMS

In papers[6,7], unsteady-state regime arising upon propagation of the stationary fundamental mode from linear to nonlinear section of a single-mode step-index waveguide was studied via numerical modeling. It was shown that the stationary solution to the paraxial nonlinear wave equation (2.9) at some distance from the end of a nonlinear waveguide has the form of a transversely stable distribution ("nonlinear mode") dependent on the field intensity, with a width smaller than that of the initial linear distribution.

The nonlinear modes are usually referred to as solutions to the nonlinear Helmholtz equation in the waveguide cross-section[8]. For their investigation, power-dispersion diagrams are commonly used that give values of critical powers and are helpful in stability analysis of the fundamental mode.[8]

We consider here step-index waveguides that are nonlinear both in the core and in the cladding, and their junctions (Fig.1): structure A, junction of waveguides that possess different nonlinear properties, in particular, linear and nonlinear waveguides of the same width, and structure B, junction of two waveguides which possess the same nonlinearity and refractive index contrast; their widths are different, however. The first structure is regular for low-intensity light beams, the second structure is inherently irregular. One-way propagation of a light beam through the structures can be reduced to the solution of the Cauchy problem with an initial field transverse distribution correspondning to the mode profile of the first waveguide of the junction.

In terms of nonlinear dynamical systems, the second waveguide of the junction can be considered as a system that is initially more or less far from its stable point. The global dynamics of the system is directly related to the spatial transfomation of the total field behind the plane of junction. In structure A, the initial linear mode transforms into a nonlinear mode of the waveguide with the same width and refractive index. In the structure B, the initial filed distribution corresponds to a nonlinear mode of the first waveguide; it differs from nonlinear mode of the second waveguide, however. The dynamics in both cases is complicated and involves nonlinear modes as well as radiation. Global dynamics of a non-integrable system usually requires numerical simulations. For the junctions, the Cauchy problem also cannot be solved analytically.

We find below a set of stationary nonlinear modes for the step-index nonlinear waveguide, investigate their stability and global dynamics. The latter is simulated numerically by the FD-BPM as a solution to the Cauchy problem for waveguide junctions under consideration.

3.1 Theoretical approach

Consider a step-index dielectric waveguide with a linear refractive index profile, $n_l(x) = n_{co}\ (|x| \leq a),\ n_{cl}\ (|x| > a)$.

If the linear refractive index contrast $\Delta n = n_{co} - n_{cl}$ is small and also if a nonlinear change to the refractive index is comparable with Δn then the back reflections from the junctions can be ignored because the reflection coefficient is of the order of $[\Delta n/(n_{co}+n_{cl})]^2$, i.e. approximately 10^{-6} for the refractive index contrast adopted here. These physical assumptions are inherent in using paraxial approach for the solution of the scalar wave equation for the stationary transverse electrical field of TE light beam propagating through the junction:

$$2i\beta\frac{\partial \tilde{E}}{\partial z} + \frac{\partial^2 \tilde{E}}{\partial x^2} + (k^2 n_l^2(x) - \beta^2)\tilde{E} + \varepsilon_0 \mid \tilde{E} \mid^2 \tilde{E} = 0 \qquad (3.1)$$

Here $\tilde{E}(x,z)$ is slowly varying amplitude of the total electrical field $E(x,z) = \tilde{E}(x,z)\exp(i\beta z - i\omega t)$, β is a parameter responsible for fast oscillations of the total field in longitudinal direction.

If β is taken as a longitudinal wavenumber of a linear mode, then we can use standard notation of linear waveguide optics[14]: $k^2 n_{co}^2 - \beta^2 = u^2 / a^2$, $(|x| \leq a)$ and $k^2 n_{cl}^2 - \beta^2 = -w^2 / a^2$, $(|x| > a)$. Here k is a vacuum wavenumber, u and w are the core and cladding parameters of a linear mode, respectively, $u^2 + w^2 = V^2$, where $V = ka\sqrt{n_{co}^2 - n_{cl}^2}$ is a linear waveguide parameter.

Eq. (3.1) can be converted into a more convenient form by setting $\xi = kx\sqrt{\Delta\varepsilon}$, $A = \sqrt{n_2 I} / \sqrt{\Delta\varepsilon}$, with $n_2 = 2n_l(x)n_K$, $\Delta\varepsilon = n_{co}^2 - n_{cl}^2$, and taking into account that $k^2 n_2 I = A^2 V^2 / a^2$:

$$2i\beta'\frac{\partial A}{\partial z} + \frac{\partial^2 A}{\partial \xi^2} + \left[g(\xi) + \mid A \mid^2\right]A = 0, \qquad (3.2)$$

where $g(\xi) = u^2 / V^2$, $|\xi|^2 \leq \xi_a$, $-w^2 / V^2$, $|\xi| > \xi_a$, $\xi_a = V = ka\sqrt{\Delta\varepsilon}$, $\beta' = \beta a^2 / V^2$.

Eq.(3.2) is the mixed type linear/nonlinear Schrödinger equation and belongs to a specific class of equations which are close to, but not exactly, integrable. The lowest-order integrals that remain constant in the system described by Eq.(3.2) as z changes are the dimensionless power:

$$P = \int_{-\infty}^{\infty} | A(\xi, z) |^2 d\xi \tag{3.3}$$

and the Hamiltonian:

$$H = \frac{1}{2\beta'} \int_{-\infty}^{\infty} (| \partial A(\xi, z) / \partial \xi |^2 - g(\xi) | A(\xi, z) |^2 - \frac{1}{2} | A(\xi, z) |^4) d\xi \tag{3.4}$$

where $\xi_2 = kX_2\sqrt{\Delta\varepsilon}$.

In numerical simulations, the time-averaged power

$$T_n(z) = \int_{-\xi_n}^{\xi_n} | A(\xi, z) |^2 d\xi \Big/ \int_{-\xi_n}^{\xi_n} | A(\xi, 0) |^2 d\xi \tag{3.5}$$

propagating in the positive z-direction, normalized to its initial value at $z = 0$ has been evaluated by integration within the waveguide core $(n = 1)$ and within the computational window $(n = 2)$.

The Hamiltonian was also evaluated by integration within the computational window:

$$H(z) = \frac{1}{2\beta'} \int_{-\xi_2}^{\xi_2} (| \partial A(\xi, z) / \partial \xi |^2 - g(\xi) | A(\xi, z) |^2 - \frac{1}{2} | A(\xi, z) |^4) d\xi \tag{3.6}$$

3.2 Nonlinear modes in planar waveguides

Stationary (i.e. for $\partial A / \partial z = 0$) localized solutions to Eq.(3.2) represent nonlinear modes in the planar waveguide and may be found in an analytical form via matching the partial solutions of Eq.(3.2) at the core/cladding boundary. The partial solutions are: Jacobi elliptical function[15] in the core and $A \sim w \cosh^{-1}\left[w(\xi - \xi^*)V\right]/V$ in the cladding (the functional dependence similar to a fundamental soliton in a uniform nonlinear medium). Here ξ^* is a parameter which depends on the boundary conditions. Contrary to the modes of a linear waveguide, the transverse profile of a nonlinear mode depends on the power in the mode.

In this paper, we have used the next numerical scheme to evaluate transverse profiles of the nonlinear modes. First, continuity conditions for $A(\xi)$ and $\partial A / \partial \xi$ at the core-cladding boundary have been applied for a known solution in the cladding, $A \sim w \cosh^{-1}\left[w(\xi - \xi^*)V\right]/V$, with the field magnitude $A(\xi_a)$. In this way, initial conditions have been obtained for the calculation of the nonlinear mode profile in the core. Then, by variation of the fitting parameter ξ^* in numerical calculation of Eq.(3.2) we have found such a value that meets the symmetric $\left((\partial A / \partial z)_{z=0} = 0\right)$ or antisymmetric $\left(A(0) = 0\right)$ condition for the mode amplitude at the waveguide axis. Examples of the transverse profiles for symmetric fundamental nonlinear mode corresponding to several values of the nonlinear mode power P are shown in Fig. 3. The mode with larger power has narrower transverse profile so that the power is more confined in the waveguide core.

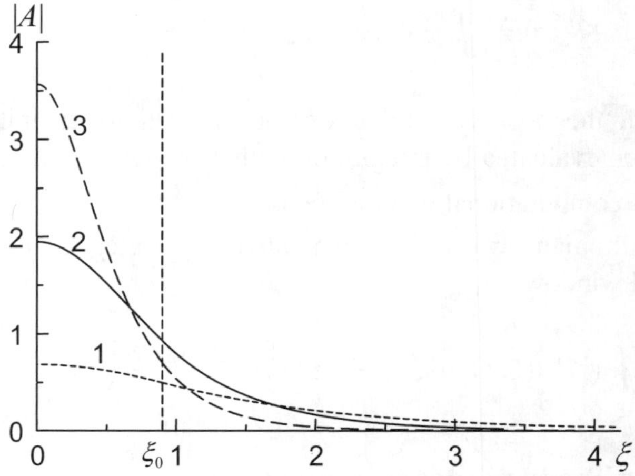

Figure 3. Transverse profiles of a symmetric nonlinear mode with P=1 (1), 5 (2), 10 (3). Vertical dashed line denotes the core/cladding boundary.

In general, the features of a nonlinear mode in a step-index planar waveguide depend on the parameter $n_K I / \Delta n$ – the ratio of the nonlinear part of the refractive index to the linear contrast of the waveguide. If $n_K I \ll \Delta n$, the transverse profile of the symmetric nonlinear mode coincides with the profile of the linear fundamental mode. If $n_K I \gg \Delta n$, the high-intensity light beam has the same transverse profile as a spatial soliton of a uniform nonlinear medium. Upon its propagation in the waveguide, the high-intensity light beam is not influenced by the core/cladding boundary.

In this paper, we are interested in structures with $n_K I \approx \Delta n$ when non-linear and waveguiding effects are comparable. For the estimation of the range of powers we are interested in, take into account that for the weakly-guiding structure $n_K I / \Delta n \approx n_2 I / \Delta\varepsilon = A^2$. Then the relationship between the dimensionless power (3.3) and the parameter $n_K I / \Delta n$ can be presented as a power dependence of A_0^2 where A_0 is the axial value of the nonlinear mode amplitude (Fig.4).

According to the results, $1 < n_K I / \Delta n < 10$ for the range of dimensionless powers which approximately fall into the interval 2<*P*<10. The latter is the range of powers that we are mostly dealing with in this paper.

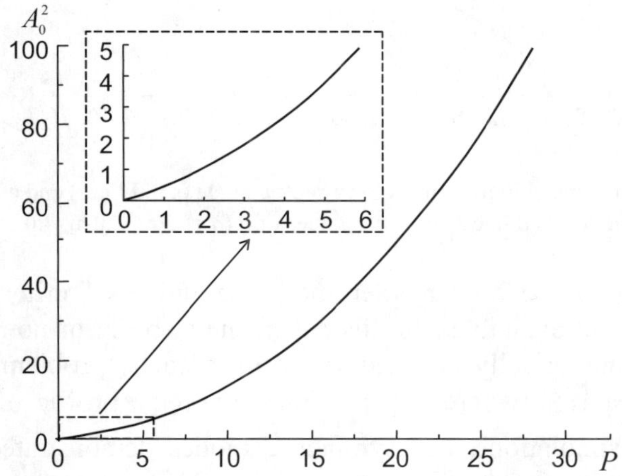

Figure 4. Axial value of $|A|^2$ depending on power of the mode, $a = 2$ μm.

For the analysis of a nonlinear mode stability which is important for a problem of nonlinear waveguide excitation, consider the power dispersion diagrams (Fig.5).

Branches corresponding to the symmetric mode (solid lines) meet the condition $dP / d\beta > 0$ so that the mode possesses global stability independently on its power[8]. Branches corresponding to anti-symmetric mode (dashed lines) show critical powers which depend on the waveguide width. The values of critical power denote the upper power limit for a single mode regime in the nonlinear waveguide with a given V (or core width).

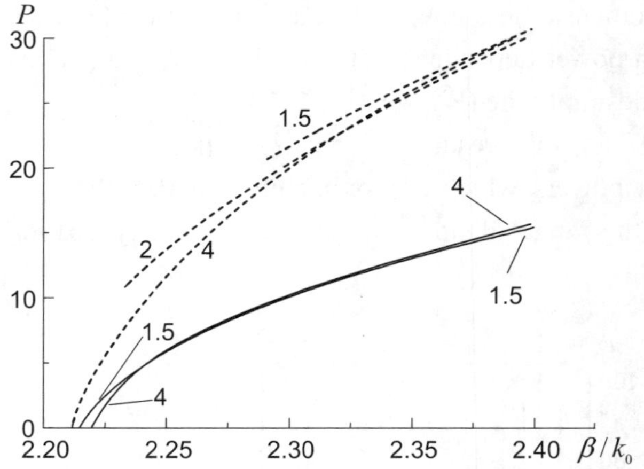

Figure 5. Power dispersion diagrams of symmetric mode (solid lines) and anti-symmetric mode (dashed lines) for some values of *a* (*a*=1.5 m, 2.0 m, 4.0 m).

Concerning to the higher-order modes, a universal criterium of their stability is not yet available[8]. In this paper, the stability of nonlinear modes was analysed numerically via solution of the Cauchy problem for Eq.(3.2) with $A(\xi,0) = A(\xi)$ where $A(\xi)$ is the transverse profile of a nonlinear mode. In the simulations, anti-symmetric modes demonstrated decay and emission of power outside the guiding region. Meanwhile, the fundamental symmetric mode was stable along any propagation distance.

As it is well known[8], stationary solutions to Eq.(3.2) occur at the extrema of the Hamiltonian for a given power. The solutions that correspond to global or local minimum of *H* for a family of solitons are stable. The representation of the output nonlinear waveguide as a nonlinear dynamical system by the Hamiltonian allows to predict, to some extent, the dynamics of the total field behind the waveguide junction.

In Fig.6, the Hamiltonian corresponding to the symmetric nonlinear mode in a regular waveguide versus the linear waveguide parameter *V* is presented for some given values of power *P*. The dependences have minima that move to smaller values of *V* with the increase of the beam power ($P = 5, V^{\min} \approx 0.7$, $P = 8, V^{\min} \approx 0.6$, $P = 10, V^{\min} \approx 0.5$).

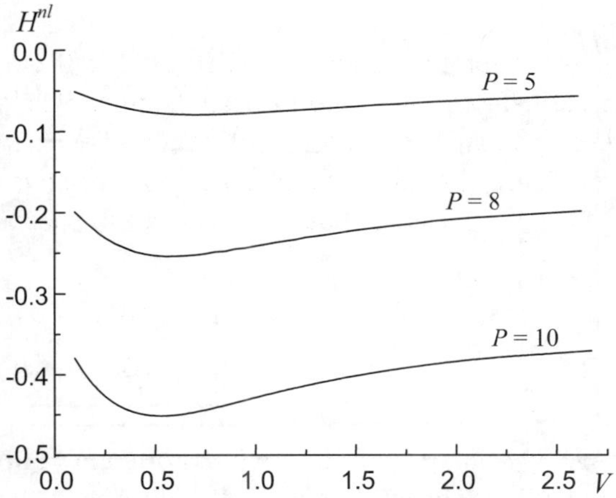

Figure 6. Hamiltonian (3.4) calculated with $A(\xi)$ corresponding to symmetric nonlinear mode versus the linear waveguide parameter V for some values of power P (3.3).

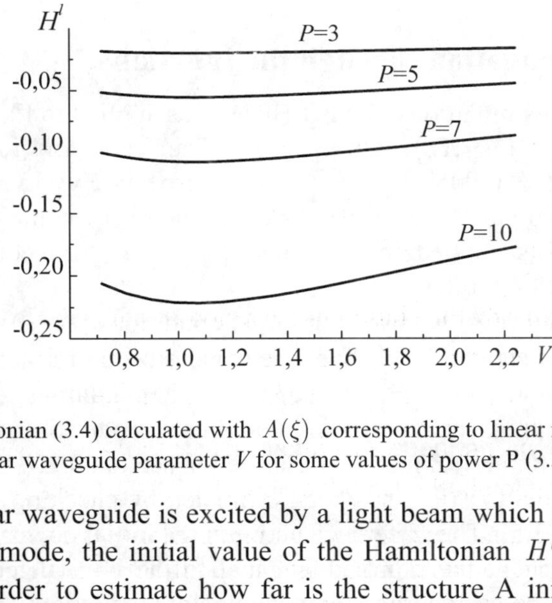

Figure 7. Hamiltonian (3.4) calculated with $A(\xi)$ corresponding to linear mode versus the linear waveguide parameter V for some values of power P (3.3).

If a nonlinear waveguide is excited by a light beam which is not matched with nonlinear mode, the initial value of the Hamiltonian $H^l(0)$ is greater than H^{nl}. In order to estimate how far is the structure A initially from its stable state, the Hamiltonian (3.4) with a linear mode profile given by (2.11) has been calculated. In Fig.7, $H^l(0)$ depending on V is presented for some values of light beam power.

The difference $H^l - H^{nl}$ shows how far is the structure A initially from its stable point (Fig.8). In the considered range of V, for powers $P < 7$, the difference is greater in narrower waveguides (curves P = 3 and P = 5 in Fig.8). For larger powers, a positive slope of the curve is also observed in some range of V (curve P = 10 in Fig.8).

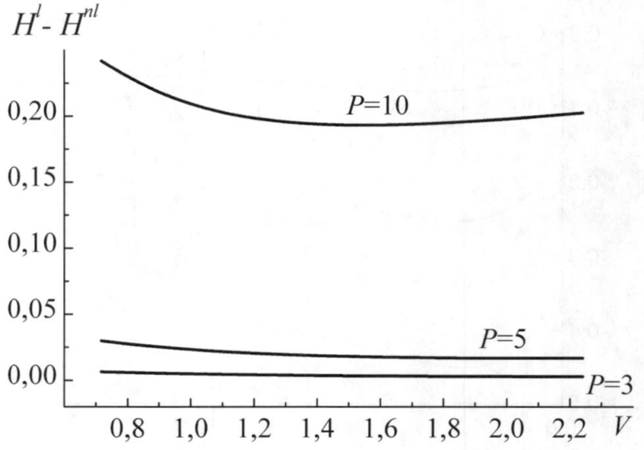

Figure 8. Difference $H^l - H^{nl}$ versus the linear waveguide parameter V for some values of power P (3.3).

3.3 Propagation through the junctions

In numerical simulations, the FD-BPM[9] was applied to Eq.(3.2). We used a transverse grid $(-\xi_b, \xi_b)$ with $\xi_b = 50$. Transverse and longitudinal grid steps were $\Delta\xi = 0.005$ and $\Delta z = 0.1$, respectively. In order to avoid reflections from the transverse grid boundaries, complex scaling of coordinates[16] was applied in the range $\xi_2 < \xi < \xi_b$ in combination with radiation conditions at the grid edges.[17]

For the structure A, we used the initial conditions $A(\xi,0) = F_0\psi(\xi)$, F_0 being the amplitude and $\psi(\xi)$ the transverse profile of a linear mode given by (2.11). For the structure B, we used the initial conditions $A(\xi,0) = A(\xi)$, $A(\xi)$ being the transverse profile of a nonlinear mode calculated for a given power P by the numerical method described in section 3.2.

Power of radiation field has been estimated by checking the total power determined by (3.5). The difference between the initial power $T_2(0) = 1$ and the power $T_2(z)$ at the point z is equal to the normalized power losses $P^{rad}(z)$. It is important to note that as the computational window has finite dimensions, some part of radiation leaks out the window behind the plane of junction so that the total power decreases (as the total power in the unbounded domain is conserved, the system remains Hamiltonian).

3.3.1 Junction of linear and nonlinear waveguides

An example of a linear mode transformation into the nonlinear mode in the structure A is shown in Fig.9. Unsteady-state regime[14] is observed that

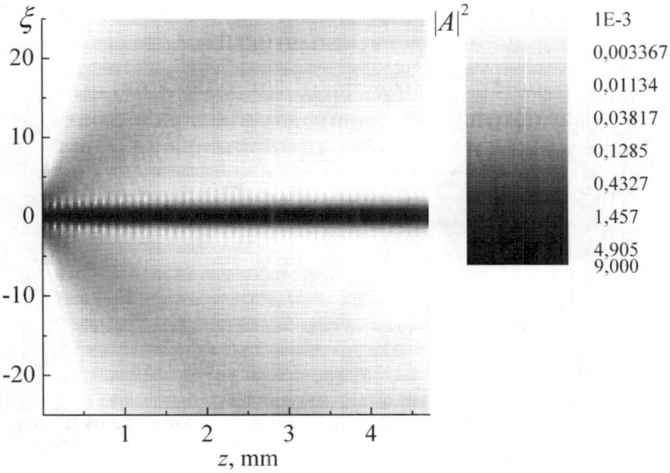

Figure 9. Unsteady-state regime in nonlinear waveguide of the structure A, a = 1.8 m, P = 7.

comprises self-focusing of the light beam in nonlinear waveguide and emission of some part of power outside the guiding region. In the waveguide core, the unsteady-state regime has a finite length and is followed by a steady-state regime where only nonlinear mode propagates in the core (Fig.9, $z > 4$ mm).

As a result of self-focusing, power propagating within the core $T_1(z)$ increases behind the region of unsteady-state regime (Fig.10). Meanwhile the total power of the light beam calculated within the computational window decreases due to emitting of radiation field outside the window.

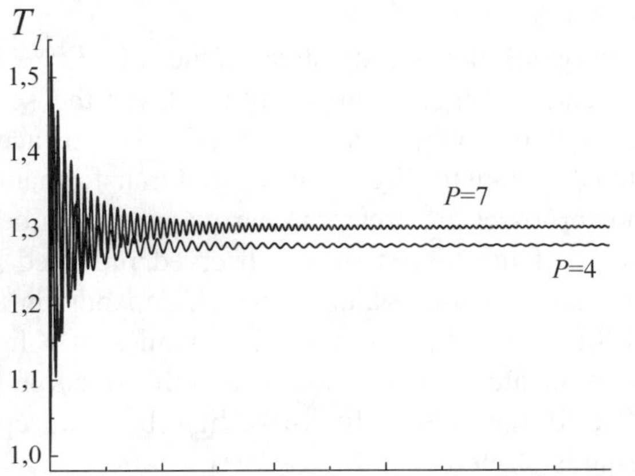

Figure 10. Longitudinal variation of the normalized power (3.5) propagating within the core of the nonlinear waveguide of the structure A for some values of power P (3.3), a = 1.8μm.

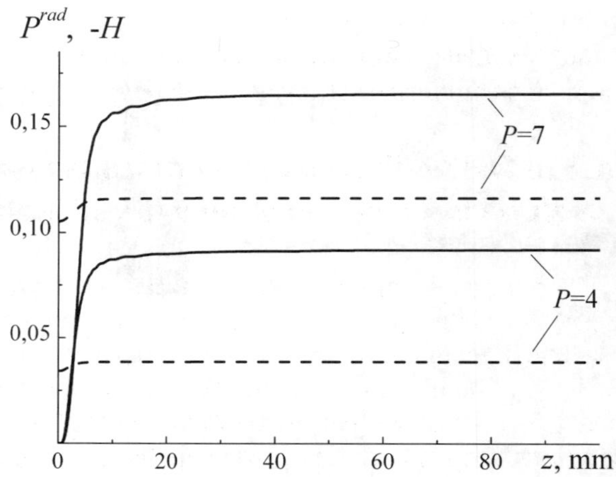

Figure 11. Longitudinal variation of the power of radiation field (solid lines) and of the Hamiltonian (3.6) (dashed lines) for some values of initial power P (3.3), $a = 1.8$ μm.

The normalized power of the radiation field $P^{rad}(z) = 1 - T_2(z)$ grows with the initial beam power P (Fig.11, solid lines). The Hamiltonian of the system decreases over the length of the unsteady-state regime (Fig.11, dashed lines) because the linear mode tends to transform into the nonlinear mode. However, the steady-state value of the Hamiltonian H^f (calculated at $z = 100$ mm) is greater than the value H^{nl} evaluated for the nonlinear mode of the same power as the initial light beam because a part of power has gone with radiation field. The difference between these values of the Hamiltonian $\Delta H = H^f - H^{nl}$ is shown in Fig.12 versus the linear waveguide parameter V.

For the comparison, the steady-state value of P^{rad} (calculated at $z = 100$ mm) depending on V is also plotted. Over the set of numerical experiments, the initial power was fixed ($P = 3$). In accordance with estimations of the Hamiltonian (Fig.8), the spatial transformation of an initial light beam and power losses are greater in narrower waveguides.

Such a behavior of the total field is observed provided that the beam power is smaller than a definite value P_s which depends on the waveguide width (for a = 1.8μm, $P_s \sim 8$). The spatial dynamics of a light beam with $P > P_s$ is more complicated because nonlinear self-effects in radiation field increase so that the formation of soliton-like light beams propagating in the waveguide cladding is observed.

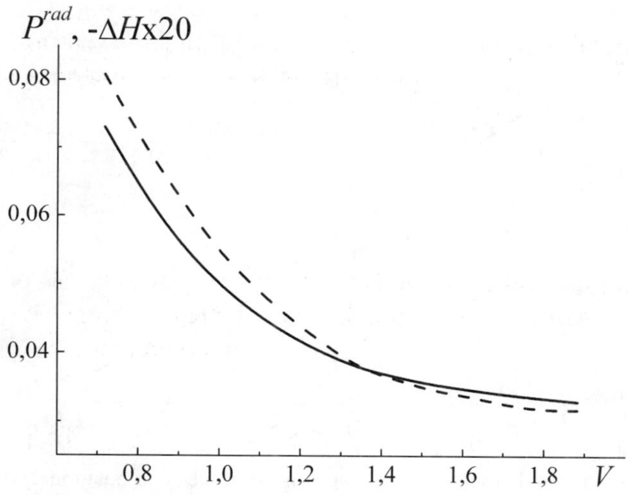

Figure 12. Power of radiation field (solid line) and variation of the Hamiltonian (3.6) (dashed line) versus the waveguide parameter, $z = 100$ mm, $P = 3$.

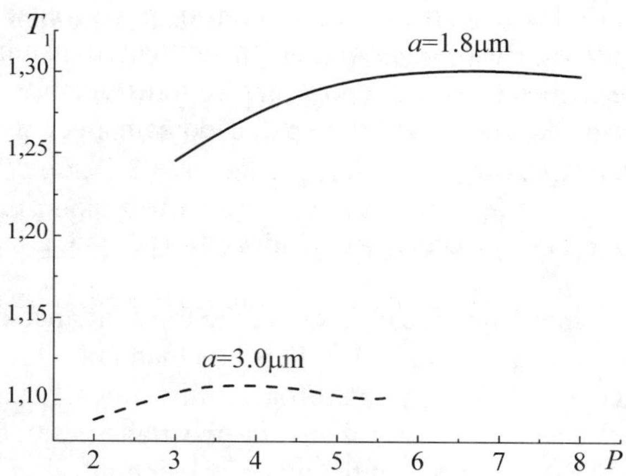

Figure 13. Normalized power (3.5) propagating within the core of the nonlinear waveguide A versus the initial power P for some values of the core half-width a, z=100 mm.

The transmittance of the structure A depending on the input power P was evaluated via calculation of $T_1(z)$ (Fig.13) and $T_2(z)$ (Fig.14). It is seen that self-focusing of the light beam in the core of nonlinear waveguide increases with input power, but rate of the increase diminishes so that for the powers $P > 7$ ($a = 1.8$ μm) and $P > 4$ ($a = 3.0$ μm) the dependence is weak. Negative slope of the curve in this range results from the mentioned above soliton-like

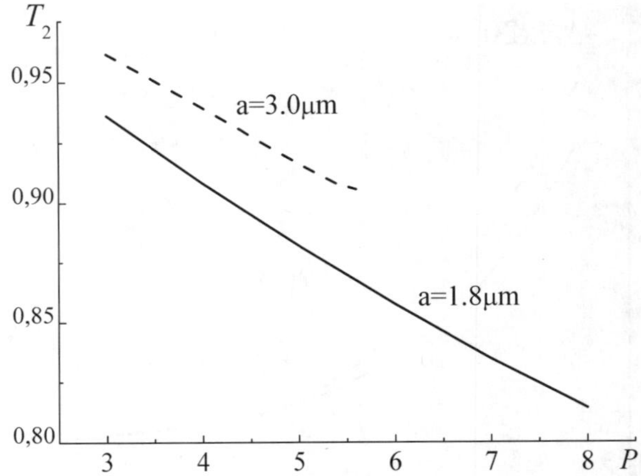

Figure 14. Normalized power (3.5) propagating within the computational window in a nonlinear waveguide A versus the initial power P for some values of the core half-width a, z = 100mm.

behavior of the radiation field. Power propagating in the positive z-direction within the computational window diminishes with the input power.

This dependence has a character close to linear in the considered range of initial powers. By this way it is demonstrated that transmittance of the structure is power dependent, and moreover, the transmittance-versus-power relation is different depending on the spatial domain over which the light beam intensity is integrated.

3.3.2 Junction of nonlinear waveguides

Contrary to the linear/nonlinear junction, the transmittance of a nonlinear step-like discontinuity (structure B, Fig.1) is less than unity for low-intensity light beams. Under the assumption of a unidirectional propagation, the transmittance can be evaluated by matching the transverse components of electrical field of TE polarization at the plane of junction:

$$F_1\psi_1 = F_2\psi_2 + A^{rad}, \tag{3.7}$$

where ψ_1 and ψ_2 are the transverse profiles of the fundamental linear mode (defined by (2.11) with the replacement $x \to \xi$) in the first and second waveguides of the junction, F_1 and F_2 are amplitudes of the modes, respectively. A^{rad} represents the radiation field which is usually expressed as an integral over the continuum of radiation modes[14]. The linear transmittance of the structure T^l can be evaluated as a ratio of the output beam power to the input beam power with the account of orthogonality conditions for linear modes of the second waveguide:

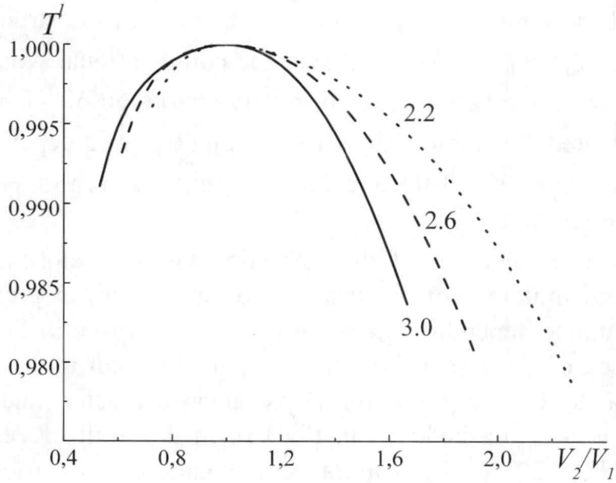

Figure 15. Transmittance of linear step-like discontinuity for some values of a_1 (a_1 = 2.2 μm, 2.6 m, 3.0 m).

$$T^l = \int_{-\infty}^{\infty} |A_2(\xi, z)|^2 d\xi \Big/ \int_{-\infty}^{\infty} |A_1(\xi, z)|^2 d\xi \tag{3.8}$$

The transmittance depends on the half-widths a_1 and a_2, not only on the ratio a_2 / a_1 (Fig.15). In the unidirectional approximation, the value of the linear transmittance is the same for light beams propagating in opposite directions through the discontinuity.

The transmittance of the nonlinear step-like discontinuity in cylindrical waveguide has been evaluated[18] under the assumption that profiles of low-intensity nonlinear modes can be approximated by profiles of linear modes. According to the results, nonlinear transmittance is less or greater than the linear one depending on waveguide parameters of the first and the second waveguides, $V_1 = ka_1(n_{co}^2 - n_{cl}^2)^{1/2}$, $V_2 = ka_2(n_{co}^2 - n_{cl}^2)^{1/2}$, respectively. However, the difference between transverse profiles of linear and nonlinear modes can be significant for the considered range of powers (Fig.8). That is why numerical modeling of nonlinear mode propagation through the discontinuity is a reasonable way to study the spatial transformation of the mode field.

The transmittance T^{nl} of the structure B,

$$T_n^{nl} = \int_{-\xi_n^2}^{\xi_n^2} |A_2(\xi, z)|^2 d\xi \Big/ \int_{-\xi_n^1}^{\xi_n^1} |A_1(\xi, z)|^2 \, d\xi, \; n = 1, 2 \tag{3.9}$$

was evaluated by intensity integration over the core of the structure (n = 1, $\xi_1^1 = ka_1\sqrt{\Delta\varepsilon}$, $\xi_1^2 = ka_2\sqrt{\Delta\varepsilon}$) and over the computational window (n = 2, $\xi_2^1 = \xi_2^2 = \xi_2 = kX_2\sqrt{\Delta\varepsilon}$). In Fig.16, relative variation of T_2^{nl} with respect to T^l is presented for some values of the input power depending on the waveguide parameter V_2 of the second waveguide, with a given half-width a_1 of the first waveguide.

If the nonlinear increase of the refractive index is close to its linear contrast, the transmittance of the structure B can be less or greater than its linear transmittance depending on a_1, a_2 (P = 3, solid line in Fig.16). If the nonlinear increase of the refractive index is greater than its linear contrast, the transmittance of the structure B is always greater than its linear transmittance (P = 5, dashed line in Fig.16). In the limit of very intensive light beams (P > 20) the transmittance is total, $T^{nl} = 1$ (dotted line in Fig.16).

Contrary to the linear/nonlinear junction, the transmittance of the structure B has a tendency to grow with input power up to unity in the limit $P(0) \gg 1$ (Fig.17). The values of the nonlinear transmittances are different for light beams propagating in opposite directions through the discontinuity.

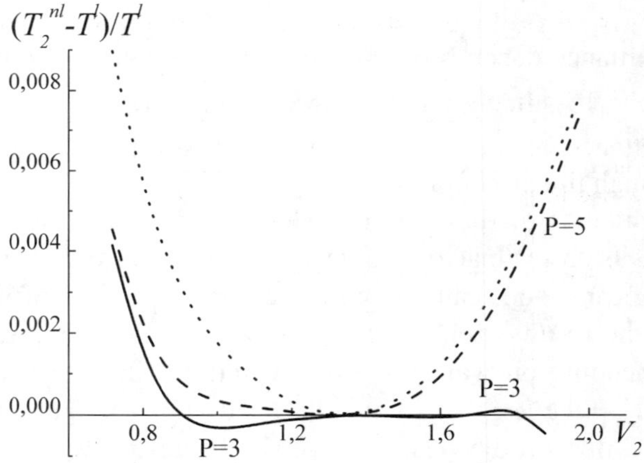

Figure 16. Transmittance of nonlinear step-like discontinuity in comparison with linear transmittance (Fig.15) for some values of initial power P (3.3), a_1 = 3.0 μm,

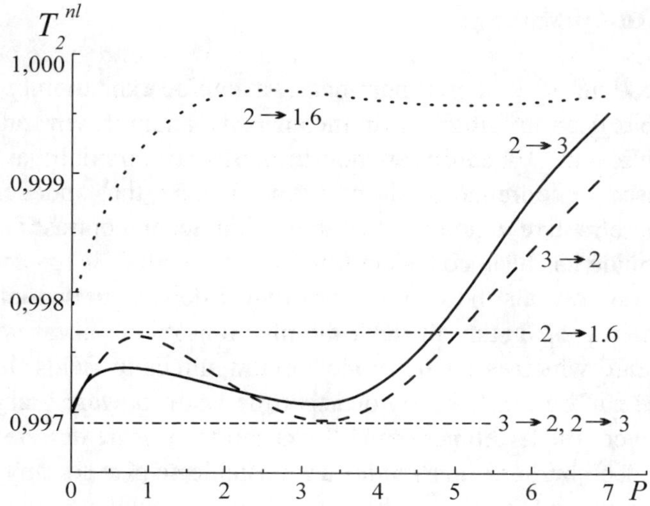

Figure 17. Transmittance of nonlinear junctions (a_1(μm)→a_2(μm)) calculated at z=100mm by intensity integration within the computational window (horizontal lines denote linear transmittance).

If the intensity is integrated within the core of the structure B, the transmittance grows or decreases with power behind the down-step structure or up-step structure, respectively (Fig.18). In the limit $P(0) \gg 1$ this transmittance tends to unity.

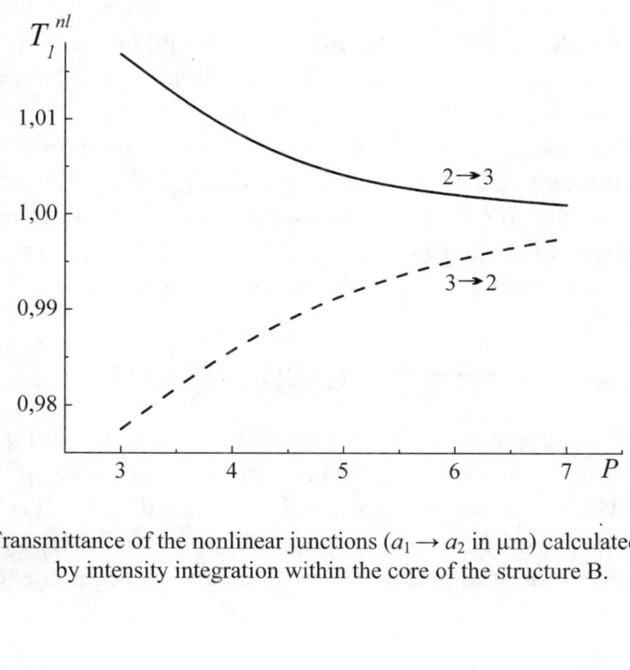

Figure 18. Transmittance of the nonlinear junctions ($a_1 \rightarrow a_2$ in μm) calculated at $z = 100$ mm by intensity integration within the core of the structure B.

3.4 Conclusions

In this section, a problem of nonlinear waveguide excitation by stationary light beam has been investigated. In the analysis, an approach traditional for nonlinear optics and based on solution to nonlinear paraxial wave equation has been used. The range of light beam powers that induce nonlinear variation of refractive index comparable with linear contrast of the step-index waveguide has been considered.

The analysis reveals the complicated power-dependent spatial dynamics of a light beam propagating through the junctions of nonlinear waveguides. In a waveguide which is single-mode for low intensity fields, higher non-linear modes can be excited provided that the beam power is large enough. Critical power of a higher mode excitation is greater in narrower waveguides. The higher-order modes are unstable so that for any transverse profile of input field, stable profile of the fundamental nonlinear mode is finally observed in the second waveguide of the junction behind a region of unsteady-state regime. Power of radiation field emitted over the region of unsteady-state regime depends on power of the input light beam. Thus, light transmission by junctions of nonlinear waveguides is power-dependent, and moreover, it is different for light beams propagating in opposite directions through the junction.

Two kinds of nonlinear junctions considered above have different functions with respect to the power of input light beam. The transmittance of the linear/nonlinear junction decreases with input power. The efficiency of the nonlinear action of the structure is greater in narrower waveguides. The transmittance of the junction of nonlinear waveguides has extremes in dependence on the input power but grows up to unity in the limit of high-intensity light beams.

All the above results can be directly used in the problem of optical pulse propagation through the junctions provided that the quasi-static approximation is feasible. As the transmittance of a waveguide junction depends on power of a stationary component of the pulse, variation of an input pulse envelope behind the junction should be observed.

4. NON-STATIONARY LIGHT BEAMS

In this section, propagation of an optical pulse in a step-index nonlinear waveguide is simulated by the numerical technique described in Section 2. As a comparative parameter, the power P^{peak} evaluated by (3.2) ($P^{\text{peak}} = P$) at the pulse peak ($t = t_c$) of the initial temporal distribution (2.10) of the pulse is used. First, we consider the problem of a nonlinear waveguide excitation by the non-stationary light beam using the quasi-static approximation. Next, role of the SS effect and the second-order GVD effect, separately, in spatiotemporal dynamics of optical pulse in step-index nonlinear waveguide is analyzed.

4.1 Quasi-static approximation: propagation through the junctions

In the quasi-static approximation, Eq.(2.9) has been solved numerically by the FD-BPM for each stationary component of temporal distribution of the light beam. Amplitudes of the stationary components were specified by the form of pulse temporal envelope. The initial pulse envelope was assumed Gaussian.

Crank-Nicholson and Douglas schemes with improved interface conditions have been applied to transform Eq. (2.9) into a finite-difference equation.

4.1.1 Junction of linear and nonlinear waveguides

In the structure A, the transverse profile of the initial pulse varies behind the junction due to the self-focusing effect. In the nonlinear waveguide of the structure, a fraction of an initial pulse power is emitted from the guiding region (Fig.19).

According to the results of Section 3.3.1 obtained for a stationary light beam, power losses in the nonlinear waveguide of the structure A grow with power. As the temporal distribution of the initial pulse in the linear waveguide is Gaussian (given by (2.11) with the pulse peak at $t = t_c$) radiation field is emitted predominantly from the central part of the pulse in the nonlinear waveguide. Then the radiation field has also a temporal envelope with a peak power at $t = t_c$.

After the radiation field has gone, the diffraction of the light beam is compensated by the self-focusing effect, so that the spatiotemporal distribution becomes stable (Fig.20).

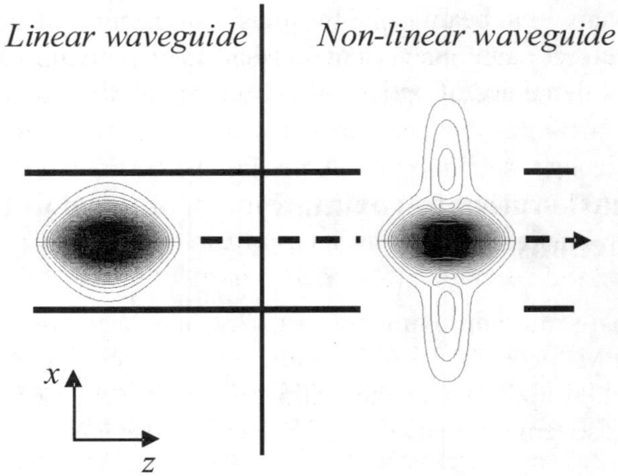

Figure 19. Pulse propagating through the structure A.

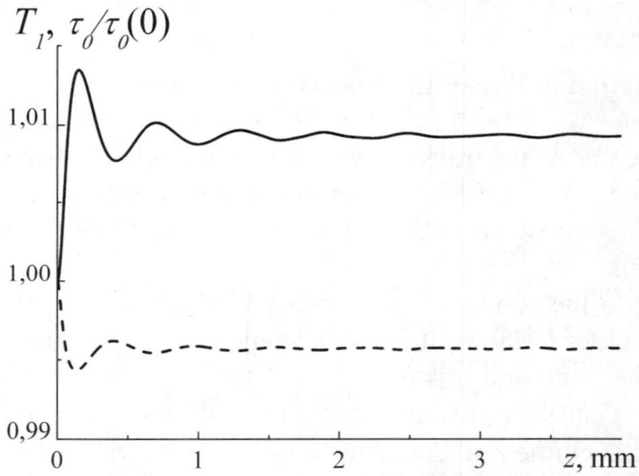

Figure 20. Longitudinal variation of the pulse energy propagating within the core T_1 (2.17) (solid line) and pulse duration at the waveguide axis τ_0 (2.12) (dashed line) normalized to its initial value, $a = 3\mu m$, $P^{peak} = 4$.

Thus, in the quasi-static approximation, the length of the unsteady-state regime in the core of the nonlinear waveguide is finite, as in the case of the stationary light beam propagation (see Fig.10, 11).

Behind the junction, power of the field propagating within the core increases due to the self-focusing effect, while the pulse duration at the waveguide axis decreases. In the quasi-static approximation, this effect does not depend on the initial pulse duration. Total losses vary with power at the pulse peak similar to the case of stationary wave propagation in the structure A, i.e. they increase with the power (Fig.21, compare with Fig.11).

As it was shown in Sec. 3.3.1, the self-focusing of stationary light beam and consecutively the power fraction propagating within the core of a nonlinear waveguide of the structure A grow with the initial beam power (Fig.10, 13). In the case of optical pulse propagation, this results in the variation of the pulse spatiotemporal distribution behind the junction so that the pulse shape and duration (2.12) depend on the transverse coordinate. In the core, the pulse peak is self-focused more than the pulse wings. The power evaluated over the computational window decreases with the input power and has the smallest value at the pulse peak due to greater emission of radiation field from this part of the pulse.

Another result of Sec. 3.3.1 is that the transmittance of the structure A is power dependent, moreover the transmittance evaluated via intensity integration within the core (T_1) and within the computational window (T_2) depends on the initial beam power differently (Fig.13,14). The pulse duration also depends on the domain $(-X_n, X_n)$ over which the moments M_i (2.15) of the temporal distribution of the pulse are integrated. The pulse duration versus X_n is presented in Fig.22.

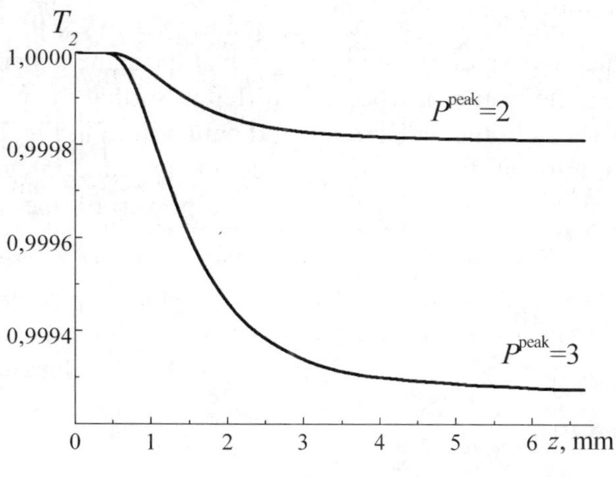

Figure 21. Longitudinal variation of the pulse energy propagating within the computational window T_2 (2.17), $a = 3.0\mu m$

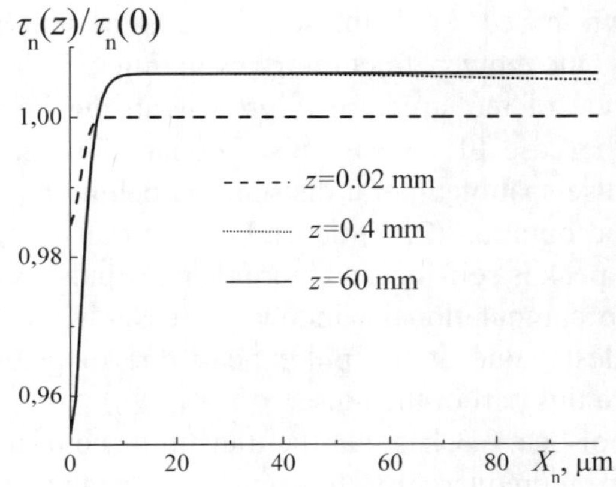

Figure 22. Normalized pulse duration calculated by intensity integration within the spatial domain $(-X_n, X_n)$ in the waveguide cross-section, depending on the dimensions of the domain, a = 3.0μm.

If the beam power is measured within the waveguide core ($X_1 = a$) or within a smaller domain, these results show compression of the pulse. This effect is the most pronounced at the waveguide axis ($X_0 = 0$). If the beam power is evaluated over a domain with $|X_n| > a$ at the same distance from the plane of junction, the result may be quite different so that the pulse duration can be greater than its initial value (solid and dotted lines in Fig.22).

The pulse duration at the nonlinear waveguide axis and integrated within the waveguide core decreases with the peak power of the initial pulse (Fig.23).

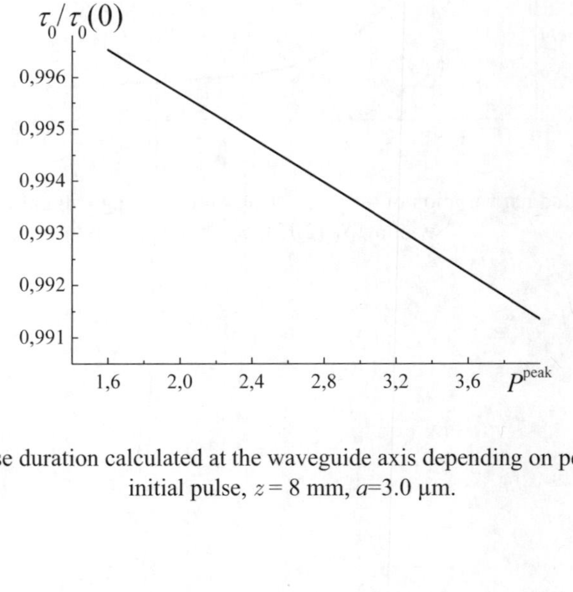

Figure 23. Pulse duration calculated at the waveguide axis depending on peak power of the initial pulse, z = 8 mm, a=3.0 μm.

This dependence is linear for the range of powers corresponding to the linear part of the transmittance-versus-power dependence in Fig.13 (dashed curve).

4.1.2 Junction of nonlinear waveguides

According to the results of Sec. 3.3.2, the transmittance of the structure B depends on the geometry of the junction and on the dimensions of the domain $(-X_n, X_n)$ in the waveguide cross-section where the light beam intensity is integrated. For this junction we use the same excitation scheme as for the structure A, so that the spatiotemporal distribution of the initial pulse is given by (2.10), (2.11) in a linear waveguide with the same width as the first waveguide of the structure B.

The relative variation of the pulse duration (2.16) calculated via integration of the light beam intensity within the core of the structure is shown in Fig.24. First, as the pulse propagates through the junction of linear and nonlinear waveguides, its duration decreases. Then, when the pulse propagates through the structure B, its duration varies in accordance with estimations of nonlinear transmittance in p.3.2.2. In the step-down or step-up discontinuity, the pulse compression or broadening is observed, respectively, provided that the beam power is integrated within the core of the structure.

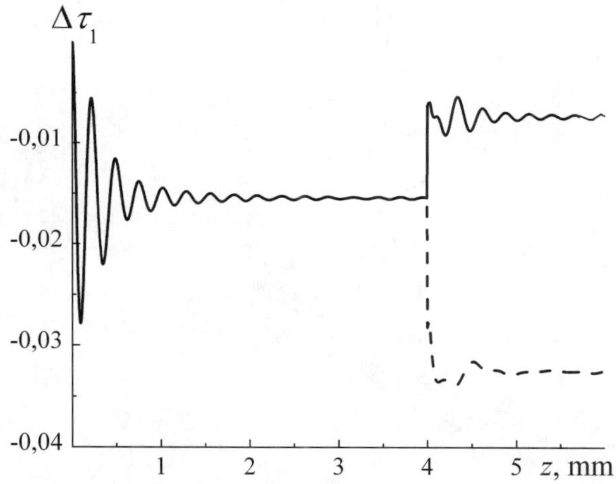

Figure 24. Longitudinal dependence of the relative variation of the pulse duration (2.16) behind the junction of nonlinear waveguides. $a_1 = 3.0$ µm , $a_2 = 4.0$ µm (solid line); $a_1 = 3.0$ µm , $a_2 = 2.0$ µm (dashed line), $P^{peak} = 3$.

4.2 Account of the self-steepening effect

In the numerical modeling of optical pulse propagation with account of the SS effect, Eq.(2.8) with $\tau_R = \varepsilon_2 = \varepsilon_3 = 0$ has been solved. The Crank-Nicholson scheme was used to transform Eq.(2.8) into a finite-difference equation.

Since group velocity of optical pulse propagating in a medium with the Kerr nonlinearity is intensity-dependent, the central part of the pulse propagates more slowly than its edges. As a result, the pulse peak shifts from the center of the computation window, and, in the region of its tail, one can observe the SS effect (Fig.25). The SS effect decreases in waveguide cladding so that temporal distribution of the pulse is close there to its initial Gaussian form.

Calculations of the pulse displacement and asymmetry by (2.12) and (2.13) (Fig.26 and Fig.27, respectively) show that the pulse group velocity increases linearly with propagation distance, while the symmetry of its leading and trailing parts is broken, with the degree of asymmetry growing rapidly with distance. For a given peak power, the degree of asymmetry grows more strongly for shorter pulses.

For a given pulse power, the SS effect depends on the transverse profile of the field and is less pronounced in waveguides with thinner cores (the dashed lines in Fig.26, 27 are plotted for a waveguide with $d = 2.4\mu$m).

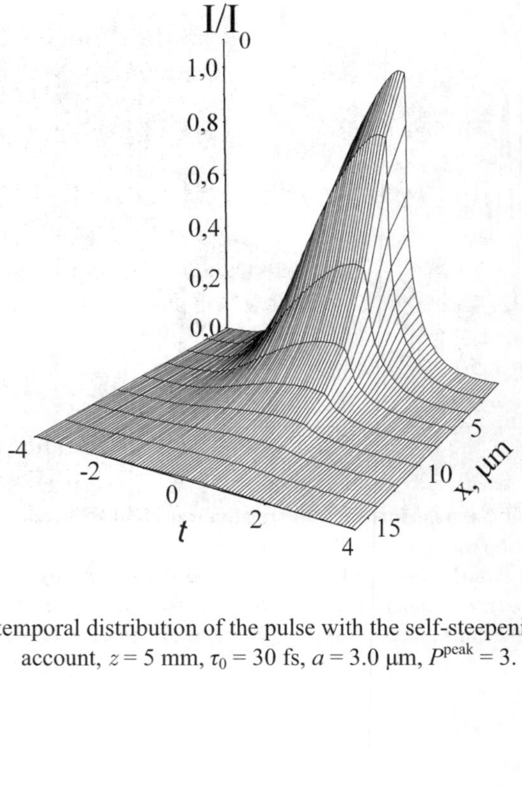

Figure 25. Spatiotemporal distribution of the pulse with the self-steepening effect taken into account, $z = 5$ mm, $\tau_0 = 30$ fs, $a = 3.0$ μm, $P^{peak} = 3$.

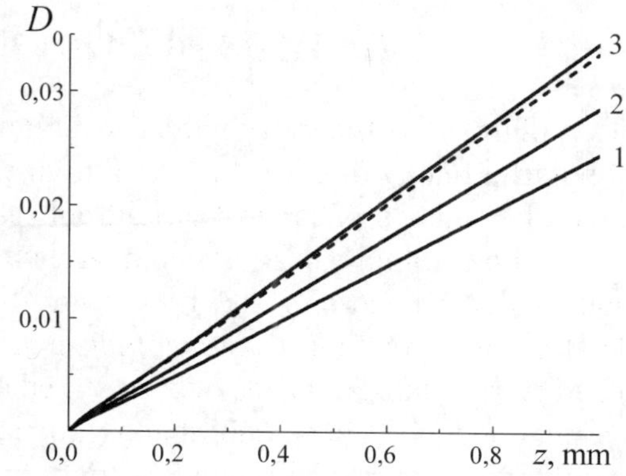

Figure 26. Longitudinal variation of the pulse displacement (2.13) at the waveguide axis in nonlinear waveguide of the structure A with respect to the center of moving coordinate system, τ_0=70fs (1), 60fs (2), 50fs (3), a=3.0μm (solid lines), a= 2.4μm (dashed line), P^{peak}=3.

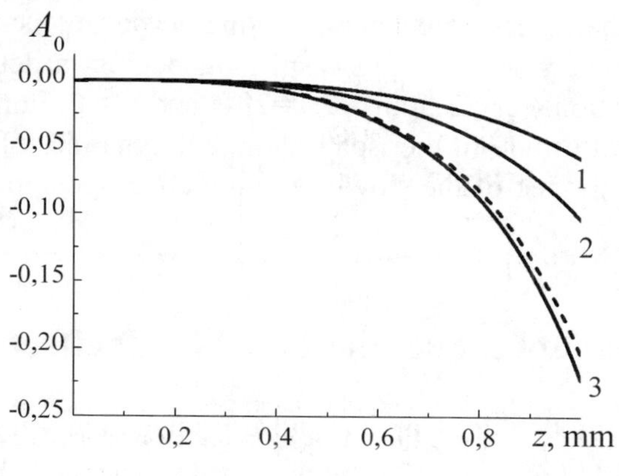

Figure 27. Longitudinal variation of the pulse asymmetry (2.14) calculated at the waveguide axis in nonlinear waveguide of the structure A, τ_0=70fs (1), 60fs (2), 50fs (3), a=3.0μm (solid lines), a=2.4μm (dashed lines), P^{peak}=3.

This weakening of the SS effect can be explained by emission of radiation field due to the SS effect predominantly from high-intensity central parts of the pulse, the emission being greater in waveguides with thinner core (see Section 3).

Power losses resulting from the SS effect are observed at the earliest stages of the unsteady-state regime in the nonlinear waveguide of the junction (Fig.28). The losses grow significantly beginning at some distance

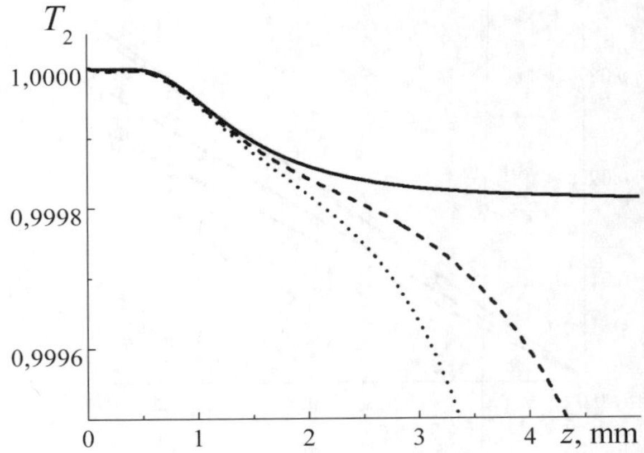

Figure 28. Longitudinal variation of power of the light beam calculated over the computational window in nonlinear waveguide of the structure A, $\tau_0(0)$=20fs (dashed line), $\tau_0(0)$= 10fs (dotted line), solid line corresponds to the quasi-static approximation.

from the junction (z = 2.5 μm in Fig.28) where the maximum at the pulse tail becomes very sharp. This steepening of the pulse tail is followed by its splitting. The point where the pulse splitting begins is determined by the initial pulse duration ($z \approx$ 2.5μm for $_0$=20fs and $z \approx$ 3.5μm for $_0$=30 fs in Fig. 28). In order to study further spatiotemporal dynamics of the pulse, other numerical methods free of the slowly varying envelope approximation have to be used.

4.3 Account of the second-order GVD effect

In numerical modeling of optical pulse propagation with account of the second-order GVD effect, Eq.(2.8) with $\tau_R = \varepsilon_2 = \varepsilon_3 = 0$ has been solved. The Crank-Nicholson scheme was used to transform Eq.(2.8) into a finite-difference equation.

Consider now how the solution obtained in the quasi-static approximation changes if the second-order group velocity dispersion is taken into account. It is seen from Fig. 29 that in a nonlinear waveguide of the structure A with the MD and the SS effects is first observed that spatial and temporal parameters of the field vary similarly to the case of the quasi-static approximation. Then, at a given power, spatiotemporal distribution varies depending on the value and sign of the dispersion coefficient k_2. The pulse duration decreases in the case of anomalous GVD ($k_2 < 0$) and increases in the case of normal GVD ($k_2 > 0$).

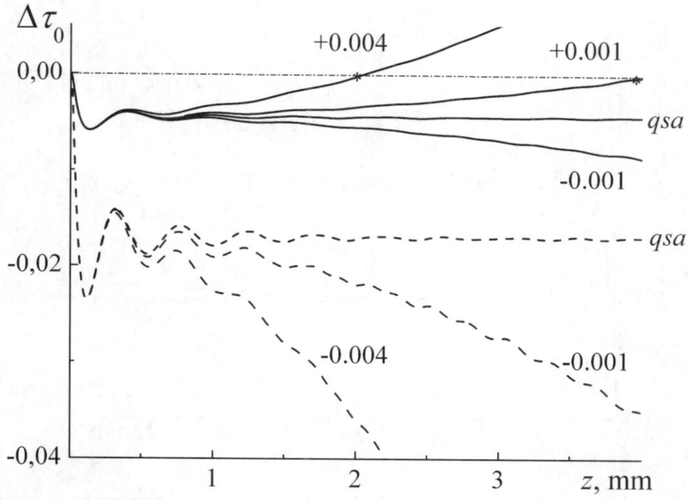

Figure 29. Longitudinal dependence of the relative variation of the pulse duration at the waveguide axis for some values of k_2 (k_2=0.001,-0.001,0.004,–0.004), in fs^2/μm. P^{peak} = 3 (solid lines), 4 (dashed lines), τ_0 = 16 fs, a = 3.0 μm; *qsa* – quasi-static approximation.

It is seen that longitudinal dependences of the variation in the pulse duration plotted for k_2 of the same modulus but different signs are approximately symmetrical with respect to the curve obtained in the quasi-static approximation. Hence, at normal GVD, the pulse duration is shorter than the initial one as a result of the self-focusing effect until, owing to GVD, this duration continuously increases to its initial value (denoted by asterisks) and then exceeds it. The size of the region within which the pulse compression is observed at $k_2 > 0$ decreases with increasing $|k_2|$. In turn, normal GVD weakens the self-focusing effect and leads to continuous decrease in power of the pulse fraction propagating in the core (Fig.30, $k_2 > 0$). On the contrary, the anomalous GVD enhances the self-focusing effect, so that the power in the core increases (Fig.30, $k_2 < 0$). At a given k_2, these effects are more pronounced for a pulse with a higher peak intensity P^{peak} (compare solid and dashed curves in Fig.30).

The relative decrease $\Delta\tau_0$ of the pulse duration depends approximately linearly on P^{peak} when $k_2 < 0$ (Fig.31). For $k_2 > 0$, at high powers (e.g. $P^{peak} > 2$ in Fig.31), the relative increase of the pulse duration is also almost linear. In this case, such propagation is possible when the pulse duration virtually does not depend on the power of the initial beam. This might result from a balance between the self-focusing effect which decreases the pulse duration (dashed line in Fig.31) and the normal GVD effect which broadens the pulse.

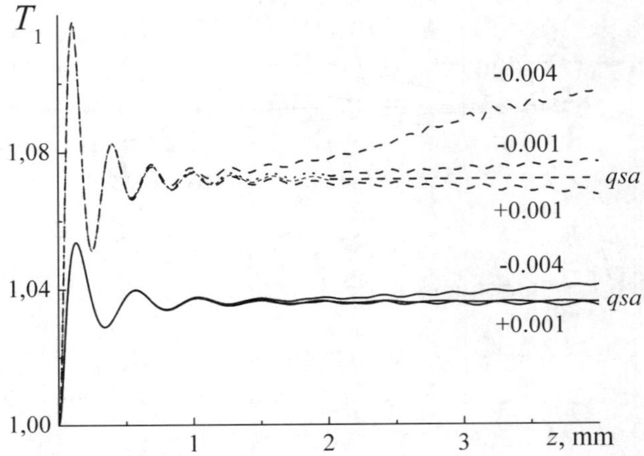

Figure 30. Longitudinal dependence of power of the pulse (2.17) propagating within the waveguide core for some values of k_2 (k_2=0.001,-0.001,0.004,–0.004), in fs^2/μm. P^{peak} = 3 (solid lines), 4 (dashed lines), τ_0 = 16fs, a = 3.0μm; *qsa* – quasi-static approximation.

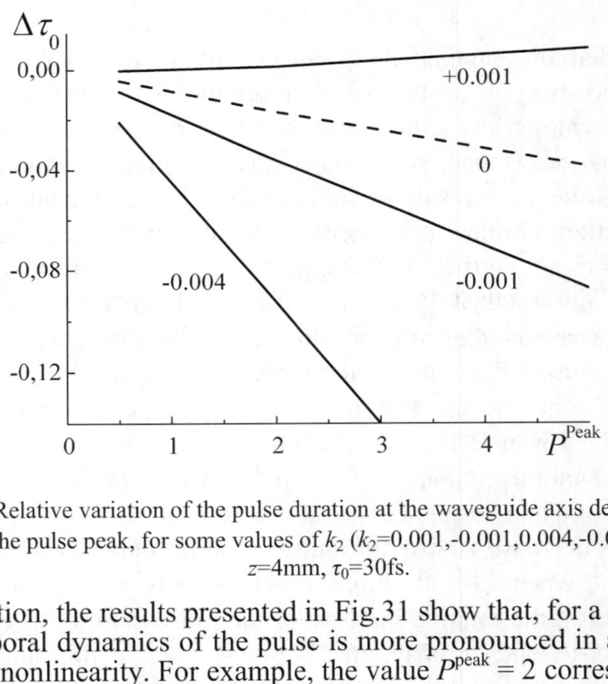

Figure 31. Relative variation of the pulse duration at the waveguide axis depending on the power at the pulse peak, for some values of k_2 (k_2=0.001,-0.001,0.004,-0.004), in fs^2/μm; z=4mm, τ_0=30fs.

In addition, the results presented in Fig.31 show that, for a given P^{peak}, the spatiotemporal dynamics of the pulse is more pronounced in a medium with the higher nonlinearity. For example, the value P^{peak} = 2 corresponds to $|A|^2$ = 1 (see Fig.4) and consequently $n_K I_0 = \Delta n \approx 0.003$ (here I_0 is the axial value of the intensity at the pulse peak). This value corresponds to the maximum admissible peak intensity $I_0 \approx$ 100 GW/cm^2 of a femtosecond pulse that propagates in a silica waveguide with $n_K \approx 10^{-16}$ cm^2/W. The higher values of P^{peak} correspond to the pulse propagation, e.g., in a chalcogenide waveguide with $n_K \approx 10^{-14}$cm^2/W. Whereas in the silica waveguide, the variation $\Delta\tau$ amounts to 1–2%, in the chalcogenide waveguide it is around 10–20%.

The character of variations in the spatiotemporal field distribution in the guiding region depending on the sign of the GVD is shown in Fig.32. The most noticeable effect of dispersion manifests itself in the waveguide core, where in the case of normal GVD, the pulse continuously broadens simultaneously with the compression of its transverse distribution (Fig.32a). In the case of anomalous GVD, a so-called "light bullet" is formed (Fig.32b), which looks like a sharp increase in the intensity at the peak of the spatiotemporal distribution.

In the region of unsteady-state regime, the radiation field is emitted from the core into the cladding, the character of the field distribution in the cladding being almost independent on the sign of the GVD (Fig.33). This is a satellite pulse propagating in the cladding with the peak power at the point $t = t_c$ the same as for the initial pulse. In the case of normal GVD the pulse formed by radiation field has greater duration in comparison with the case of an anomalous GVD.

It is noteworthy that in this case, in contrast to the quasi-static propagation, the region of unsteady-state regime is not finite because the balance between diffraction and nonlinear refraction in dispersive medium is disturbed by the same effect which was discussed concerning to the SS effect in Sec. 4.2. Specifically, due to continuous variation in the temporal distribution of the pulse, the transverse distribution also changes and a fraction of the field is emitting continuously from the guiding region.

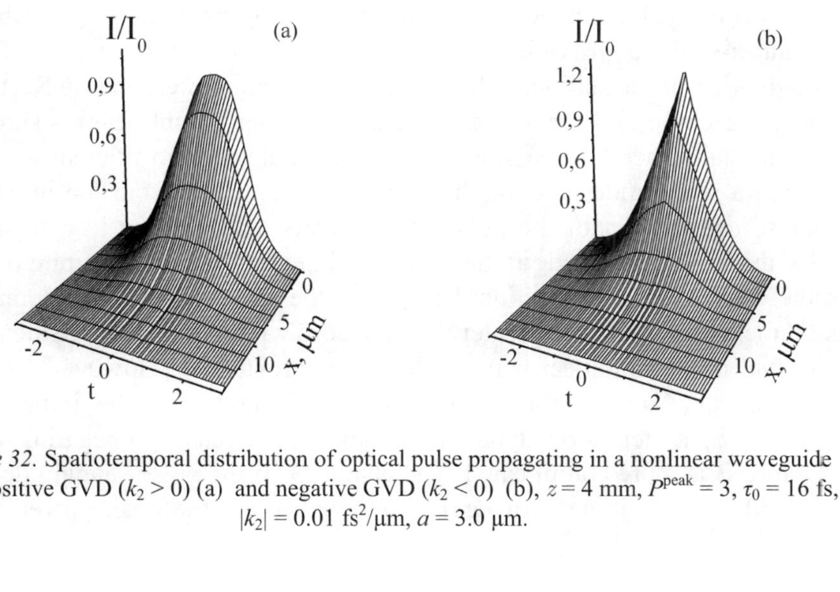

Figure 32. Spatiotemporal distribution of optical pulse propagating in a nonlinear waveguide with positive GVD ($k_2 > 0$) (a) and negative GVD ($k_2 < 0$) (b), $z = 4$ mm, $P^{peak} = 3$, $\tau_0 = 16$ fs, $|k_2| = 0.01$ fs^2/μm, $a = 3.0$ μm.

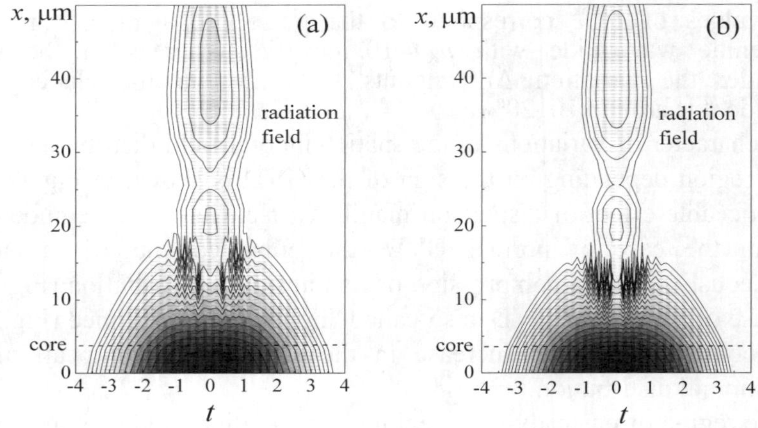

Figure 33. Emission of the radiation field from the guiding region resulting from the second-order GVD effect (contour plot, logarithmic scale), $k_2 > 0$ (a) $k_2 < 0$ (b), $z = 4$mm, $P^{peak} = 3$, $\tau_0 = 16$ fs, $|k_2| = 0.01$ fs^2/μm, $a = 3.0$ μm.

4.4 Conclusions

In this section, a problem of a nonlinear waveguide excitation by a non-stationary light beam has been investigated. Optical pulse propagation through the junctions of nonlinear waveguides was simulated by the numerical technique described in Section 2.

In general, joint action of a nonlinear refraction, self-phase modulation and GVD influences the pulse propagation. If the diffraction length of the light beam is much shorter than the characteristic length of pulse variation owing to the second-order GVD and SS effects considered above which influence the pulse envelope, the quasi-static approximation is feasible. In the quasi-static approximation, the analysis is based on the results of investigation of the stationary light beam propagation presented in Section 3. The spatiotemporal distribution of the non-stationary light beam is shown to become stable behind a region of unsteady-state regime after propagation through a waveguide junction. The pulse shape differs from the initial one because central part of the pulse is self-focused more than its wings. The pulse duration calculated via the intensity integration within a finite domain in the waveguide cross-section depends on the dimensions of the domain. If the dimensions are much greater than the core width, the pulse broadening is observed in the nonlinear waveguide of the structure A. However the pulse duration calculated at the waveguide axis is less than the initial pulse duration. Spatiotemporal dynamics of an optical pulse propagating in the structure B is more complicated. The character of the pulse shaping depends on widths of the joined waveguides as well as on the peak power of the initial pulse. If the peak power is large enough the pulse compression is observed behind the junction independently on the structure geometry provided that the intensity is integrated over a spatial domain with dimensions much greater than the core width. If the intensity is integrated over the core of the structure B, the pulse duration decreases or increases, respectively, in the step-down or step-up discontiniuty.

If the diffraction length of the light beam is comparable with the characteristic length of pulse envelope variation owing to the second-order GVD and SS effects, these effects are to be taken into consideration together with the self-focusing effect. In this case, the negative GVD enhances and the positive GVD weakens the self-focusing effect. The opposite action of the positive GVD effect and the self-focusing effect can be used for the stabilization of optical pulse duration with respect to variations of the pulse power.

In general, as the variation of the temporal profile of the non-stationary light beam due to the SS effect or the second-order GVD effect is continuous, emission of radiation field from the guiding region is also continuous upon propagation of the pulse. This emission prevents formation of a spatiotemporal soliton in the step-index guiding structures.

It is important to note that feasibility of the numerical technique used in this work is limited. In addition to limitations discussed in Sections 2, 3 it has been proved that sharp variations in temporal and spatial profiles of the non-stationary light beam cannot be simulated in the frames of the paraxial approach. That is why the method of modeling presented here can be applied to pulse durations greater than 10 fs.

5. CONCLUSION

In this paper, spatiotemporal dynamics of non-stationary light beam propagating through the junctions of step-index nonlinear waveguides have been investigated.

Theoretical approach based on solution to the (2D+T) parabolic wave equation traditional for methods of nonlinear optics has been used. This approach is feasible if linear refractive index contrast in the waveguide is small and comparable with the nonlinear part of the refractive index. Then the back-reflected field can be ignored.

The results presented in this work demonstrate that the spatiotemporal phenomena arising upon propagation of high-intensity optical pulses in step-index optical waveguides are significant under some conditions and should be accounted for in modeling and design of integrated optical devices. In miniature devices (waveguide gratings with the period comparable to the wavelength, etc.), even if the quasi-static approximation might not be feasible, the results of the quasi-static self-focusing can be applicable. In a waveguide segments longer than the diffraction length, the influence of the MD and SS effects should be significant.

These results can be used for comparative analysis in the elaboration of nonlinear implementations of more rigorous computational methods, free of the slowly varying envelope approximation.

All the specific phenomena described above result from the emission of radiation field outside the guiding region. This effect has already been demonstrated as useful for pulse compression[18] and all-optical power limiting.[19] It can be also applied for pulse shaping and optical gating. The variations of spatiotemporal envelope of non-stationary light beam resulting from the self-effects in nonlinear waveguides are to be accounted for in some novel laser technologies, for example, in waveguide writing by high-intensity femtosecond pulses.

6. REFERENCES

1. S.A. Akhmanov, V.A. Vysloukh, A.S. Chirkin, *Optics of Femtosecond Laser Pulses* (American Institute of Physics, NY 1992).
2. G.P. Agrawal, *Nonlinear fiber optics* (Academic Press, Inc., Boston, MA, 1989).
3. A.A. Zozulya, S.A. Diddams, A.G.Van Engen, T.S. Clement, "Propagation dynamics of intense femtosecond pulses: multiple splitting, coalescence and continuum generation", *Phys.Rev.Lett.* **82**, 1430-1433 (1999).
4. K. Mori, H. Takara, and S. Kawanishi, "Analysis and design of supercontinuum pulse generation in a single-mode optical fiber", *J.Opt.Soc.Am.* **18**, 1780-1792 (2001).
5. G. Fibich, W. Ren, X-P. Wang, "Numerical simulations of self-focusing of ultrafast laser pulses", *Phys. Rev. E* **67,** 056603 (2003).
6. E.A.Romanova, L.A.Melnikov, "Optical pulse propagation in an irregular waveguide with spatially distributed Kerr-like non-linearity", *Optics and Spectroscopy* **95**, 268-275 (2003).
7. E.A. Romanova, L.A. Melnikov, "Spatiotemporal dynamics of femtosecond pulses in non-linear optical waveguides with material dispersion", *Optics and Spectroscopy* **96**, 90-95 (2004).
8. N.N. Akhmediev, "Spatial solitons in Kerr and Kerr-like media", *Opt. and Quantum Electron.* **30**, 535-569 (1998).
9. W.P. Huang (Ed.), *Methods for modelling and simulation of guided-wave optoelectronic devices: waves and interactions* (PIERS 11, EMW Publishing, Cambridge, MA, 1995).
10. P.G. Suchoski, Jr., and V. Ramaswamy, "Exact numerical technique for the analysis of step discontinuities and tapers in optical dielectric waveguides", *J.Opt.Soc.Am.A.* **3**, 194-203 (1986).
11. J.M. Ortega, W.G. Poole, Jr.: *An introduction to numerical methods for differential equations* (Pitman Publishing Inc., 1981).
12. J. Yamauchi, J. Shibayama, and H. Nakano, "Application of the generalized Douglas scheme to optical waveguide analysis", *Opt. Quantum Electron.* **31**, 675-687 (1999).

13. Y.P. Chiou, Y.C. Chiang, H.C. Chang, "Improved three-point formulas considering the interface conditions in the finite-difference analysis of step-index optical devices", *J. Lightwave Technol.* **18**, 243-251 (2000).
14. A.W. Snyder and J.D. Love, *Optical Waveguide Theory* (Chapman & Hall, London, England, 1983).
15. H.W. Schurmann, V.S. Serov and Yu.V. Shestopalov, "TE-polarized waves guided by a lossless nonlinear three-layer structure", *Phys.Rev.E* **58**, 1040 – 1050 (1998).
16. C.W. McCurdy and C.K. Stroud, "Eliminating wavepacket reflection from grid boundaries using complex coordinate contours", *Computer Phys. Commun.* **63**, 323-330 (1991).
17. G.R. Hadley, "Transparent boundary conditions for the BPM", *IEEE J. Quantum Electron.* **28**, 363-370 (1992).
18. L.A. Melnikov, E.A. Romanova, E.V. Bekker, "Non-linear transmission of a single-mode optical fiber with a sharp change of core diameter", *Optics and Spectroscopy* **89**, 761-765 (2000).
19. E.V. Bekker, E.A. Romanova, L.A. Melnikov, T.M. Benson, P. Sewell, "All-optical power limiting in waveguides with periodically distributed Kerr-like nonlinearity", *Appl. Phys.* B **73**, 531-534 (2001).

EFFICIENT FREQUENCY-DOUBLING OF FEMTOSECOND PULSES IN WAVEGUIDE AND BULK NONLINEAR CRYSTALS

Design, fabrication, theory and experiment

Ben Agate[1], Edik U. Rafailov[2], Wilson Sibbett[1], Solomon M. Saltiel[3], Kaloian Koynov[4], Mikael Tiihonen[5], Shunhua Wang[5], Fredrik Laurell[5], Philip Battle[6], Tim Fry[6], Tony Roberts[6] and Elizabeth Noonan[6]

[1]School of Physics and Astronomy, University of St Andrews, St Andrews, KY16 9SS, Scotland, U.K.; [2]Division of Electronic Engineering and Physics, University of Dundee, Dundee, DD1 4HN, Scotland, U.K.; [3]Faculty of Physics, University of Sofia, 5 J. Bourchier Boulevard, BG-1164, Sofia, Bulgaria; [4]Max Planck Institute for Polymer Research Ackermannweg 10, D-55128 Mainz, Germany; [5]Department of Physics, Royal Institute of Technology, Roslagstullsbacken 21, SE-10691 Stockholm, Sweden; [6]AdvR Inc., Suite K, 910 Technology Boulevard, Bozemon, Montana 59718, USA

Abstract: We investigate efficient frequency-doubling of low energy femtosecond pulses in bulk and waveguide nonlinear crystals, thereby demonstrating how to achieve a compact and portable ultrafast blue light source. Using a femtosecond Cr:LiSAF laser (fundamental wavelength 860 nm), we examine the relative merits of the process of second harmonic generation (SHG) using bulk potassium niobate, bulk aperiodically-poled KTP, periodically-poled and aperiodically-poled KTP waveguide crystals. While SHG conversion efficiencies up to 37% were achieved using the waveguides, non-traditional strong focusing in the bulk samples yielded efficiencies of 30%. We have developed several theoretical models to accurately describe the temporal and spectral properties of the generated blue light, as well as the observed saturation behavior of the conversion process in the waveguide structures.

Key words: Diode-pumped lasers; frequency conversion; modelocked lasers; waveguides; periodic structures, quasi-phasematching, femtosecond pulses.

S. Janz et al. (eds.), Frontiers in Planar Lightwave Circuit Technology, 189–227.

1. INTRODUCTION

The recent progress in photobiology research, such as microscopy, optical micromanipulation and bio-medical imaging continues to justify the need for compact, low-cost visible and near-infrared lasers that demonstrate true portability and practicality. The development and exploitation of photonics-based techniques that are applicable to biology and medicine is of particular interest. Laser sources at the shorter blue and ultraviolet wavelengths provide clear advantages over infrared lasers in allowing for stronger beam focusing, enhanced resolution in multi-dimensional imaging techniques[1] and high-resolution spectroscopy[2]. By using ultrashort-pulse lasers in preference to continuous-wave sources, it is possible to investigate ultrafast biological processes[3], increase the resolution of microscopy[4,5], and amplify subtle signals in nonlinear and multi-photon techniques[6-8]. Femtosecond pulses in the blue spectral region have already been utilized in the study of protein dynamics[9].

Straightforward frequency-doubling of near-infrared femtosecond lasers remains one of the simplest and most efficient access routes into the blue spectral region, due mainly to the inherently high peak intensities of the ultrashort laser pulses. While conventional titanium-sapphire lasers have long been the market leader in providing femtosecond pulses in the near-infrared, their impractical size, cost and power requirements render such sources impractical for space-limited on-site applications in bio-medical and clinical laboratories. Consequently, we have recently demonstrated how small-scale femtosecond Cr:LiSAF lasers have much to offer in combining superior operational efficiencies with impressive design flexibility[10].

In this paper, we begin by describing second harmonic generation (SHG) in the femtosecond regime. We then present several theoretical models which accurately describe efficient SHG of femtosecond pulses in both bulk and waveguide nonlinear crystal structures, and detail the design and fabrication processes of these samples. We describe how a compact Cr:LiSAF laser design can be used to demonstrate an attractively simple approach to developing a portable ultrafast blue light source[11]. With this femtosecond Cr:LiSAF laser (operating at around 850 nm), we then experimentally access the blue spectral region with a series of SHG investigations using four types of nonlinear crystal. The performance of aperiodically-poled potassium titanyl phosphate (appKTP) in bulk and waveguide form is assessed in comparison to bulk potassium niobate ($KNbO_3$)[12] and periodically-poled KTP (ppKTP) waveguide[13]. Each nonlinear crystal is placed in a very simple extra-cavity single-pass arrangement at room temperature, requiring minimal wavelength and temperature stabilization. The ultrashort-pulse blue light generated from all four nonlinear crystals is either femtosecond or picosecond in duration. This investigation provides the means to compare and contrast the relative merits of bulk and waveguide nonlinear materials, as well as periodically- and aperiodically-poled structures. Observations show that, while superior infrared-to-blue SHG efficiencies are achieved with bulk $KNbO_3$ and the ppKTP waveguide, aperiodically-poled structures permit higher blue peak powers, shorter blue pulses and broader tunability in the blue spectral region. To support our experimental findings, we analyze the crystal performances against our mathematical models, before concluding with a discussion and summary.

2. FREQUENCY-DOUBLING OF FEMTOSECOND PULSES

2.1 Background

The theoretical study of second harmonic generation (SHG) using focused Gaussian beams by Boyd and Kleinman[14] has long been a reliable resource for those studying frequency conversion processes. However, the Boyd-Kleinman theory applies only to cw beams, and cannot be relied upon to correctly describe harmonic generation using ultrashort (femtosecond) pulses. We have recently published a theoretical model[15] that describes SHG of femtosecond pulses in bulk nonlinear crystals, by taking into account the associated critical effects of group velocity mismatch (GVM). Our model explains successfully the experimentally observed behavior of SHG in the femtosecond regime[12,16], in contrast to the somewhat inaccurate predictions of Boyd and Kleinman. We have also adapted this model to describe SHG of femtosecond pulses in waveguide nonlinear crystals[11,13], which accurately describes the observed saturation of the SHG conversion process.

The efficiency of any parametric process is subject to limitations imposed by GVM, which describes a temporal walk-off between the interacting beams. This walk-off arises from a mismatch in the group velocities, and becomes particularly significant in the ultrashort pulse regime. Second harmonic generation under conditions of large GVM is characterized by a nonstationary length, L_{nst}, defined by $L_{nst} = \tau/\alpha$, where τ is the time duration of the fundamental pulses, and the GVM parameter, $\alpha = 1/v_2 - 1/v_1$, where v_2 and v_1 are the group velocities of the second harmonic (SH) and fundamental wave respectively. The nonstationary length, L_{nst}, is the distance at which two initially overlapped pulses at different wavelengths become separated by a time equal to τ. In the ultrashort pulse limit (i.e. for $L_{nst} << L$, where L is the length of the nonlinear media) and for the case of an unfocused fundamental beam, the generated SH pulses are longer in time by a factor L/L_{nst}. In contrast to the well known quadratic dependence for the frequency doubling of cw waves, this process depends linearly on the length of the nonlinear media, L. As a result of this deleterious temporal broadening effect, nonlinear media are often chosen such that $L \approx L_{nst}$. However, while this may ensure generated SH pulses having durations close to that of the fundamental pulses, the interaction length is reduced and the SHG efficiency is compromised.

The natural way to increase the efficiency of such a frequency conversion process is to use a focused fundamental beam (or, alternatively, a waveguide structure). An established theory of SHG using focused cw beams[14] predicts, for negligible birefringence walk-off, an optimal focusing condition which is expressed by the ratio $L/b = 2.83$, where b is the confocal parameter ($b = k_1 w_{01}^2$, where w_{01} and k_1 are the focal spot radius and the wave vector of the fundamental wave respectively). However, this theory applies only to the long-pulse or cw case, where GVM is negligible ($L_{nst} >> L$). Our recently published theoretical model[15] defines, for the first time, the optimum focusing conditions for SHG using focused beams in the ultrashort-pulse regime, where GVM is significant (i.e. where $L \geq L_{nst}$).

Despite the limitations imposed on the length of the nonlinear media by the unwanted effects of GVM, several experimental papers on frequency doubling in $KNbO_3$[12,16-19], LBO[20] and BBO[20] have recently demonstrated that efficient SHG is possible by focusing femtosecond (120-200 fs) pulses in "thick" nonlinear media where the ratio $L/L_{nst} > 20$. These experiments have demonstrated that a conversion efficiency exceeding 60% is possible and, as we have shown[12], the duration of the generated near-transform-limited SH pulses remained within the femtosecond regime, increasing only 2-3 times with respect to the fundamental duration. In these experiments the optimal ratio L/b was found to be in the region of 10, which is far from the known Boyd and Kleinman ratio of $L/b = 2.83$. To our knowledge, no other theoretical investigations exist that can predict the optimal focusing for SHG under conditions of large GVM, as we discuss here. In the reported work of Weiner and Yu[16,19] a simple model is proposed that predicts well the efficiency of the SHG process in bulk nonlinear crystals, as well as identifying that an increase in SHG efficiency relates to an increase in the L/b ratio. Their model, however, does not predict the existence of optimal values for L/b and phase mismatch as obtained in both experiments discussed above. Also, it cannot describe the evolution of the SH temporal shape inside the crystal and does not give an optimal focusing position inside the nonlinear media.

Our theoretical model for SHG in bulk nonlinear crystals, which describes the process of SHG under conditions of large GVM, assumes an undepleted fundamental beam, and that both the fundamental and SH beam have Gaussian transverse distributions. The results of the model are later compared with experimental data from SHG experiments using focused femtosecond pulses in the bulk nonlinear crystals of potassium niobate ($KNbO_3$) and aperiodically-poled potassium titanyl phosphate (appKTP). This model is presented in section 2.2.

In the case of SHG in waveguide nonlinear crystals, we describe a theoretical model which accounts for the temporal behavior of the interacting pulses and the possible z-dependence of the phasematching condition. The model also describes the observed saturation and subsequent decrease in SHG conversion efficiency in the waveguide samples, as a result of two-photon absorption (TPA) of the second harmonic (SH) wave. The results of this model are later compared with experimental data from SHG experiments using femtosecond pulses in the waveguide nonlinear crystals of periodically-poled potassium titanyl phosphate (ppKTP) and appKTP. This model is presented in section 2.3.

2.2 Theoretical model to describe SHG of focused femtosecond pulses in periodically-poled and aperiodically-poled bulk nonlinear crystals

As discussed above, we have recently reported a theoretical model that defines the optimal conditions for SHG using focused beams in the ultrashort-pulse regime. The model[15], verified by experiment[12], is suitable for cases when birefringence walk-off can be neglected, such as SHG with non-critical phasematching (as with $KNbO_3$) and SHG in a homogeneous quasi-phasematched structure[21]. It provides information on the efficiency of the SHG process, the pulse duration of the SH pulses (figure 1a), and the modification of the phasematching tuning curves. It also allows for optimization of the SHG process by selecting the optimal phase mismatch, focusing strength (figure 2) and position of focusing within the crystal.

Here we will show that our previous model[15] can be extended to describe the process of SHG with focused beams in aperiodically-poled nonlinear media (i.e. those with inhomogeneous mismatch).

While inhomogeneous mismatch can result from a non-constant temperature variation along the sample length, we will instead consider the typical case of a $\chi^{(2)}$ crystal with linearly chirped quasi-phasematched (QPM) grating. The crystal is a nonlinear medium with a locally varying phasematching parameter. For a linearly chirped quasi-phasematched (QPM) grating the mismatch phase factor, $\Phi(z)$, which controls the second harmonic generation process at point z can be presented in the form

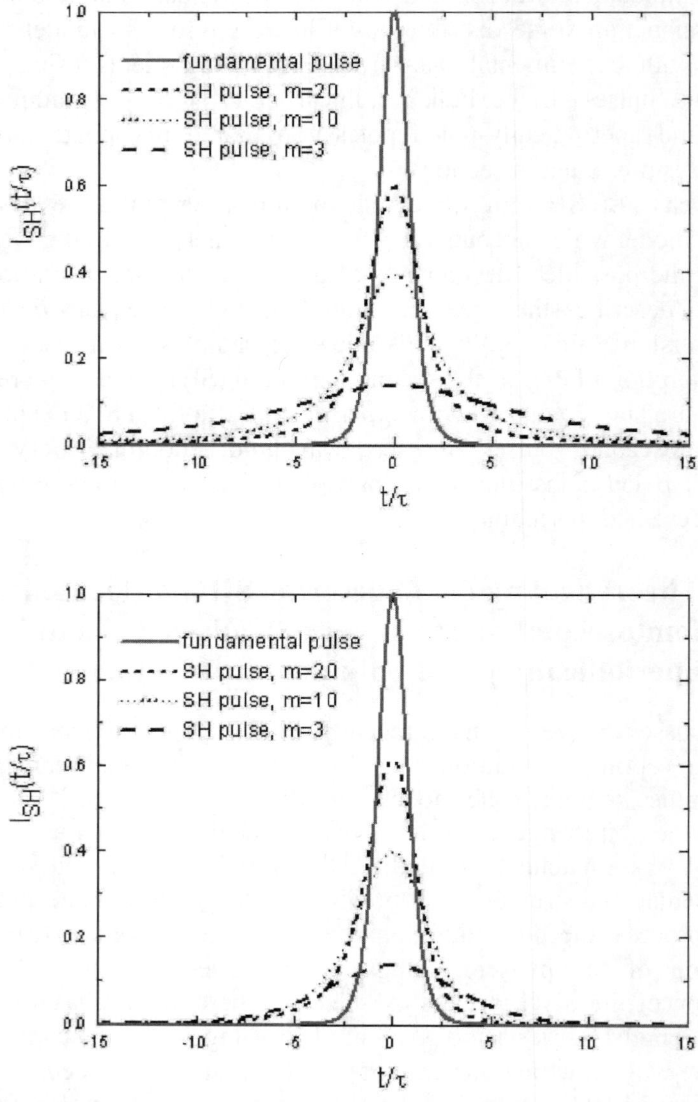

Figure 1. Temporal profiles and relative intensities of the generated SH pulses calculated for several values of focusing strength, *m* = *L/b*, and ratio $L/L_{nst} = 40$ for: (a) uniform (periodically-poled) nonlinear media[15]; and (b) linearly chirped (aperiodically-poled) quasi-phasematched grating with $\Delta\Lambda = 0.2\ \mu m$ ($D_g = 10.3\ mm^{-2}$). The fundamental pulse shape (grey line) is shown for comparison.

$\Phi(z) = \Delta k_0 z + D_g z^2$, which allows the localized wavevector imbalance to be presented as $\Delta k(z) = d\Phi(z)/dz = \Delta k_0 + 2D_g z$, where D_g is calculated from $D_g = \pi\Delta\Lambda/(\Lambda_0^2 L)$ and $\Delta k_0 = k_2 - 2k_1 - 2\pi/\Lambda_0$. The quantity $\Delta\Lambda = \Lambda(z = L) - \Lambda(z = 0)$ defines the change of the grating period across the sample length, L, and $\Lambda_0 = [\Lambda(z = L) + \Lambda(z = 0)]/2$ is the mean grating period. It is useful to note that for sufficiently long samples (as is the case for both $KNbO_3$ and appKTP samples used in our experiments, described in sections 7.1 and 7.2), the variation of the grating period across the sample, $\Delta\Lambda$, defines the wavelength acceptance bandwidth, $\Delta\lambda_{fund}$, of the SHG process in the case of unfocused or weakly focused beams:

$$\Delta\lambda_{fund} = \left| \frac{2\pi\Delta\Lambda}{\Lambda_0^2} \left(\frac{d\Delta k_0}{d\lambda} \right)^{-1} \right| \tag{1}$$

The parabolic equations derived in a slowly varying envelope approximation that describe the second harmonic generation (SHG) of ultrashort pulses in media with locally inhomogeneous wave-vector mismatch, have the form:

$$\left(\frac{\partial}{\partial z} + \frac{i}{2k_1}\Delta_\perp + \frac{1}{v_1}\frac{\partial}{\partial t} \right) A_1 = 0 \tag{2}$$

$$\left(\frac{\partial}{\partial z} + \frac{i}{2k_2}\Delta_\perp + \frac{1}{v_2}\frac{\partial}{\partial t} \right) A_2 = -i\sigma_2 A_1^2 \exp(i\Phi(z)) \tag{3}$$

where A_1 and A_2 denote the complex amplitudes of the fundamental and the second harmonic wave respectively, and are functions of three spatial coordinates and one temporal coordinate, $A_j = A_j(x, y, z, t)$. $\Delta_\perp$ stands for the operator $\partial^2/\partial x^2 + \partial^2/\partial y^2$. The nonlinear coupling coefficient, σ_2, is calculated as $\sigma_2 = 2\pi d_{\text{eff, SHG}}/(\lambda_1 n_2)$, where the magnitude of $d_{\text{eff, SHG}}$ depends on the method of phase-matching and the type of the nonlinear medium that is used. The depletion of the fundamental beam, birefringence walk-off and absorption losses for the both interacting waves are neglected.

Following the previously described method[15] we obtain an expression for the SH amplitude, $S(L, p)$, at the output of a nonlinear media with linearly chirped (aperiodically-poled) quasi-phase-matched (QPM) grating:

$$S(L, p) = -i\sigma_2 A_o^2 b H_{tr}(m, \mu, v, \gamma, p) \tag{4}$$

with

$$H_{tr}(m,\mu,\nu,\gamma,p)=\frac{1}{2}\int_{-m(1+\mu)}^{m(1-\mu)}\frac{du}{1-iu}T\left(\frac{p}{\tau}+\gamma u\right)^2\exp(i\nu u+iD_g b^2u^2/4) \quad (5)$$

where $\nu=\frac{\Delta k_o b}{2}=\frac{\Delta k_o L}{2m}$; $m=\frac{L}{b}$; $\gamma=\frac{\alpha b}{2\tau}=\frac{1}{2m}\frac{L}{L_{nst}}$ and μ indicates the position of the focused spot inside the crystal: $\mu=0$ corresponds to center of the crystal; $\mu=-1$ to the focus at the input face; $\mu=+1$ to the focus at the output face. In addition, p is the local time, α is the GVM parameter, and τ is related to the full-width at half-maximum (FWHM) fundamental pulse duration, τ_0, by $\tau=\tau_0/1.76$.

The energy of the second harmonic (SH) is found by integrating the SH intensity over space and time while assuming Gaussian spatial and hyperbolic secant temporal profiles. The normalized efficiency, η_0 (%pJ^{-1}), of the SHG process is then calculated by:

$$\eta_o=\frac{16\pi^2 d_{eff}^2}{3\lambda^3 c\varepsilon_o n_2 n_1}\frac{Lh_{tr}}{\tau}10^{-10} \quad (6)$$

where n_2 and n_1 are respectively the refractive indices at the second harmonic and fundamental wavelengths, and h_{tr} is a transient focusing factor calculated from

$$h_{tr}(m,\mu,\nu,\gamma)=\frac{3}{4m}\int_{-\infty}^{+\infty}\left|H_{tr}(m,\mu,\nu,\gamma,p')\right|^2 dp' \quad (7)$$

The absolute efficiency, η (%), in the absence of depletion of the fundamental wave is defined by $\eta=\eta_0 W_{fund}$, where W_{fund} is the fundamental pulse energy in pJ. At higher pump intensities, when the depletion of the fundamental is weak but essential, the corrected value for the absolute efficiency, η (%), can be found by[20]:

$$\eta=\frac{100W_{fund}\eta_o}{100+W_{fund}\eta_o} \quad (8)$$

By controlling the parameter D_g, defined above as $D_g=\pi\Delta\Lambda/(\Lambda_0^2 L)$, we can apply this model to accurately describe the SHG process in linearly chirped (aperiodically-poled) QPM structures with focused beams. $D_g=\Delta\Lambda=0$ corresponds to homogeneous (periodically-poled) QPM gratings, and for this case the model provides the results described previously[15]. $D_g \neq 0$; $\Delta\Lambda \neq 0$ corresponds to inhomogeneous (aperiodically-poled) QPM structures, and the results of the model in describing the generated SH pulse shape are presented in figure 1b. Figure 2 provides theoretical curves which define the optimal SHG focusing conditions for different values of $\Delta\Lambda$.

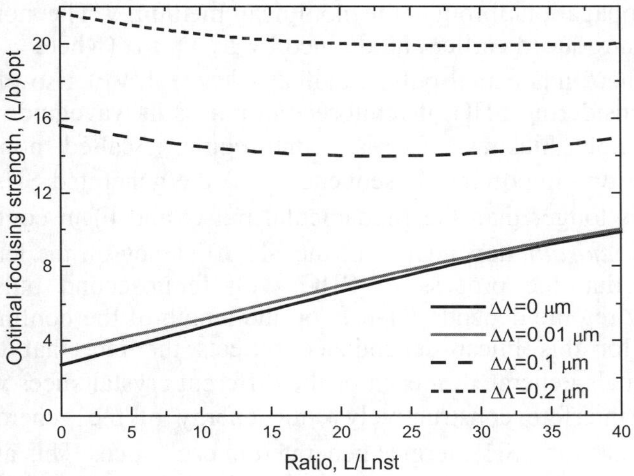

Figure 2. Dependence of optimal focusing strength, $m_{opt} = (L/b)_{opt}$, on the ratio, L/L_{nst} for different values of $\Delta\Lambda$. Crystal length is $L = 4$ mm. The efficiency of the SHG process can be maximized by selecting the correct focusing lens for a given ratio L/L_{nst}.

As can be seen from the comparison of figure 1a and figure 1b, the SH pulse duration is shorter when the SHG process takes place in linearly chirped QPM structures. The output pulses also have certain amount of phase chirp, which can be exploited to achieve further pulse shortening[22,23]. Figure 1 perfectly illustrates the advantage of using relatively thick nonlinear crystals with focused beams: the generated SH intensity is higher and the pulses are shorter in the case of tight focusing (i.e. larger $m = L/b$).

As illustrated in figure 2 the strong dependence of the optimum focusing strength, $m_{opt} = (L/b)_{opt}$, on the ratio L/L_{nst} (which is typical for materials with constant Δk) no longer applies when dealing with aperiodically-poled structures (those with a linear change of Δk in z).

We believe that our model can be extended even further to accurately describe other nonlinear optical interactions such as sum and difference frequency mixing, as well as higher-order harmonics generation.

2.3 Theoretical model to describe SHG of femtosecond pulses in waveguide nonlinear crystals

As mentioned earlier, second harmonic generation using femtosecond pulses is characterized by group velocity mismatch (GVM), which accounts for the temporal walk-off between the fundamental and second harmonic pulses in propagation through the nonlinear medium. The nonstationary length, L_{nst}, introduced earlier and defined by $L_{nst} = \tau/\alpha$ (where τ is the time duration of the fundamental pulses, and $\alpha = 1/v_2 - 1/v_1$) also plays a vital role when considering SHG of femtosecond pulses in waveguide structures. When the sample thickness, $L >> L_{nst}$, the regime is called 'nonstationary' and there are two important consequences: a) the generated SH pulses are L/L_{nst} times longer than the fundamental pulse; and b) in contrast to the well known *quadratic* dependence of the SH efficiency on the length of the nonlinear media, the process of SHG with femtosecond pulses in the nonstationary regime depends *linearly* on the length of the nonlinear media. The reason for this linear dependence reflects the fact that the second harmonic signals generated in each of the different crystal slices with length L_{nst} do not interfere constructively - the energy of the generated SH is simply the sum of the SH energies generated in each slice. This allows us to use a very simple formula to describe the process of SHG in the nonstationary regime ($L >> L_{nst}$), which applies to both SHG in bulk media with unfocused beams, and to SHG in waveguided crystals.

In deriving the formula we require, let us first recall that the amplitude of phasematched SHG in a sample with $L = L_{nst}$ in an undepleted approximation is described by the differential equation:

$$\frac{dA_2}{dz} = -i\sigma_2 A_1 A_1 \qquad (9)$$

The solution of Eq. (9) gives for the power of the SH wave:

$$P_2 = \frac{2\sigma_2^2}{\varepsilon_o nc} \frac{P_1^2}{A_{eff}} L_{nst}^2 \qquad (10)$$

where P_1 is the power of the fundamental wave, and A_{eff} is the effective cross section of the waveguide. Rewriting this expression in terms of energy we obtain an expression for the SHG efficiency:

$$\eta = \frac{W_2}{W_{fund}} = \frac{2\sigma_2^2}{\varepsilon_o nc} \frac{W_{fund}}{A_{eff}\tau} L_{nst}^2 \tag{11}$$

where W_2 and W_{fund} are respectively the SH and fundamental pulse energies. If now the waveguide length is several times longer than L_{nst}, then the output energy is scaled by the factor L / L_{nst}:

$$\eta_{nd} = \frac{2\sigma_2^2}{\varepsilon_o nc} \frac{W_{fund} L_{nst}^2}{A_{eff}\tau} \frac{L}{L_{nst}} = \frac{8\pi^2 \left| d_{eff}^{(2)} \right|^2}{\varepsilon_o c n^3 \lambda^2} \frac{W_{func} L}{\alpha A_{eff}} \tag{12}$$

For bulk samples the effective cross-section, $A_{eff} = \pi w_{01}^2$. It can be seen from Eq. (12) that SHG energy conversion in the nonstationary regime ($L >> L_{nst}$) depends linearly on the length of nonlinear media and does not depend on the pulse duration. Equation (12) can be recovered from the results of Weiner[16] if we take weak focusing approximation of their Eq. (4).

We remind the reader that Eq. (12) is derived under conditions where the effects of pump depletion are neglected. In much the same way as it was described in section 2.2, we can expand the application of Eq. (12) for the range of higher pump intensities for which the depletion of the fundamental is weak but essential. The corrected value for the SHG efficiency, η_{cor}, which accounts for the effects of pump depletion is[20]:

$$\eta_{cor} = \frac{\eta_{nd}}{1 + \eta_{nd}} \tag{13}$$

We will later use Eq. (13) for analyzing the experiments of SHG in ppKTP and appKTP waveguides (described in section 7). For now, we compare the analytical solution of Eq. (13) with the direct numerical solution of the differential equations in Eq. (14), which account for the temporal behavior of the interacting pulses, the nonlinear losses of the SH wave, and the possible z-dependence of the phasematching condition:

$$\left(\frac{\partial}{\partial z}+\frac{1}{v_1}\frac{\partial}{\partial t}\right)A_1=-i\sigma_1 A_2 A_1^* \exp\left[-i\Phi(z)\right]$$
$$\left(\frac{\partial}{\partial z}+\frac{1}{v_2}\frac{\partial}{\partial t}\right)A_2=-i\sigma_2 A_1^2 \exp\left[i\Phi(z)\right]-\gamma A_2^2 A_2^* \quad (14)$$

where A_1 and A_2 are the complex amplitudes of the fundamental and SH waves respectively and are functions of (z, t). The nonlinear coupling coefficients, $\sigma_{1,2}$, are calculated as $\sigma_{1,2}=2\pi d_{eff,SHG}/(\lambda n_{1,2})$, where the magnitude of $d_{eff,SHG}$ for QPM waveguides is $d_{eff,SHG}=(2/\pi)d_{33}\beta$, and β is an overlap integral and is < 1. The parameter, $\gamma=(3\pi/4\lambda_2 n_2)\,\mathrm{Im}(\chi^{(3)})$ where $\mathrm{Im}(\chi^{(3)})$ is the imaginary part of the third-order nonlinear susceptibility[24]. The role of this term in the second equation of Eq. (14) is to account for the nonlinear losses observed in the experiments with both ppKTP and appKTP. The phase factor in system Eq. (14) is defined as $\Phi(z)=\Delta k_0 z$ for the ppKTP waveguide, and as before $\Phi(z)=\Delta k_0 z+D_g z^2$ for the appKTP waveguide.

In figure 3 we compare the analytical solution of Eq. (12) and Eq. (13) with the numerical solution of Eq. (14) for the SHG conversion efficiency for three different pulse profiles (hyperbolic secant, Gaussian and rectangular). It can be seen that the simplified model perfectly describes the process of SHG when $L>5L_{nst}$. Further, the analytical model of Eq. (12) and Eq. (13), valid for samples much longer the nonstationary length L_{nst}, does not depend on the pulse shape.

The effects of two-photon absorption (TPA) of the SH wave are included in the second equation of Eq. (14), in the form of the parameter $\gamma=(3\pi/4\lambda_2 n_2)\,\mathrm{Im}(\chi^{(3)})$, in order to explain the imbalance of the conservation of energy at the waveguide output. Experimental observations[13] have shown that, for input fundamental pulse energies of around 60 pJ, the ratio of $(P_{1,out}+P_{2,out})/P_{1,in}$ reached 75%, where P_1 and P_2 are respectively the average powers of the fundamental and SH beams. As the exact value of $\mathrm{Im}(\chi^{(3)})$ for KTP at 430 nm is, to our knowledge, not available in the literature, we have used these measured nonlinear-type losses in energy to set the value of the parameter, γ, in the system of Eq. (14).

The theoretical curves obtained from the solution of Eq. (14), with $\gamma=0.03L\sigma^2$, are shown later in figure 18. If we take $d_{eff,SHG}$ = 3.6 pm/V, the value of γ used in the calculation corresponds to a TPA coefficient of 3.8 cm/GW, which is close to the reported TPA coefficient of $KNbO_3$ (3.2 cm/GW)[25].

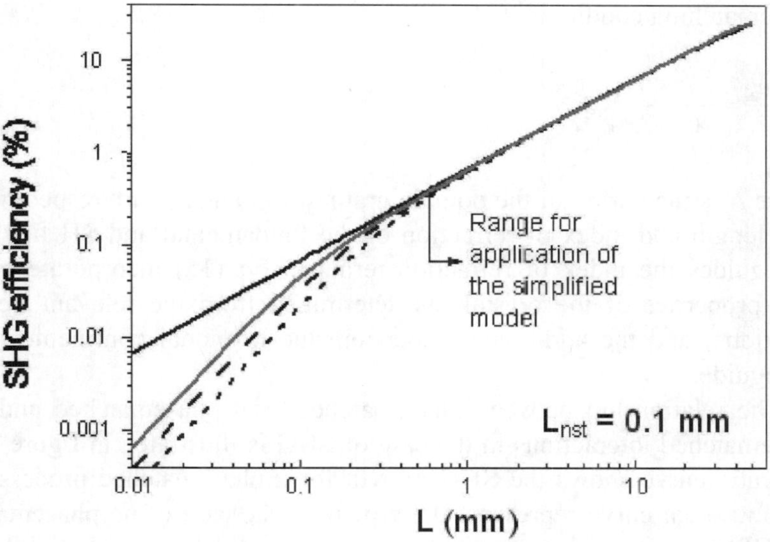

Figure 3. Comparison of simplified analytical solution for η_{cor} (black line) given by Eq. (13), and numerical solution of Eq. (14). Depletion effects are included for three different temporal pulse shapes: secant hyperbolic (gray line), Gaussian (dash line) and rectangular (short dash line). Nonlinear losses are neglected ($\gamma = 0$). Calculations are performed for a KTP waveguide with cross-sectional dimensions of 2 × 2 μm ($d_{eff} = 4.6$ pm/V). Input pulse energy is 1 pJ.

3. FABRICATION OF WAVEGUIDE CRYSTALS

Efficient nonlinear optical frequency conversion requires phasematching, large pump intensities, long interaction lengths and materials with large nonlinear optical coefficients. The limitations imposed on nonlinear optical mixing processes due to the low peak powers found in most quasi-cw and cw lasers can be overcome by creating waveguides in the nonlinear material.

The waveguide confines the beam, allowing for a high intensity over long interaction lengths. This can lead to a significant increase in the optical conversion efficiency, thus making guided wave nonlinear optical conversion ideal for applications requiring cw or low peak power quasi-cw lasers. Furthermore, efficient optical mixing throughout the entire transparency range of the crystal can be achieved by using a quasi-phasematched (QPM) process.

In a QPM nonlinear optical process, the waveguide is segmented into regions with alternating anti-parallel ferroelectric domains. For SHG, the period of the domain grating for first order QPM is determined by the quasi-phasematching condition[26,27]:

$$\frac{n_2}{\lambda_2} = \frac{2n_1}{\lambda_1} + \frac{1}{\Lambda} \tag{15}$$

where Λ is the period of the domain grating, $\lambda_{1,2}$ and $n_{1,2}$ are respectively the wavelength and index of refraction of the fundamental and SH fields. For waveguides the index of refraction terms in Eq. (15) incorporate both the bulk properties of the crystal (as determined from the relevant Sellmeier equations) and the additional dispersion due to modal confinement in the waveguide.

The relationship between phasematched, non phasematched and quasi-phasematched interactions in the case of SHG is illustrated in figure 4. The quadratic curve shows the SHG growth for a phasematched process, while the sinusoidal curve represents the growth in the case of no phasematching. In a QPM process, when the fundamental and second harmonic fields are get out of phase, the periodically poled regions flip the phase of the second harmonic field allowing the SHG field to grow as it propagates through the crystal. In contrast to the usual birefringent phasematching, quasi-phasematched mixing processes can be exploited over the entire transparency range of the crystal.

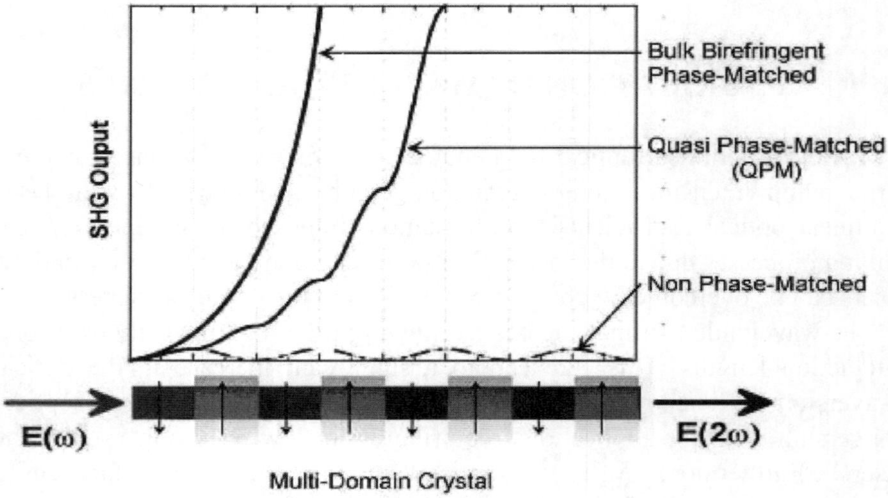

Figure 4. The graph shows the SHG output as a function of distance through a nonlinear crystal for various phase matching conditions.

The waveguides used in the experiments described in section 7 were fabricated using photolithography to transfer a mask containing the waveguide patterns to a KTP wafer and then using an ion-exchange process to embed the waveguides. The exact steps to fabricate the waveguides in a KTP wafer are outlined in figure 5 and are described in detail below.

The initial step is to layout the waveguide pattern using the *AutoCAD* software. The waveguide pattern is segmented, with the period of the segmentation equal to the required QPM period for SHG (e.g. for conversion of 850 nm the grating period is ~4 μm).

The waveguide pattern is transferred to the KTP substrate using photolithography as shown in the first four steps in figure 5. The first step in the photolithography process is to spin on a thin layer of photoresist that evenly covers the surface of a *z*-cut KTP substrate. The photomask, with waveguide design, is imaged onto the photoresist using a projection

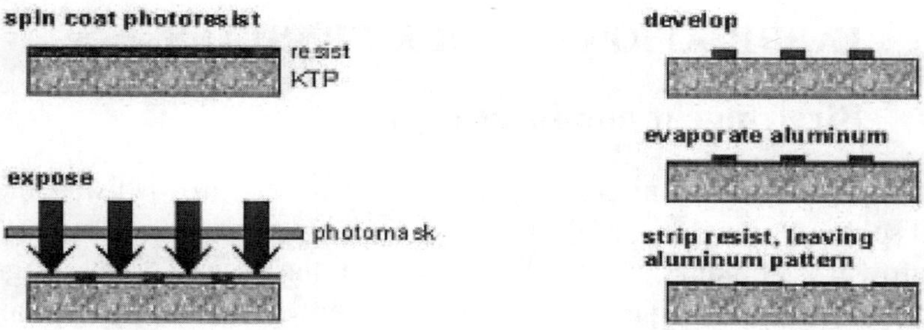

Figure 5. The five steps required to transfer the waveguide pattern to the KTP substrate.

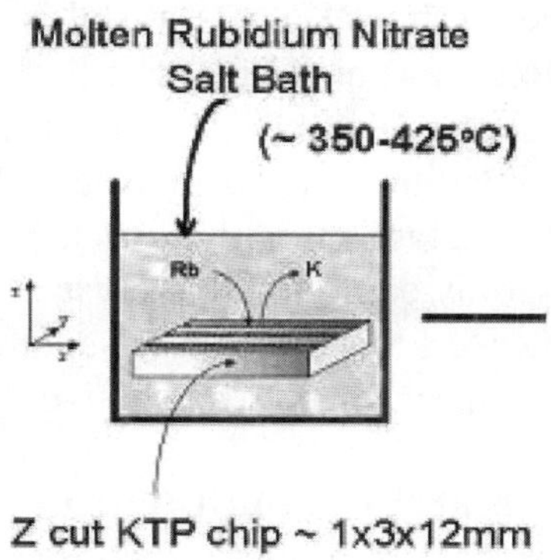

Figure 6. The illustration depicts the ion-exchange process that occurs in the oven.

lithography system. A developing process removes the exposed photoresist so only the unexposed regions of the KTP are covered. Next, a thin aluminum or SiO_2 coating is deposited onto the wafer. Submerging the wafer in acetone removes the remaining resist along with any aluminum (but not the SiO_2) in the exposed area creating a waveguide pattern.

The KTP chips are individually immersed in a molten bath of a mixture of $RbNO_3$ and $Ba(NO_3)_2$. Within this bath, the Rb ions diffuse into the unmasked portions of the KTP chip, while the K ions diffuse out of the substrate and into the bath, shown in the illustration in figure 6. In the diffused regions, the rubidium ions increase the index of refraction relative to the undiffused KTP and thus form the optical waveguide. Note that due to the presence of barium, there is an increase in the index of refraction and the ferroelectric domain in the diffused region is reversed and hence the term chemical poling is used for this process.

4. FABRICATION OF BULK CRYSTALS

4.1 Birefringent nonlinear crystals

The bulk potassium niobate ($KNbO_3$) crystal was grown using the top seeded solution growth (TSSG) method, which is one of the commonly used 'flux growth' techniques. In the TSSG method, the desired crystal is grown from a solvent containing the dissolved constituent components. Crystals are grown at a temperature below their melting point which prevents decomposition before melting or any unwanted phase transitions.

$KNbO_3$ is a ferroelectric crystal of the 'perovskite' family, which transforms from cubic to tetragonal, tetragonal to orthorhombic, and orthorhombic to rhombohedral phase at temperatures of 418°C, 203°C, and -50°C respectively. Once the bulk $KNbO_3$ crystal is slowly cooled down to room temperature to avoid unwanted cracks, it is then cut into blocks and poled by applying an electric field at a temperature just below the phase transition (~200°C) in order to generate a single domain. The required efficient phasematching interaction within such a nonlinear crystal has traditionally been accessed by exploiting the crystal's birefringence (as with $KNbO_3$). In general, however, this birefringent approach severely limits the flexibility in choice of nonlinear material and active wavelength.

4.2 Quasi-phasematched nonlinear crystals

The alternative is quasi-phasematching (QPM), where the desired crystal can be designed for a specific wavelength and fundamental spectral bandwidth with an appropriate periodic structure. For the case of incident femtosecond pulses, characterized by significantly wider spectral bandwidths than continuous-wave sources, this is of particular interest as the spectral shape can be matched to the periodic structure. In addition, pulse shortening (or expansion) is possible by selectively arranging the different frequency converting regions. Finally a higher conversion efficiency can be obtained with QPM because the strongest nonlinear component, d_{33}, can be accessed, which is not the case for birefringent phasematching. Effective QPM has been demonstrated in many crystals, including $LiNbO_3$[28], $LiTaO_3$[29], RTP[30] and RTA[31], but we will concentrate here on KTP[32].

Periodically-poled KTP can be conveniently used at room temperature and can withstand high intensities without showing any sign of damage. The samples used in this work were made from flux-grown KTP, and the periodic structure was fabricated by electric-field poling according to the technology developed by Karlsson *et al*[32]. Flux-grown KTP can show substantial composition variations even though the optical properties are

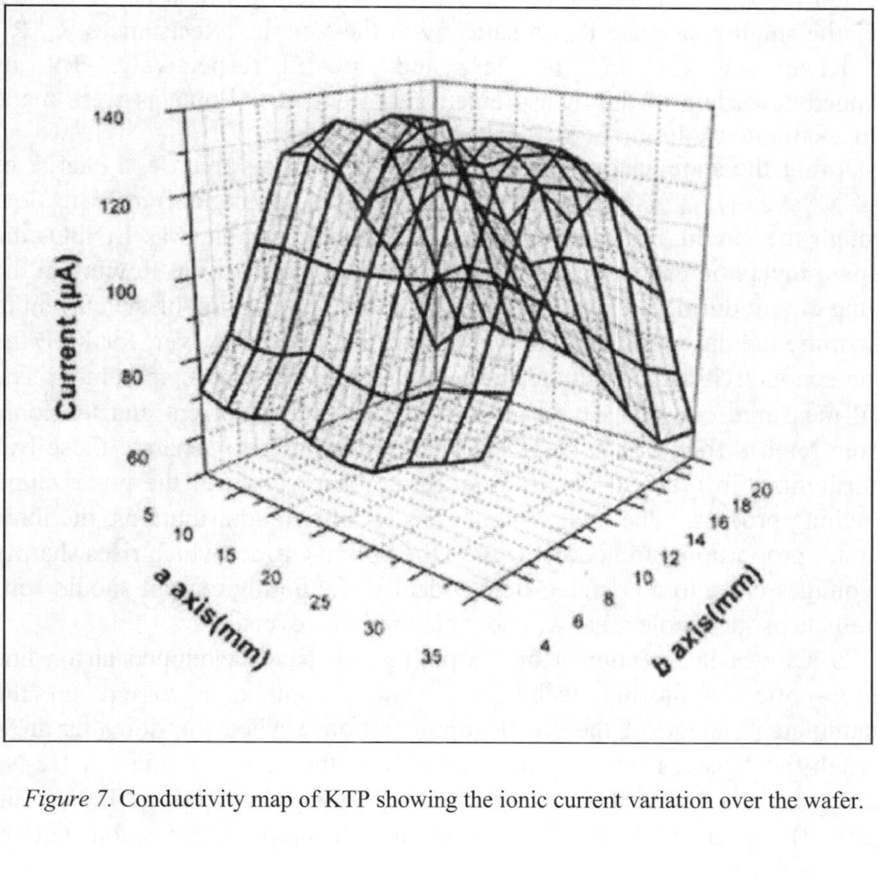

Figure 7. Conductivity map of KTP showing the ionic current variation over the wafer.

seen as identical. The composition variation, which is observed as a conductivity variation when an electric field is applied over the sample, can be detrimental for periodic poling. It is commonly seen that a sample with large degrees of such variations will result in three areas of composition, over poled, periodically-poled and unpoled regions. A typical conductivity map for a flux grown KTP wafer is seen in figure 7. As this figure illustrates, the change in conductivity (which can be as large as a factor of two) occurs mainly along the b-axis, whilst it is more or less constant along the a-axis.

The electric circuit used for the periodic poling as well as for the conductivity measurements is shown in figure 8. A high-voltage amplifier (*Trek 20/20C*) amplifies a waveform generated by an arbitrary signal generator (*Agilent 33120A*). The serial resistance of resistor R_1 limits the current passing through the sample during poling. The voltage over the sample is determined through the V_1, which is the voltage over R_3 and which forms a voltage divider parallel to the sample. By properly selecting the resistance R_2, the voltage is measured by an oscilloscope. The current through the sample is determined from the voltage V_2, which is the voltage over the small resistance R_4, in series with the sample. Resistances R_1, R_2, and R_3 are set as 15 kΩ, 100 MΩ, and 1.6 MΩ, respectively. R_4 was changed according to the measurement requirements. Contacts were made with a saturated solution of KCl.

During the spontaneous polarization reversal in an area, *A*, a charge of $Q = 2P_s A = \int i \cdot dt$ is deposited, which causes the poling current, *i*, to flow through the circuit. In low-conductivity ferroelectrics such as $LiNbO_3$ the domain inversion can be controlled by monitoring the current flowing in the poling circuit during the charge transport, and an integration of the current to determine the deposited charge over a specific area. However, for KTP the ionic conductivity is substantial in the *c*-direction at room temperature. The total measured current is then the sum of the poling current and the ionic current, and difficulties have arisen when attempting to separate these two contributions in order to obtain an accurate observation of the polarization switching process. The total current consists of two contributions, the ionic current, proportional to voltage U^2 and the poling current which rises sharply at voltages close to a coercive field. Ideally, the poling current should self-terminate as the whole area, A, under electrode is reversed.

To achieve better control of the poling, we have developed an on-line electro-optic monitoring technique[30]. The technique is based on the accumulated change of the electro-optic response when the domains grow through the crystal from the patterned side to the opposite side. A He-Ne beam polarized 45° to the z-axis is launched along the x-axis of the crystal (figure 9). When an electric field is applied, the output polarization state of

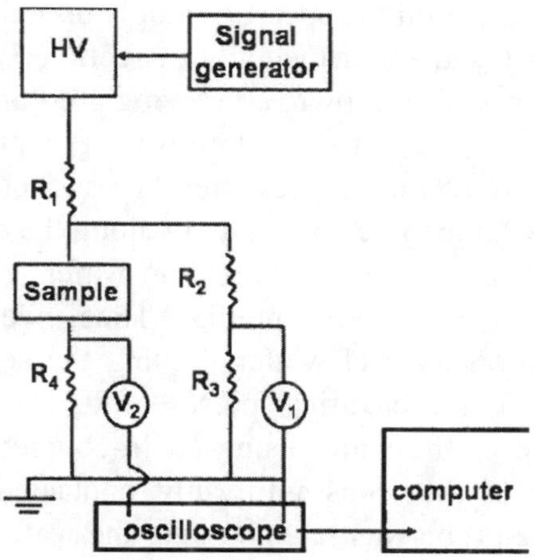

Figure 8. The electrical circuit used for all the polarization-switching experiments, as well as for the conductivity measurements.

the He–Ne beam will be changed due to the electro-optic effect. This results in a time dependent variation of the polarization state during the rise and the fall of the pulse. During the rest of the pulse, when the field is constant, the polarization state will be time dependent only if the sign of the electro-optic effect is reversed, i.e. the crystal is being poled.

The change of polarization state can be observed by measuring the intensity of the He-Ne beam through a polarizer orthogonal to the initial polarization. The intensity at the detector will then show a periodic oscillation during poling, which will stop when the poling starts (figure 9). When the poling parameters are chosen carefully, and when the crystal is homogeneous, only the area under the electrodes are poled, and no over-poling is obtained.

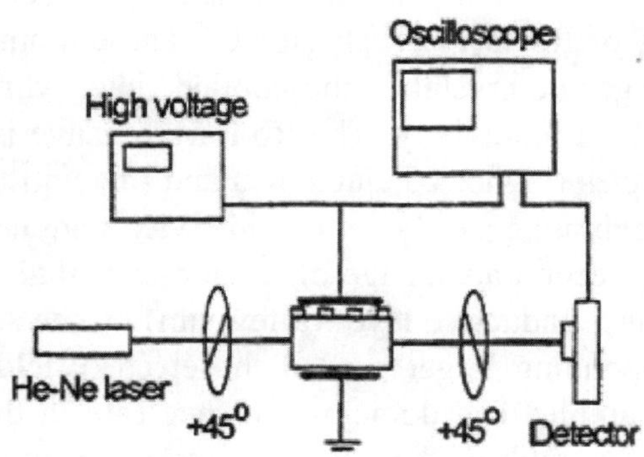

Figure 9. Schematic for monitoring the poling process[30].

Let us now concentrate on the specific sample used in this work. For the bulk appKTP crystal (used experimentally in section 7), the commercially-obtained KTP wafer was flux-grown, *c*-cut, single domain, 1 mm thick, and with the *c*-faces polished to an optical finish. For practical reasons, the wafers were cut into 10×5 mm^2 pieces after the conductivity measurements.

In flux-grown KTP, the ionic conductivity along the *c*-axis can vary by as much as an order of magnitude over a single wafer[33]. This was found to severely complicate the periodic poling. Thus, we first mapped the conductivity distribution on each wafer. Using the set-up of figure 8, 5-millisecond positive square electrical pulses of 1.5 kV in magnitude were applied to the c^--face of the wafer using an In-covered probe of 1 mm in diameter, while the c^+-face was uniformly contacted using a saturated solution of KCl. The probe was scanned over the wafer surface, measuring the ionic current through the resistor R_4, which was set to 10 kΩ.

A typical measured ionic-current map of a KTP wafer, which illustrates the conductivity distribution, is seen in figure 7. The measured current of 100 μA yields an absolute value of the conductivity at 1 mm^2 of about 8.5×10^{-7} S/cm. It shows a parabolic variation along the b-axis, increasing by a factor of 2 from to the edges to the center, whereas it remains almost constant along the *a*-axis. The concentration of K^+ vacancies, which is the main contributor to the ionic conductivity, is very sensitive to crystal growth temperature. Thus, the variations in conductivity over the KTP wafer are most probably due to a temperature gradient during crystal growth that results in a spatially varying stoichiometry. In this work, samples with low conductivity from the edge of the wafer whose conductivities are relatively low (7×10^{-7} S/cm) were chosen. To get even more homogeneous conductivity and to get good nucleation for domain reversal we ion-exchanged the sample in pure $RbNO_3$ for 6 hrs at 350°C. Rb is then diffusing into the surface layer of the crystal replacing K and filling K-vacancy positions[32]. The reduction in conductivity is a consequence of the larger ionic radius of Rb in the KTP lattice. These atoms will hence drift more slowly through the crystal in the applied field, which is the same as having a lower ionic conductivity. The Rb ion exchange is faster in regions of higher K-vacancies (higher conductivity) and this will then help to make the material more homogeneous in conductivity and hence improve the periodic poling. A second advantage of the ion-exchange is that the crystal will have a thin low conductive layer (a few μm) on the surface which then will take a corresponding larger part of the electric field when voltage is applied over the sample. The domains will nucleate in this layer and then propagate through the bulk of the sample to the opposite side again helping creation of a periodic domain structure of high quality.

The sample was patterned with periodic photoresist using conventional photolithography. In the poling the photoresist works as an isolator on the sample surface and the area under the mask will stay un-inverted while the areas under the openings are the regions that will be inverted during periodic poling. The back-side of the sample was coated with an Al film to prevent depletion of K from the sample during the following E-field poling.

The sample, with patterned area of 4×4 mm^2 was contacted with $KCl:H_2O$ liquid electrodes, and two 2.2 kV pulses (5 ms long) were applied. The sample was then tested in an SHG experiment and the back metal electrode and the photoresist was removed.

In order to write the periodic structure onto the KTP crystal, a photolithographic mask was constructed. Conventional masks have a single-grating period, typically ranging from 3.4 μm to 3.6 μm for first-order QPM applications, which is patterned over the whole length of the crystal[28-32]. In recent years, paralleled with improvement of e-beam and UV-laser lithography technologies, novel grating structures have appeared, such as multi-gratings, Fibonacci sequenced gratings, and aperiodic (chirped) gratings[34-36]. For the case of aperiodically-poled gratings, a linear chirp is deployed over a mask with length L, having a starting period of Λ_s and an ending period of Λ_f. The chirp parameter, α, is then simply calculated from $\alpha = (\Lambda_s - \Lambda_f)/L$. In our case the manufactured mask length had a length of $L = 4$ mm, with a starting and ending period of $\Lambda_s = 4.1$ μm and $\Lambda_f = 4.3$ μm, respectively, and a chirp-parameter of $\alpha = 0.05$. These parameters are valid for an idealized linear chirp. However, when manufacturing the actual grating the limiting factor is the resolution of the lithographic process; for our chirped grating a resolution of 0.05 μm was used.

5. PORTABLE FEMTOSECOND INFRARED LASER

Our recent investigations into efficient frequency-doubling of femtosecond pulses in waveguide and bulk nonlinear crystals have allowed us to propose the design of a portable ultrafast blue light source[11], which takes advantage of a previously reported ultra-compact femtosecond Cr:LiSAF laser design[10,37-40]. The extremely low pump-threshold conditions (~22 mW to sustain modelocking) associated with this Cr:LiSAF laser permit the use of inexpensive (~US$40) single-narrow-stripe AlGaInP red laser diodes (50-60 mW) as a pump source[41]. Their modest electrical drive requirements (~100 mA current, ~200 mW electrical power) mean that these diodes require no active electrical cooling and can easily be powered for a number of hours by regular 1.5 V penlight (AA) batteries. With a maximum pump power from two pairs of pump laser diodes of ~200 mW, the

Cr:LiSAF laser crystal does not require any cooling. As a result, relatively inexpensive femtosecond Cr:LiSAF lasers with optical output powers up to 45 mW, robustly modelocked with a semiconductor saturable absorber mirror (SESAM), are achievable from entirely self-contained and portable units with dimensions of 22 cm × 28 cm. [A version of this laser design incorporating one pair of pump laser diodes is shown in figure 10] A second pair of pump laser diodes can easily be incorporated onto the baseplate to maximize operational performance (figure 12).

The modest electrical drive requirements of these diodes, and the resulting option to power the laser with standard penlight (AA) batteries, allow these Cr:LiSAF lasers to boast an impressive electrical-to-optical efficiency of over 4 %, which until recently[42] was the highest reported overall system efficiency of any femtosecond laser source. The amplitude stability of the laser output was observed to be very stable with a measured fluctuation of less than 1% for periods in excess of 1 h. These measurements were made on a laser that was not enclosed and located in a lab that was not temperature-controlled. In a more enclosed and controlled local environment we would expect the amplitude fluctuations of this laser to be extremely small. While the output powers achievable from these lasers have been limited by the available power from the AlGaInP red laser pump diodes, there are already strong indications that commercial access to higher-power suitable diode lasers is imminent.

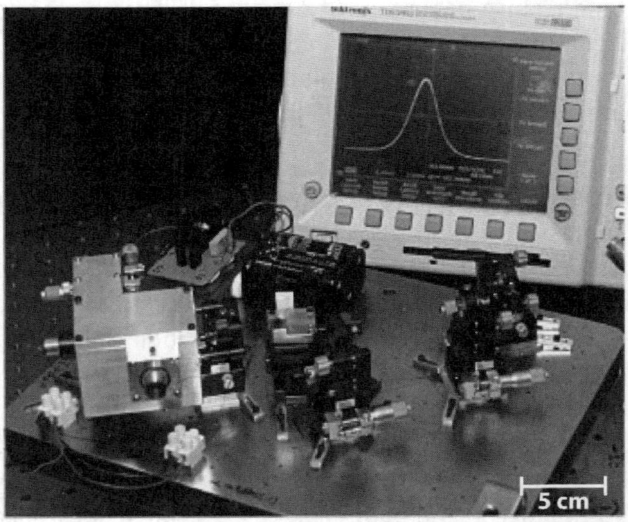

Figure 10. Photograph of ultra-compact, portable femtosecond infrared laser pumped by two inexpensive (~US$40) single-narrow-stripe AlGaInP diodes and powered by standard penlight (AA) batteries[10]. An oscilloscope in the background records an intensity autocorrelation of the femtosecond pulses.

6. SIMPLIFIED SECOND HARMONIC GENERATION SCHEME

By incorporating a straightforward but highly efficient frequency-doubling scheme into the compact femtosecond infrared laser of figure 10, a portable ultrafast blue light source can be achieved. To generate blue light (~425 nm), all that is required is a single extracavity lens to focus the ~850 nm light from a Cr:LiSAF laser through a nonlinear frequency-doubling crystal in a single pass configuration (figure 11). A half-wave plate (HWP) may also be required to select the appropriate linear polarization.

The phasematching conditions of all four doubling crystals investigated in this paper comfortably allow efficient and optimal blue-light generation at room temperature. In addition, the phasematching acceptance bandwidths (which define the accuracy to which the crucial parameters of nonlinear crystal temperature, fundamental wavelength and incident angle must be maintained to successfully optimize the SHG process) are sufficiently relaxed to require minimal attention once the process has been optimized. Each nonlinear crystal is mounted on a basic thermoelectric cooler (TEC) which provides some current-controlled temperature stabilization should it be required.

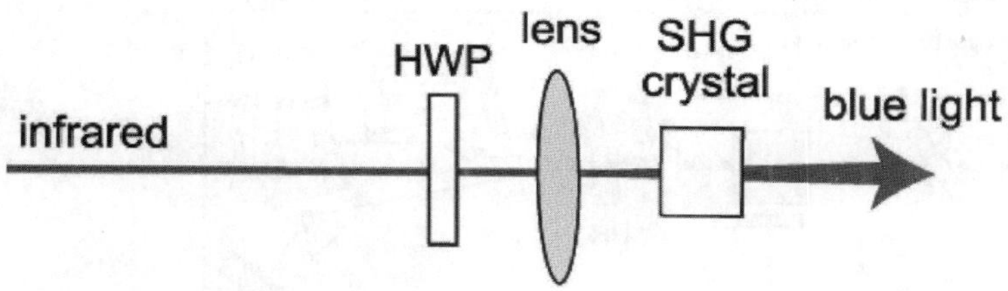

Figure 11. Simplified efficient blue generation scheme. A single pass of pulsed infrared light through an extracavity lens and nonlinear crystal is all that is required. A half-wave plate (HWP) may also be included if necessary.

7. EXPERIMENTAL PERFORMANCE OF WAVEGUIDE AND BULK NONLINEAR CRYSTALS

Investigations into the relative performance of the four nonlinear crystal types were carried out using the femtosecond Cr:LiSAF laser illustrated in figure 12. Although this laser configuration differs slightly from the laser configuration of figure 10 (with an extra pair of pump laser diodes for increased output power[10], and an intracavity prism to allow wavelength tuning), these modifications do not compromise the small scale or potential portability of the laser construction.

With four single-narrow-stripe pump laser diodes (two providing up to 60 mW at 685 nm, and two providing up to 50 mW at 660 nm) the laser was capable of generating 120-210 fs pulses at a repetition-rate of 330 MHz and average output powers up to 45 mW. By simply tilting the angle of the output coupler, selection of the central operating wavelength was also possible between 825 nm and 875 nm (defined by the reflectivity bandwidth of the SESAM mirror). The infrared output beam was strongly linearly polarized (due to the Brewster surfaces of the Cr:LiSAF laser crystal) and the beam quality was measured to have an $M^2 = 1.1$. The stability of the pulsed output over a number of hours was observed to be excellent, together with negligible power degradation.

Each nonlinear crystal was assessed in an extracavity single-pass arrangement at room temperature (figure 11). A half-wave plate was required for certain nonlinear crystals to provide the correct linear polarization. The relative performance of the four nonlinear crystals is summarized below in sections 7.1 to 7.4.

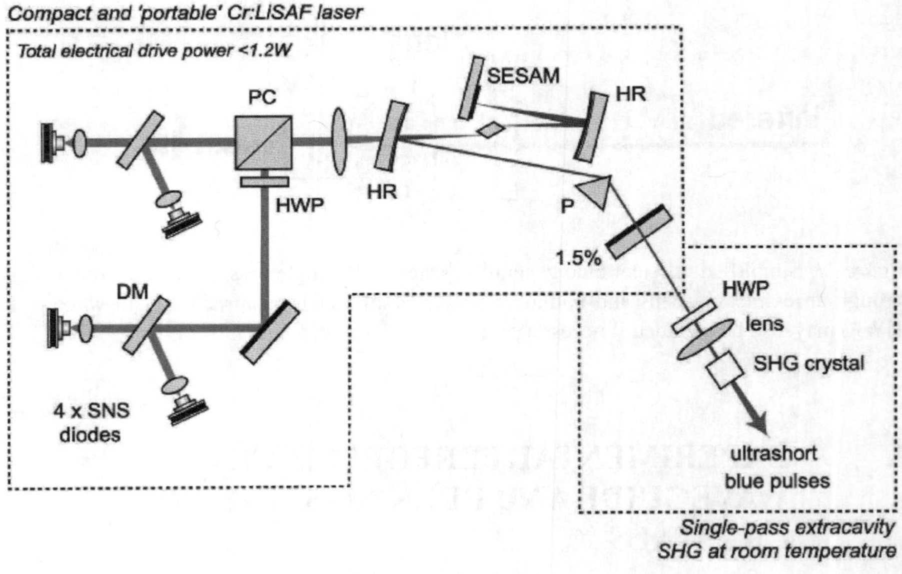

Figure 12. Laser configuration for SHG experiments, incorporating four single-narrow-stripe (SNS) red laser diodes and a prism (P) for wavelength tuning (DM: dichroic mirror; HWP: half-wave plate; PC: polarization cube; HR: high reflector; 1.5 %: output coupler).

7.1 Bulk $KNbO_3$

Bulk potassium niobate ($KNbO_3$) is well suited to our needs, because birefringent type-I non-critical phasematching (NCPM) can be exploited for highly efficient SHG of ~850 nm at room temperature[12,16,17]. This NCPM avoids any spatial walk-off between the fundamental and second harmonic beams, as well as maximizing the angular acceptance of the phasematching process.

Using an experimentally-optimized focusing lens (f = 15 mm; spot radius, w = 4.3 μm) and a 3 mm $KNbO_3$ crystal (cut for NCPM at 22°C and 858 nm; AR-coated), up to 11.8 mW of blue average power with a spectral width up to $\Delta\lambda_{SH}$ = 1.4 nm at 429 nm was generated with only 44.6 mW of incident fundamental. The maximum observed SHG conversion efficiency was as high as 30 %. The overall efficiency of the electrical-to-blue process was over 1 %, and the blue pulses were measured by autocorrelation to be ~500 fs in duration[12].

The measured full-width-at-half-maximum (FWHM) wavelength and temperature acceptance bandwidths were 2.7 nm and 5 °C respectively. The beam quality parameter of the generated femtosecond blue beam was observed to be M^2 = 1.8. We believe the quality of the blue beam deteriorated slightly from the fundamental beam (M^2= 1.1) as a result of the strong focusing. Because exact phasematching conditions are satisfied only at the beam centre, peripheral rays propagate under conditions of slight mismatch and as such accumulate phase distortions.

7.2 Waveguide ppKTP

Potassium titanyl phosphate (KTP) is another suitable nonlinear crystal for SHG, given that it can be waveguided and periodically-poled to readily satisfy quasi-phasematching (QPM) conditions at room temperature. This waveguide ppKTP crystal, periodically-poled for SHG of ~850 nm and not AR-coated, had cross-sectional dimensions of ~4×4 μm in diameter and was 11 mm in length (including a 3 mm Bragg grating section[43] which did not affect the performance of the Cr:LiSAF pump laser).

With an appropriate aspheric lens (f = 6.2 mm) to optimize the coupling of the fundamental light into the waveguide, up to 5.6 mW of average output blue power with a spectral width of $\Delta\lambda_{SH}$ = 0.6 nm at 424 nm was achieved for 27 mW of incident fundamental. Accounting for a coupling efficiency of 70%, the associated internal SHG conversion efficiency within the waveguide was calculated to be as high as 37 %[13]. With the Cr:LiSAF laser requiring no more than 1.2 W of electrical drive, this corresponds to an overall electrical-to-blue system efficiency of 0.5 %. Evidence of a saturation and subsequent decrease in overall efficiency of the SHG process (figure 18) has been attributed to two-photon absorption of the second harmonic (SH) wave[13]. The superiority of the ppKTP waveguide at low pulse energies is evident when comparing the slope efficiency per unit length of this experimental data (6.9 %$pJ^{-1}cm^{-1}$) with other similar SHG experiments using bulk $KNbO_3$ (1.0 %$pJ^{-1}cm^{-1}$ [16], 0.73 %$pJ^{-1}cm^{-1}$ [12]) and waveguide ppKTP (0.15 %$pJ^{-1}cm^{-1}$ [44]).

A broad temperature acceptance bandwidth of ΔT~30 °C (FWHM) centered at 18 °C easily permitted room-temperature operation. The characterization of fundamental pulses entering ($\Delta\tau$~170 fs; $\Delta\nu\Delta\tau$~0.32) and leaving ($\Delta\tau$~195 fs; $\Delta\nu\Delta\tau$~0.37) illustrated that these pulses were dispersed only slightly on propagation through the waveguide.

7.3 Bulk appKTP

While periodically-poled materials are typically characterized by very narrow spectral acceptance bandwidths (< 1 nm), aperiodically-poled structures (characterized by a linear gradient in grating period) have the advantage of providing sufficiently broad spectral acceptance bandwidths to utilize more of the spectrum associated with picosecond[45] and femtosecond[46] pulses. They can also simultaneously provide some pulse compression of sufficiently pre-chirped incident pulses[22,23].

The bulk appKTP crystal was 4 mm in length, and was not AR-coated. The grating periods varied from 4.1 μm to 4.3 μm in order to provide a fundamental bandwidth of 7 nm, for SHG at room temperature centered at 851 nm. With an experimentally-optimized focusing lens (f = 15 mm; w = 4.3 μm), 3.2 mW of blue average power was produced at 429 nm with $\Delta\lambda_{SH}$ up to 1.4 nm, from 27 mW of incident fundamental. This corresponds to an SHG conversion efficiency of 11.8 %, and an overall electrical-to-optical system efficiency of 0.3 %. The wavelength acceptance bandwidth was measured to be 4.5 nm. In attempting to measure the temperature acceptance bandwidth, a negligible deterioration in SHG efficiency was observed when adjusting the bulk appKTP crystal temperature between 10°C and 50°C.

Although the absolute efficiency of the SHG process is lower than that achieved from the $KNbO_3$ crystal, figure 13a illustrates that, either with suitable pre-chirping of the fundamental pulses or recompression of the generated blue pulses, the bulk appKTP crystal is able to provide ultrashort blue pulses with higher peak powers, P_{pk} [$P_{pk} = E_p/\Delta\tau_{SH}$, E_p is the pulse energy and $\Delta\tau_{SH}$ is the second harmonic pulse duration]. Insufficient power was available to measure $\Delta\tau_{SH}$ from the bulk appKTP crystal, but $\Delta\tau_{SH}$ for our experimental conditions was calculated to be ~370 fs, with ~270 fs

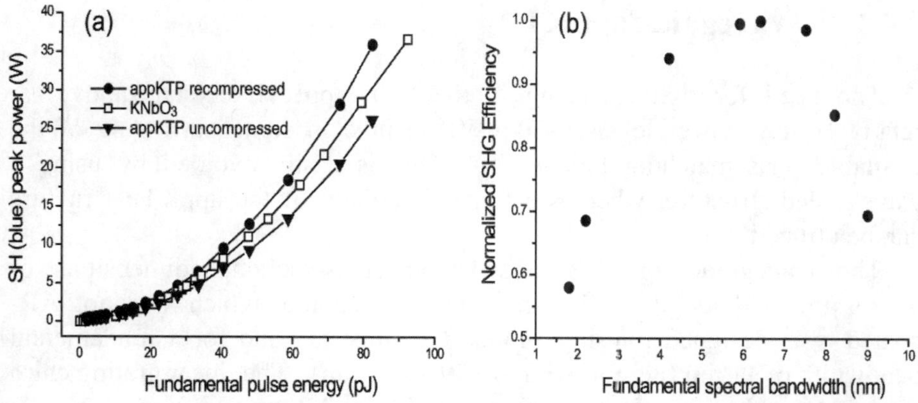

Figure 13. (a) Relative blue pulse peak powers from bulk appKTP and bulk KNbO3 crystals. (b) Dependence of SHG efficiency on fundamental spectral bandwidth of femtosecond Cr:LiSAF laser.

possible with suitable compression. The data for the $KNbO_3$ crystal used in figure 13a was calculated using measured values of $\Delta\tau = 540$ fs and $\Delta\nu\Delta\tau = 0.39$[12].

The bulk appKTP crystal is the least impressive of the four nonlinear crystals in terms of absolute SHG efficiency (figure 20). While the nonlinear coefficient, d_{eff}, of KTP (7.8 pmV^{-1}) is less than that of $KNbO_3$ (12.5 pmV^{-1}), an additional explanation is evident from figure 13b, which is measured for constant fundamental pulse energy. This clear dependence of SHG efficiency on fundamental spectral bandwidth (broadened by the insertion of more glass from the intracavity prism in figure 12) is surprising, given the fact that the bulk appKTP crystal was designed for broad fundamental pulses ($\Delta\lambda$ ~7 nm). However, the single-pass SHG scheme of figure 11 allows tight focusing of the fundamental beam in the centre of abulk nonlinear crystal in order to generate sufficiently high intensities for efficient nonlinear conversion. Although the aperiodic structure in the bulk appKTP crystal varied from 4.1 μm to 4.3 μm over a length of 4 mm, the associated confocal parameter (estimated to be the distance over which SHG takes place) is only 0.3 mm. Therefore, in the case of such tight focusing, the fundamental field will not experience the full gradient of the aperiodic structure. As such, the effective fundamental bandwidth is narrowed. The optimum 7 nm bandwidth indicated by figure 13b represents an expected expansion of the wavelength response due to the broadband fundamental.

7.4 Waveguide appKTP

The need for tight focusing in the bulk appKTP crystal (above) at relatively low power levels (~30 mW) resulted in some narrowing of the available phasematching bandwidth. This is easily avoided by using a waveguided structure, where the full aperiodicity of the appKTP structure can be utilized.

The waveguide appKTP crystal had cross-sectional dimensions of ~4 × 4 μm, and length of 12 mm. The waveguide, which was not AR-coated, was designed for single-mode transmission, and for a fundamental bandwidth of 11 nm (i.e. for SHG of 839 – 851 nm). The linear grating chirp is represented by a change in grating period from 3.8 μm to 4.0 μm.

With a suitable coupling lens of 6.2 mm focal length, up to 5.4 mW of blue average power at around 422 nm was obtained from 24.8 mW of fundamental. The internal SHG efficiency reached a maximum of 32 %, with an overall system electrical-to-blue efficiency of 0.4 %. The effects of utilizing the full aperiodicity of the appKTP waveguide structure was evident when observing the generated blue spectral bandwidths. Whereas the bulk appKTP crystal provided blue spectra with bandwidths of around 1 nm, broader blue spectra up to ~2.5 nm were observed (figure 14b), which supported shorter duration blue pulses. The structured oscillatory nature of the blue spectral profile, also recently observed elsewhere[47], is discussed in section 8.2.

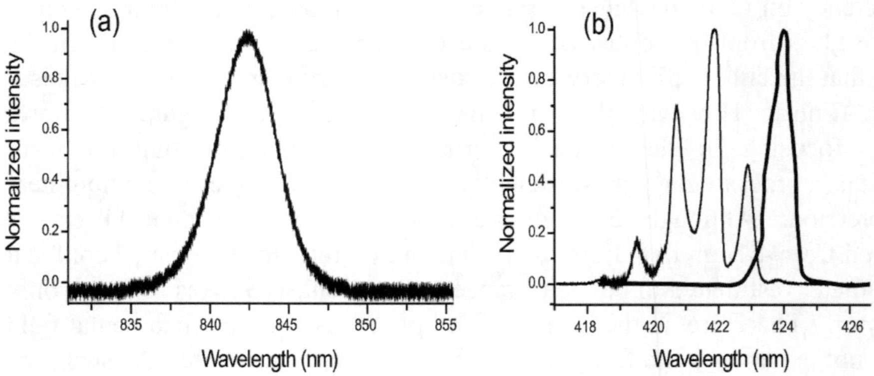

Figure 14.(a) Spectrum of fundamental pulses transmitted through the appKTP waveguide crystal (Δλ ~5 nm). (b) Typical structured spectra of the second-harmonic blue pulses leaving the appKTP waveguide, with a FWHM spectral bandwidth of up to approximately 2.5 nm (*thin line*); and typical unstructured spectra of the second-harmonic blue pulses leaving the appKTP bulk crystal (thick line).

Tunability of the blue light from 418 nm to 429 nm was also possible, via simple tuning of the Cr:LiSAF pump laser. This 11 nm tuning range coincides with the fundamental acceptance bandwidth of 11 nm for which the appKTP waveguide was designed.

8. COMAPARISON OF EXPERIMENTAL RESULTS WITH THEORETICAL MODELS

8.1 Bulk nonlinear optical crystals

In modeling the SHG performance of the bulk nonlinear optical crystals, we have assumed the incident fundamental pulses are characterized by a hyperbolic secant temporal profile, and therefore $\tau = \tau_{\text{fund}}(\text{FWHM})/1.76$. For the case of $KNbO_3$, the maximum optical-to-optical SHG efficiency achieved was 30%. The corresponding value of L/L_{nst} under these conditions, with a GVM parameter, $\alpha_{KNbO3} = 1.2$ ps/mm, is $L/L_{nst} = 30$. The values of beam waist, w_{01}, and confocal parameter, b, are calculated from conventional ABCD matrices using the measured beam diameter, d, before the focusing lens. The experimental uncertainties in the determination of w_{01} and b are 5% and 10% respectively.

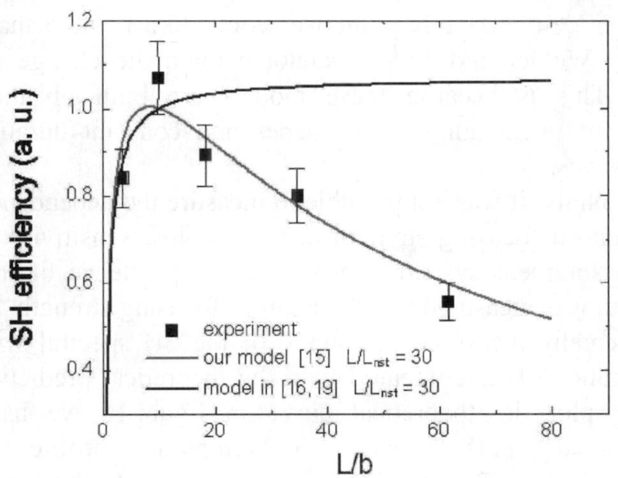

Figure 15. Experimental values of SHG efficiency using the bulk $KNbO_3$ crystal (data points), and the predictions of our model (gray curve)[15] and another model (black curve)[16,19] for L/L_{nst} = 30, as a function of the focusing strength, $m = L/b$. The theoretical curves are normalized to their values for L/b = 10, which was the experimental optimal focusing strength[12]. The maximum experimental point refers to a SHG efficiency of 30%.

In figure 15 we show the experimental values of normalized SHG efficiency as a function of the focusing strength, $m = L/b$. The experimental data is plotted together with theoretical predictions from our model described here, as well as the model published in references[16,19]. Both models assume the ratio $L/L_{nst} = 30$, and both theoretical curves are normalized for $L/b = 10$. We see from figure 15 that our model describes correctly the existence of a maximum in the dependence of SHG efficiency on focusing strength. To determine whether our model predicts the absolute efficiency, we have made use of the initial part (< 20%) of the experimental data. Working with different published values for the second order nonlinearity constant, d_{32}, of $KNbO_3$[48-50] we find that the corrected value of $d_{32} = 12.5$ pm/V published in reference[49] gives the best agreement not only with the experiment reported here, but also with the experimental work published previously[16,19].

Fitting the experimental data from the $KNbO_3$ experiment with Eq. (8), we find an experimental value of $\eta_{slope} = 0.30$ %pJ^{-1}, whereas the theoretical value from Eq. (6) is $\eta_{slope} = 0.32$ %pJ^{-1} (using $d_{eff} = 12.5$ pmV^{-1}, $\tau_0 = 210$ fs and, from the model[15], a calculated value of $h_{tr} = 0.137$). The agreement is illustrated in figure 17. We consider this agreement between experiment and theory to be very good, taking into account that the theoretical values are calculated from the empirically-determined τ_{fund} and b which have a 10% error margin. The agreement for the lengthening of the SH pulse, τ_{SH}/τ_{fund}, is also excellent, as can be seen from $(\tau_{SH}/\tau_{fund})_{EXP} = 2.6$ and $(\tau_{SH}/\tau_{fund})_{THEORY} = 2.4$. At this point we would like to note that the model described by Weiner and Yu[16,19] cannot explain the change in SH pulse durations. This is because these models are built on the restricting assumptions of unchanging pulse shapes and constant duration for both interacting waves.

Experimentally, it was not possible to measure the dependence of the SH pulse duration on focusing strength due to the low sensitivity of the pulse duration measurement system. Instead, the dependence of the SH pulse spectral width was measured as a function of focusing strength. In figure 16 the experimentally measured narrowing of the SH spectral width with an increase of ratio b/L is compared with the theoretical prediction from our model. To plot the theoretical curve in figure 16 we have assumed transform-limited pulses with a temporal profile given by $\Delta f_{fund}/\Delta f_{sh} = \tau_{sh}/\tau_{fund}$. Due to the presence of GVM, the increase in the confocal parameter, b, leads to a broadening of the SH pulse, and therefore to a narrowing of the SH spectra. On the other hand, for very small values of the confocal parameter, b, the SHG process is stationary and the transformation of the spectra behaves as if in the long-pulse limit. We can see that the model presented here also describes the spectra of the generated

SH pulses, although the presumption that the SH pulses have a hyperbolic secant temporal profile is quite approximate. Correct accounting of the SH pulse shape would improve the accuracy of the model even further. Significantly, the model reported in references[16,19] is not able to describe such a dependence for the reasons discussed above. Although our analysis has assumed fundamental pulses with no frequency chirp, it is possible to extend the present model to describe SHG with chirped fundamental pulses.

In another study of SHG in the femtosecond regime, an experiment to determine the optimal focal position within a bulk $KNbO_3$ crystal has been carried out[17]. It was shown that, at low powers below the saturation regime, the optimal position of the focal spot is indeed in the center of the crystal as predicted by our present model.

Fitting the experimental data for the bulk appKTP crystal to Eq. (8), we obtain the experimental value $\eta_0 = 0.16$ %pJ^{-1}. From Eq. (6) we calculate the theoretical value of $\eta_0 = 0.16$ %pJ^{-1} (using $d_{eff} = 7.8$ pmV^{-1}, $\tau_0 = 210$ fs and a calculated value of $h_{tr} = 0.087$). Once again, figure 17 illustrates the accuracy of the model to experimental observations.

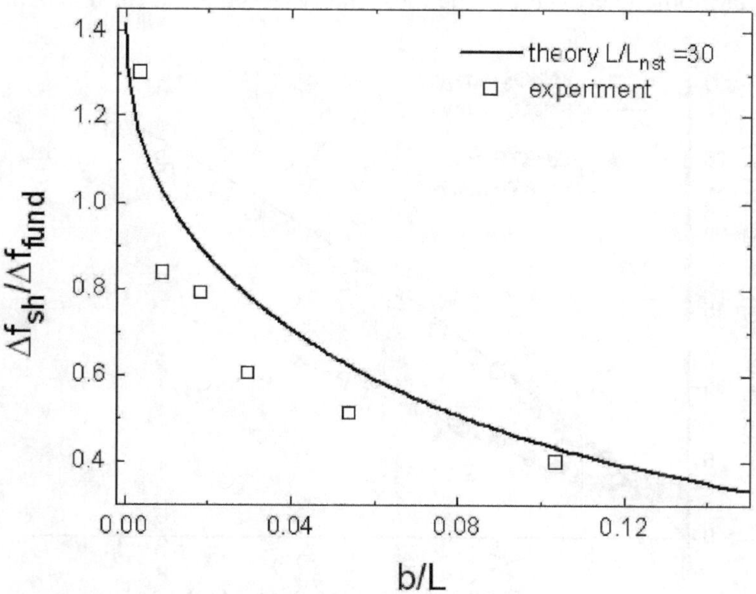

Figure 16. Experimentally measured evolution of the SH pulse spectral width relative to the fundamental spectral width for the bulk $KNbO_3$ crystal (data points), and the theoretical prediction of our model calculated for $L/L_{nst} = 30$, as a function of the ratio b/L[15].

It is worth repeating that the relatively low efficiency for the appKTP crystal is due to the fact that d_{eff}(KTP) < d_{eff} ($KNbO_3$). Performing the same assessment with lithium niobate ($LiNbO_3$) should yield up to four times the efficiency, because d_{eff} ($LiNbO_3$) = 17.6 pmV^{-1}. Unfortunately, insufficient power was available to measure the duration of the blue pulses from the bulk appKTP crystal. However, our calculations show that the generated blue pulses would be characterized by an uncompensated duration of 370 fs. These pulses could be compressed to around 270 fs in order to access higher peak powers.

8.2 Waveguide nonlinear optical crystals

When using the waveguide ppKTP crystal experimentally, the dependence of internal SHG efficiency on input power is characterized by a maximum efficiency of 37 %. A further increase in fundamental pulse energy then leads to a saturation and subsequent decrease in the efficiency of the SHG process (figure 18). This behavior was also observed in the waveguide appKTP crystal (figure 18), and has been reported elsewhere[44,47]. As we have suggested previously[13], two-photon absorption (TPA) of the second-harmonic (SH) wave is the most likely explanation for this behavior.

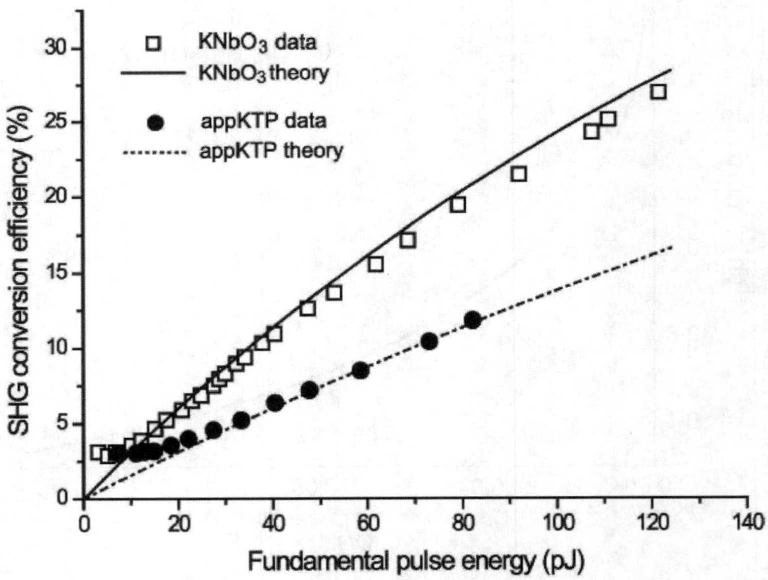

Figure 17. Theoretical predictions versus experimental data for the performance of the bulk nonlinear crystals, $KNbO_3$ and appKTP.

In figure 18 we also provide theoretical curves for the performance of the waveguide crystals, which account for the observed effects of TPA. The horizontal scales are normalized to match the initial slopes of the experimental curves. It is clear that the presence of TPA causes the slight decrease in SHG efficiency once saturation of the SHG process has begun. The fact that the measured efficiency is less than that predicted by the theory shows that the model can be improved further. This could be achieved by accounting for other effects such as nonlinear losses of the SH wave due to presence of the fundamental wave, self-phase and cross-phase modulation and blue-light induced red absorption (BLIRA)[25,51]. Thermally-induced phase mismatch may also be present as the blue wavelength is close to the absorption edge of the crystal.

Theoretical output spectra for the fundamental and SH waves for the appKTP waveguide are shown in figure 19. It can be seen that they closely represent the observed experimental oscillatory nature of the SH spectra, and the smooth profile of the fundamental output, as shown in figure 14. Our calculations predict that the oscillations in figure 19b become more exaggerated as we tune away from the perfect phasematching condition. In addition, the calculated acceptance bandwidth for our waveguide appKTP crystal of $\Delta\lambda_{fund} = 12.6$ nm is in excellent agreement with the tests carried out on the sample after fabrication.

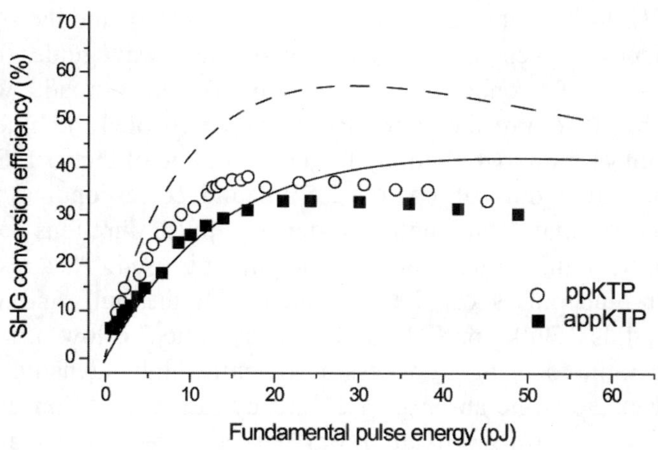

Figure 18. Saturation of SHG efficiency in the waveguide (ppKTP and appKTP) experiments (point data). Theoretical predictions incorporating the effects of two-photon absorption for the SH wave and group velocity mismatch are also shown for the ppKTP (dashed line) and appKTP waveguides (solid line).

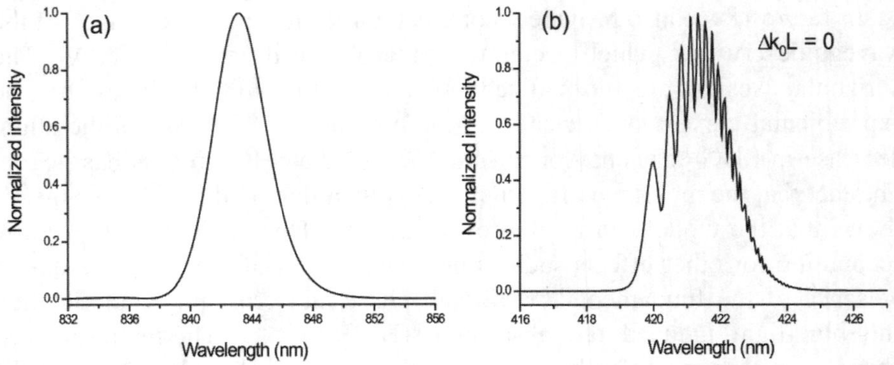

Figure 19. Theoretically calculated output spectra for (a) the fundamental and (b) second harmonic wave. See figure 14 for experimental spectra.

9. DISCUSSION AND SUMMARY

Of the four nonlinear crystal types investigated in this paper (figure 20), bulk $KNbO_3$ was superior in terms of generated blue average power (11.8 mW) and generated beam quality. $KNbO_3$ is also less susceptible to the observed saturation and subsequent decrease in SHG efficiency observed in both waveguide crystals.

The appKTP waveguide provided similar maximum internal SHG efficiency (32 %) to that of a ppKTP waveguide (37 %), and the steep rise in SHG efficiency proves the effectiveness of such waveguides under low pump-pulse-energy conditions. In addition, the broad wavelength acceptance bandwidth of the appKTP waveguide enabled the blue pulses to be tuned from 418 nm to 429 nm by frequency tuning of the Cr:LiSAF laser.

Blue spectral widths of up to ~2.5 nm and corresponding theoretical analyses confirm that significantly shorter blue pulse durations are obtained when aperiodic poling rather than periodic poling is used. As a result, higher peak power blue pulses can be generated with the bulk and waveguide appKTP crystals. Bulk appKTP is not ideally suited to low-power sources that require tight focusing to access sufficiently high intensities, although another advantage is the absence of efficiency saturation. Conveniently, all four crystals perform well at room temperature such that minimal wavelength and temperature stabilizations are required.

The theoretical model of SHG with focused beams in both homogeneous and linearly-chirped aperiodically-poled structures predicts very well the results from the bulk $KNbO_3$ and bulk appKTP crystals. The model of SHG in the two waveguide crystals explains qualitatively the saturation and subsequent decrease of the SHG efficiency, but requires further refinement to explain more fully the experimentally-observed absolute efficiency.

The advantages of using bulk, waveguide, periodically-poled or aperiodically-poled media can be exploited with the intended application in mind. The relative performance of the four nonlinear crystals is summarized in Table 1.

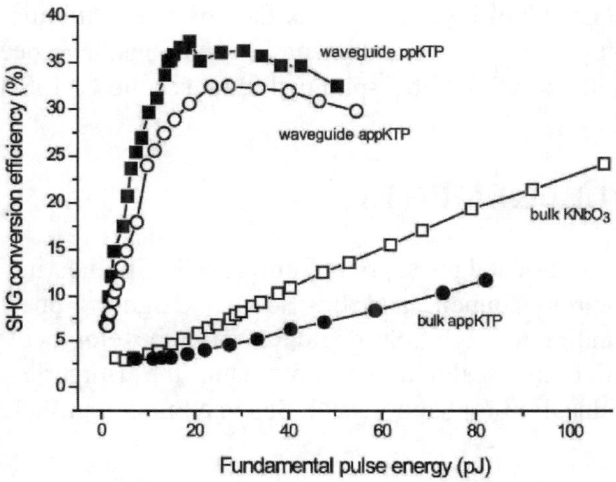

Figure 20. Relative experimental performance of the four nonlinear crystals investigated in this paper as a means to efficiently and practically generate ultrafast blue light from a compact and portable femtosecond Cr:LiSAF laser.

Table 1. Summary of the performance of the four nonlinear crystals

	bulk $KNbO_3$	waveguide ppKTP	bulk appKTP	waveguide appKTP
L	0.3	0.8	0.4	1.2
α	1.2	1.2	1.2	1.2
b	0.31	-	0.25	-
f	15	-	15	-
P_{SH}	11.8	5.6	3.2	5.4
η_{SHG}	30	37	12	32
η_{slope}	1	6.9	0.4	2.3
Ep_{SH}	36	17	10	16
$\Delta\lambda_{SH}$	<1.4	0.6	<1.4	~2.5

L: nonlinear crystal length (cm); *α*: GVM parameter (ps.mm^{-1}); *b*: confocal parameter (mm) ; *f*: focal length (mm); P_{SH} : blue average power (mW); η_{SHG}: maximum SHG conversion efficiency (%); η_{slope}: SHG slope efficiency per unit crystal length (%pJ^{-1}cm^{-1}); Ep_{SH}: blue pulse energy (pJ); $\Delta\lambda_{SH}$: blue spectral bandwidth (nm).

10. CONCLUSIONS

We have performed a thorough experimental and theoretical investigation into efficient frequency-doubling of low energy femtosecond pulses. Evaluation of periodically-poled and aperiodically-poled bulk and waveguide structures has been presented. Several theoretical models accurately describe the temporal and spectral properties of the generated second harmonic (blue) light, as well as the observed saturation behavior of the conversion process. All experimental evaluations have been carried out with a potentially portable, diode-pumped femtosecond Cr:LiSAF laser.

ACKNOWLEDGEMENTS

Ben Agate acknowledges support from the UK Engineering and Physical Sciences Research Council, and the Scottish Higher Education Funding Council for funding through a Strategic Research Development Grant award. Solomon Saltiel and Kaloian Koynov thank the Bulgarian Ministry of Science and Education for support with Science Fund grant F-1201/2002.

REFERENCES

1. P.N. Prasad, in: *Introduction to Biophotonics* (Wiley & Sons, Hoboken, 2003) pp.203-254.
2. H. Schnitzler, U. Frohlich, T.K.W. Boley, A.E.M. Clemen, J. Mlynek, A. Peters, and S. Schiller, All-solid-state tunable continuous-wave ultraviolet source with high purity and frequency stability, *Applied Optics* **41**(33), 7000-7005 (2002).
3. R.S. Knox, Ultrashort processes and biology, *Journal of Photochemistry and Photobiology* **49,** 81-88 (1999).
4. W. Rudolph, P. Dorn, X. Liu, N. Vretenar, and R. Stock, Microscopy with femtosecond laser pulses: applications in engineering, physics and biomedicine, *Applied Surface Science* **208**, 327-332 (2003).
5. S.W. Hell, P.E. Hanninen, J. Salo, A. Kuusisto, E. Soini, T. Wilson, and J.B. Tan, Pulsed and cw confocal microscopy - a comparison of resolution and contrast, *Optics Communications* **113**(1-3), 144-152 (1994).
6. B. Agate, C.T.A. Brown, W. Sibbett, and K. Dholakia, Femtosecond optical tweezers for *in-situ* control of two-photon fluorescence, *Optics Express* **12(**13), 3011-3017 (2004).
7. P.J. Campagnola, M.D. Wei, A. Lewis, and L.M. Loew, High-resolution nonlinear optical imaging of live cells by second harmonic generation, *Biophysical Journal* **77**(6), 3341-3349 (1999).
8. E. Bordenave, E. Abraham, G. Jonusauskas, J. Oberle, and C. Rulliere, Longitudinal imaging in biological tissues with a single laser shot correlation system, *Optics Express* **10**(1), 35-40 (2002).
9. L. Zhu, P. Li, J.T. Sage., and P.M. Champion, Femtosecond dynamics of heme-proteins, *Journal of Luminescence* **60**(1), 503-506 (1994).

10. B. Agate, B. Stormont, A.J. Kemp, C.T.A. Brown, U. Keller, and W. Sibbett, Simplified cavity designs for efficient and compact femtosecond Cr:LiSAF lasers, *Optics Communications* **205**, 207-213 (2002).
11. B. Agate, E.U. Rafailov, W. Sibbett, S.M. Saltiel, K. Koynov, M. Tiihonen, S. Wang, F. Laurell, P. Battle, T. Fry, and E. Noonan, Portable ultrafast blue light sources designed with frequency doubling in KTP and $KNbO_3$, *IEEE Journal of Selected Topics in Quantum Electronics: Short Wavelength and EUV Lasers* (November/December 2004).
12. B. Agate, A. J. Kemp, C.T.A. Brown, and W. Sibbett, Efficient, high repetition-rate femtosecond blue source using a compact Cr:LiSAF laser, *Optics Express* **10**(16), 824-831 (2002).
13. B. Agate, E.U. Rafailov, W. Sibbett, S.M. Saltiel, P. Battle, T. Fry, and E. Noonan, Highly efficient blue-light generation from a compact diode-pumped femtosecond laser by use of a periodically poled KTP waveguide crystal, *Optics Letters* **28**(20), 1963-1965 (2003).
14. G. D. Boyd and D.A. Kleinman, Parametric interactions of focused Gaussian light beams, *Journal of Applied Physics* **39**, 3897-3641 (1968).
15. S. M. Saltiel, K. Koynov, B. Agate, and W. Sibbett, Second-harmonic generation with focused beams under conditions of large group-velocity mismatch, *Journal of the Optical Society of America B* **21**(3), 591-598 (2004).
16. A. M. Weiner, A. M. Kan'an, and D. E. Leaird, High-efficiency blue generation by frequency doubling of femtosecond pulses in a thick nonlinear crystal, *Optics Letters* **23**(18), 1441-1443 (1998).
17. D. Gunzun, Y. Li, and M. Xiao, Blue light generation in single-pass frequency doubling of femtosecond pulses in $KNbO_3$, *Optics Communications* **180**, 367-371 (2000).
18. Y.Q. Li, D. Guzun, G. Salamo, and M. Xiao, High-efficiency blue-light generation by frequency doubling of picosecond pulses in a thick $KNbO_3$ crystal, *Journal of the Optical Society of America B* **20**(6), 1285-1289 (2003).
19. S. Yu and A.M. Weiner, Phase-matching temperature shifts in blue generation by frequency doubling of femtosecond pulses in $KNbO_3$, *Journal of the Optical Society of America B* **16**(8), 1300-1304 (1999).
20. H. Wang and A.M. Weiner, Efficiency of short-pulse type-I second-harmonic generation with simultaneous spatial walk-off, temporal walk-off, and pump depletion, *IEEE Journal of Quantum Electronics* **39**(12), 1600-1618 (2003).
21. M. M. Fejer, G.A. Magel, D.H. Jundt, and R.L. Byer, Quasi-phase-matched second harmonic generation - tuning and tolerances, *IEEE Journal of Quantum Electronics* **28**, 2631-2654 (1992).
22. P. Loza-Alvarez, D.T. Reid, P. Faller, M. Ebrahimzadeh, W. Sibbett, H. Karlsson, and F. Laurell, Simultaneous femtosecond-pulse compression and second-harmonic generation in aperiodically poled $KTiOPO_4$, *Optics Letters* **24**(15), 1071-1073 (1999).
23. M.A. Arbore, O. Marco, and M.M. Fejer, Pulse compression during second-harmonic generation in aperiodic quasi-phase-matching gratings, *Optics Letters* **22**(12), 865-867 (1997).
24. R. DeSalvo, M. Sheik-Bahae, A.A. Said, D.J. Hagan, and E.W. Van Stryland, Z-scan measurements of the anisotropy of nonlinear refraction and absorption in crystals, *Optics Letters* **18**(3), 194 (1993).
25. A.D. Ludlow, H.M. Nelson, and S.D. Bergeson, Two-photon absorption in potassium niobate, *Journal of the Optical Society of America B* **18**(12), 1813-1820 (2001).
26. M. M. Fejer. Quasi-phasematched for frequency conversion and all optical signal processing (short course #1260) in *Conference on Lasers and Electro-Optics,* Baltimore (2001).
27. A. Yariv, *Quantum Electronics* (4ed, Wiley & Sons,1997)

28. M. Yamada, N. Nada, M. Saitoh, and K. Watanabe, First-order quasi-phase matched $LiNbO_3$ wave-guide periodically poled by applying an external-field for efficient blue second-harmonic generation, *Applied Physics Letters* **62**(5), 435-436 (1993).
29. S.N. Zhu, Y.Y. Zhu, Z.J. Yang, H.F. Wang, Z.Y. Zhang, J.F. Hong, C.Z. Ge, and N.B. Ming, Second-harmonic generation of blue-light in bulk periodically poled $LiTaO_3$, *Applied Physics Letters* **67**(3), 320-322 (1995).
30. H. Karlsson, F. Laurell, and L.K. Cheng, Periodic poling of $RbTiOPO_4$ for quasi-phase matched blue light generation, *Applied Physics Letters* **74**(11), 1519-1521 (1999).
31 H. Karlsson, F. Laurell, P. Henriksson, and G. Arvidsson, "Frequency doubling in periodically poled $RbTiOAsO_4$, *Electronics Letters* **32**, 556 (1996).
32. H. Karlsson and F. Laurell, Electric field poling of flux grown $KTiOPO_4$, *Applied Physics Letters* **71**(24), 3474-3476 (1997).
33. C. Canalias, J. Hirohashi, V. Pasiskevicius, and F. Laurell, Polarization switching characteristics of flux grown $KTiOPO_4$ and $RbTiOPO_4$ at room temperature, *Applied Physics Letters* (submitted), (2005).
34. K. Fradkin-Kashi and A. Arie, "Multiple-wavelength quasi-phase-matched nonlinear interactions, *IEEE Journal of Quantum Electronics* **35** 1649 (1999).
35. L.E. Myers, R.C. Eckardt, M.M. Fejer, R.L. Byer, and W.R. Bosenberg, "Multigrating quasi-phase-matched optical parametric oscillator in periodically poled $LiNbO_3$, *Optics Letters* **21**, 591 (1996).
36. K.A. Tillman, D.T. Reid, D. Artigas, J. Hellstrom, V. Pasiskevicius, and F. Laurell, Low-threshold, high-repetition-frequency femtosecond optical parametric oscillator based on chirped-pulse frequency conversion, Journal *of the Optical Society of America B* **20** 1309 (2003).
37. J.-M. Hopkins, G.J. Valentine, B. Agate, A. J. Kemp, U. Keller, and W. Sibbett, Highly compact and efficient femtosecond lasers, IEEE Journal of Quantum Electronics **38**(4), 360-368 (2002).
38. J.-M. Hopkins, G.J. Valentine, W. Sibbett, J.A. der Au, F. Morier-Genoud, U. Keller, and A. Valster, Efficient, low-noise, SESAM-based femtosecond Cr^{3+}:$LiSrAlF_6$ laser, *Optics Communications* **154**(1-3), 54-58 (1998).
39. W. Sibbett, B. Agate, C.T.A. Brown, A.A. Lagatsky, C.G. Leburn, B. Stormont, and E.U. Rafailov, Compact Femtosecond Oscillators (invited), in: *Ultrafast Optics IV*, edited by F. Krausz, (Springer Verlag 2004) pp. 3-21.
40. C.T.A. Brown, M.A. Cataluna, A.A. Lagatsky, E.U. Rafailov, B. Agate, C.G. Leburn, and W. Sibbett, Compact laser-diode-based femtosecond sources, *New Journal of Physics* **6**, 175 (2004).
41. G.J. Valentine, J.-M. Hopkins, P. Loza-Alvarez, G.T. Kennedy, W. Sibbett, D. Burns, and A. Valster, Ultralow-pump-threshold, femtosecond Cr^{3+}:$LiSrAlF_6$ laser pumped by a single narrow-stripe AlGaInP laser diode, *Optics Letters* **22**(21), 1639-1641 (1997).
42. A.A. Lagatsky, E.U. Rafailov, C.G. Leburn, C.T.A. Brown, N. Xiang, O.G. Okhotnikov, and W. Sibbett, Highly efficient femtosecond Yb:KYW laser pumped by a single narrow-stripe laser diode, *Electronics Letters* **39**(15), 1108-1110 (2003).
43. E.U. Rafailov, D.J.L. Birkin, W. Sibbett, P. Battle, T. Fry, and D. Mohatt, Efficient direct frequency conversion of a non-resonant injection-seeded laser diode using a periodically-poled KTP waveguide crystal, *Optics Letters* **26**(24), 1961-1962 (2001).
44. Y.S. Wang, V. Petrov, Y.J. Ding, Y. Zheng, J.B. Khurgin, and W.P. Risk, Ultrafast generation of blue light by efficient second-harmonic generation in periodically-poled bulk and waveguide potassium titanyl phosphate, *Applied Physics Letters* **73**(7), 873-875 (1998).

45. D.J.L.B. Birkin, E.U. Rafailov, G.S. Sokolovskii, W. Sibbett, G.W. Ross, P.G.R. Smith, and D.C. Hanna, 3.6mW blue light by direct frequency-doubling of a diode laser using an aperiodically-poled lithium niobate crystal, *Applied Physics Letters* **78**(21), 3172-3174 (2001).
46. D. Artigas and D.T. Reid, Efficient femtosecond optical parametric oscillators based on aperiodically poled nonlinear crystals, *Optics Letters* **27**(10), 851-853 (2002).
47. I.A. Begishev, M. Kalashnikov, V. Karpov, P. Nickles, H. Schonnagel, I.A. Kulagin, and T. Usmanov, Limitation of second-harmonic generation of femtosecond Ti:sapphire laser pulses, *Journal of the Optical Society of America B* **21**(2), 318-322 (2004).
48. I. Biaggio, P. Kerkoc, L.S. Wu, P. Gunter, and B. Zysset, Refractive-indexes of orthorhombic $KNbO_3$: Phase-matching configurations for nonlinear-optical interactions, *Journal of the Optical Society of America B* **9**(4), 507-517 (1992).
49. I. Shoji, T. Kondo, A. Kitamoto, M. Shirane, and R. Ito, Absolute scale of second-order nonlinear-optical coefficients, *Journal of the Optical Society of America B* **14**(9), 2268-2294 (1997).
50. G.G. Gurzadian, V.G. Dmitriev, and D.N. Nikogosyan, *Handbook of nonlinear optical crystals* (3ed), (Springer Verlag, New York, 1999).
51. L.E. Busse, L. Goldberg, M.R. Surette, and G. Mizell, Absorption Losses in MgO-doped and undoped potassium niobate, *Journal of Applied Physics* **75**(2), 1102-1110 (1994).

H:$LiNbO_3$ AND H:$LiTaO_3$ PLANAR OPTICAL WAVEGUIDES: FORMATION AND CHARACTERIZATION

I. Savova, I. Savatinova
Institute of Solid State Physics, 72 Tzarigradsko chaussee Blvd., 1784 Sofia, Bulgaria

Abstract: This paper summarizes the results of our systematic study of PE and APE planar waveguides in $LiNbO_3$ and $LiTaO_3$. We focused on the optical and structural characterization of PE layers formed on Z-cut substrates. The refractive index change was measured and the propagation losses were estimated. Raman spectroscopy was used as a method providing direct information about the phonon spectrum. The latter was related to the structure and the properties of the protonated waveguides.

Key words: $LiNbO_3$, $LiTaO_3$, proton exchange, Raman spectra

1. INTRODUCTION

$LiNbO_3$ is a widely used ferroelectric crystal with various applications in the nonlinear optics and integrated optics (IO). Another attractive material for IO devices is $LiTaO_3$. Its electrooptic (EO) and nonlinear (NL) coefficients are comparable to those of $LiNbO_3$, and its photorefractive damage threshold is more than an order of magnitude higher than that of $LiNbO_3$ in the visible range.

Several methods can be used for waveguide fabrication in $LiNbO_3$. Among them, titanium in-diffusion and proton exchange (PE)[1] are the most popular ones since they lead to the formation of well-confined and low-loss layers. PE is mainly applied because it results in a considerable decrease of the photorefractive effect in $LiNbO_3$. However, waveguides obtained by pure PE have reduced EO and NL coefficients and usually a post exchange annealing (APE) is required for restoration of the EO activity.

S. Janz et al. (eds.), Frontiers in Planar Lightwave Circuit Technology, 229–234

Currently, the APE technique is used almost exclusively for formation of waveguides in $LiTaO_3$ as well[2], because it is simple and can be performed at temperatures below the Curie point.

Various crystalline phases can be formed in $H_xLi_{1-x}NbO_3$ and $H_xLi_{1-x}TaO_3$ systems. The crystalline structure depends on both the exchange and annealing conditions. Seven crystallographic phases (α, κ_1, κ_2, β_1, β_2, β_3, β_4) were reported for PE $LiNbO_3$ in[3], and six ones (α, κ, β_1, β_2, β_3, β_4) for PE $LiTaO_3$ in[4]. More than one phase may be present in a layer. In addition, the formation of phases, stable at high temperature and metastable at room temperature, was also reported[5].

This paper summarizes the results of our study of PE and APE wave-guides in $LiNbO_3$ and $LiTaO_3$. We focused on the optical and structural characterization of PE layers formed on Z-cut substrates. The refractive index change was measured and the propagation losses were estimated. Raman spectroscopy was used as a method providing direct information about the phonon spectrum. The latter was related to the structure and chemical bonds of a given crystalline phase. Such information may be useful for correct identification of both phase composition and the microscopic mechanisms responsible for the observed variation of the properties from phase to phase.

2. EXPERIMENT

PE layers were formed in Z-cut $LiNbO_3$ and $LiTaO_3$ congruent crystals by using a variety of proton sources, such as benzoic acid – pure (C_6H_5COOH) or diluted with lithium benzoate (C_6H_6COOLi); pyro-phosphoric acid ($H_4P_2O_7$), etc. The plates were immersed in the acid melt for various periods of time (up to 8 h in the case of $LiNbO_3$ and up to 42 h for $LiTaO_3$ substrates) at a temperature in the range of 200 – 240°C. Some of the samples were subsequently annealed at temperatures up to 420°C for periods of time varying from 10 min to 2 h.

The lateral surfaces of the samples were polished to allow micro-Raman profiling measurements.

The waveguides were optically characterized at $\lambda = 632.8$ nm. The effective mode indices were determined by the m-line technique, based on a standard two prism coupling set-up. The refractive index depth profiles of the fabricated waveguides were reconstructed by the means of the inverse WKB procedure[6].

The propagation losses and their dependence on the effective mode index were estimated with the scattering detection method[7] at $\lambda = 632.8$ nm.

The Raman spectra were registered either in backscattering or in right angle geometry. In the former case, a micro Raman set-up was used with a 100× objective (NA = 0.95). In the latter case, a waveguide Raman spectroscopy was utilized in order to confine the exciting beam within the layer (a prism coupling technique was used). A $\lambda = 488$ nm or $\lambda = 514.5$ nm radiation of an Ar^+ laser were used for excitation.

3. RESULTS AND DISCUSSION

Covering a wide range of fabrication and annealing parameters, we obtained PE layers in different $H_xLi_{1-x}NbO_3$ and $H_xLi_{1-x}TaO_3$ phases. The waveguides were multimode, with refractive index changes, thicknesses and number of optical modes depending on the proton concentration. The PE process leads to an increase in the extraordinary refractive index only, the maximal measured increase being $\Delta n_e = 0.12$ for PE $LiNbO_3$, and $\Delta n_e = 0.02$ for PE $LiTaO_3$. Figures 1 and 2 summarize the corresponding refractive index profiles calculated from the effective index measurements.

The as-exchanged layers exhibited good guiding properties with propagation losses typically less than 1 dB/cm for the higher order modes ($m = 0, 1, 2$). However, prolonged annealing at high temperatures led to significant increase of the propagation losses. For example, for a $H_xLi_{1x}NbO_3$ sample exchanged for 1 h at 200°C in C_6H_5COOH and subsequently annealed at temperatures up to 375°C for 14 h, the value of the attenuation coefficient increased to 4–6 dB/cm.

Since each $H_xLi_{1-x}NbO_3$ and $H_xLi_{1-x}TaO_3$ phase is characterized with specific crystal lattice parameters and chemical bonding, we carried out a systematic research on the relationship between the proton concentration and the Raman spectra for various phases. The spectra of PE $LiNbO_3$ and $LiTaO_3$ layers with various proton concentration are presented in Figs. 3 and 4, respectively. The variations of the Raman spectra for both types of layers confirmed the formation of different phases. However, Raman spectroscopy was not sensitive to all previously reported phases[3]. For example, we were not able to register any difference between the Raman spectra of κ_1 and κ_2 phases of $H_xLi_{1-x}NbO_3$. Similar results were reported by Rams et al[8]. From Figs. 3 and 4 it can be assumed that the proton exchange leads to the following structural changes:

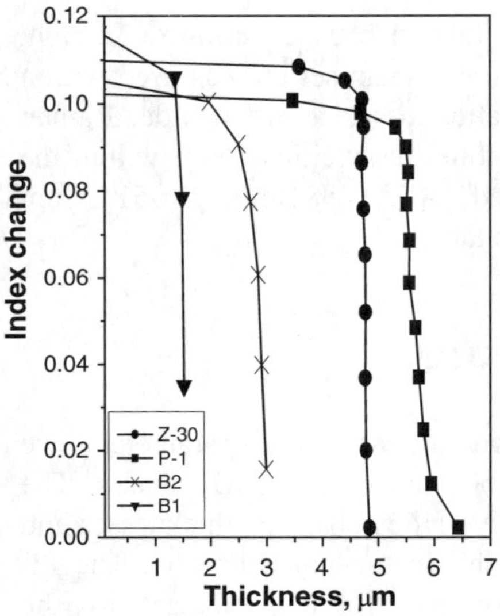

Figure 1. Refractive index profiles of $H_xLi_{1-x}NbO_3$ planar waveguides:
Z-30: PE in C_6H_6COOH at 225°C for 8 h;
P-1: PE in $H_4P_2O_7$ at 240°C for 8 h; APE at 320°C for 2.5 h;
B2: PE in C_6H_6COOH at 240°C for 7 h; APE at 260°C for 1 h;
B1: PE in C_6H_6COOH + 1% C_6H_6COOLi at 233°C for 8 h.

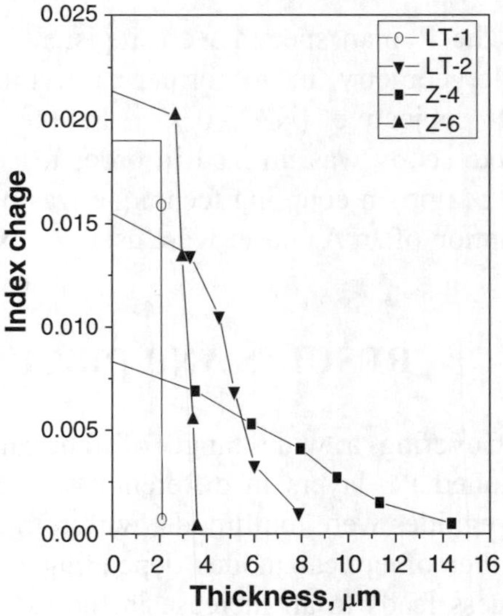

Figure 2. Refractive index profiles of $H_xLi_{1-x}TaO_3$ planar waveguides:
LT-1: PE in C_6H_6COOH at 240°C for 31 h; APE at 400°C for 2 h;
LT-2: PE in C_6H_6COOH at 240°C for 8 h; APE at 265°C for 1 h and at 295°C for 1.5 h;
Z-4: PE in C_6H_6COOH at 240°C for 8 h; APE consecutively at 265, 295, 320, 350, 400, 420°C for 1 h;
Z-6: PE in C_6H_6COOH at 225°C for 42 h; APE at 320°C for 1 h.

– At low H^+ concentration (in the α phase range for $H_xLi_{1-x}NbO_3$ and in the α and κ phase range for $H_xLi_{1-x}TaO_3$) the amount of protons is not enough to cause any considerable structural changes and the $H_xLi_{1-x}MO_3$ (M=Nb, Ta) spectra are similar to those of the pure crystals, only the FWHM of the bands is larger because of disruption of the translational symmetry due to the incorporation of protons. New bands at 65 cm^{-1} and in the range 500 to 700 cm^{-1} due to the formation of Nb(Ta)-OH complexes are also observed. An interesting peculiarity of the $H_xLi_{1-x}TaO_3$ layers is that the PE lifts the directional dependence of the oblique E phonons and degenerates their LO-TO components – effect similar to the paraelectric phase transition exhibited by ferroelectrics.

– With increase of the proton concentration (in the range of $\kappa_{1,2}$ phase $H_xLi_{1-x}NbO_3$ and $\beta_{1,2}$ phase $H_xLi_{1-x}TaO_3$), the protons are chaotically distributed in the crystal lattice, the registered spectra are diffused and the bands can hardly be separated. The bands at 65 cm^{-1} and in the range 500 to 700 cm^{-1} are among the most intensive spectral features.

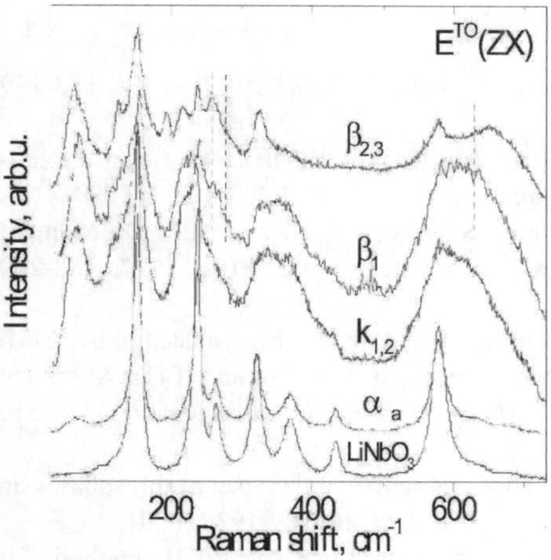

Figure 3. Raman spectra of $H_xLi_{1-x}NbO_3$ waveguides with various proton concentration.

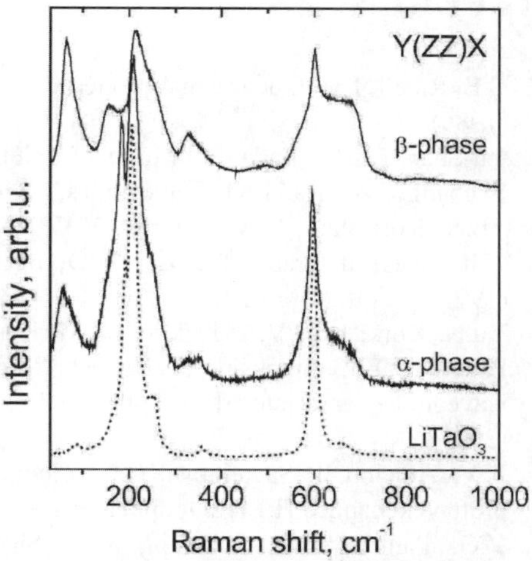

Figure 4. Raman spectra of $H_xLi_{1-x}TaO_3$ layers with various proton concentration.

– With further increase of the H^+ concentration (in β_i phase range for $H_xLi_{1-x}NbO_3$) many new bands were observed. The fact that the low concentration boundary of the β phases is approximately $x = 0.5$ leads to the assumption for some kind of ordering of Li^+ and H^+, as reported[1]. On one hand, it can be assumed that the protons form a (nearly) ordered sub-lattice. Such a structure would have a phonon spectrum different from that of a pure $LiNbO_3$, see[9]. On the other hand, the PE probably leads to a reduction of the crystal symmetry, i.e. due to the incorporation of H^+, the two Li sites in the unit cell may become non-equivalent. In such case, the symmetry would be reduced from C_{3v} to C_3. As a result, the number of molecules per unit cell would remain the same, but new bands would appear in the vibration spectrum.

In PE $LiTaO_3$, due to the lower diffusion coefficients, we could not fabricate waveguides with proton concentration, for which presence of new bands is characteristic for the Raman spectra.

For high proton concentration (in the β phase range for both $H_xLi_{1-x}NbO_3$ and $H_xLi_{1-x}TaO_3$), formation of phases stable at high temperature (200 to 240°C) and metastable at room temperature, was also be detected by Raman spectroscopy. A change in the low frequency (<1000 cm^{-1}) Raman spectra was observed for $H_xLi_{1-x}NbO_3$. For both $H_xLi_{1-x}NbO_3$ and $H_xLi_{1-x}TaO_3$ layers a change in the position and the form of the OH stretching band was detected.

In summary, the optical and structural properties of PE and APE planar waveguides in $LiNbO_3$ and $LiTaO_3$ were studied. Raman spectroscopy was used to monitor the formation of various phases.

REFERENCES

1. C.E. Rice, "The structure and properties of $Li_{1-x}H_xNbO_3$," *J. Solid St. Chem.* **64,** 188-199 (1986).
2. Yu-Shan Li, K. Tada, T. Murai, T. Yuhara, "Electrooptic Coefficient r_{33} in Proton-Exchanged z-Cut $LiTaO_3$ Waveguides,", *Japan. J. Appl. Phys.* **28**, L263–L265 (1989).
3. Yu.N. Korkishko, V.A. Fedorov, V.V. Nosikov, S.M. Kostritskii and M.P. De Micheli: "The phase diagram of $H_xLi_{1-x}NbO_3$ optical waveguides", Proc. SPIE **2997,** 188-200 (1997).
4. Yu.N. Korkishko, V.A. Fedorov, S.M. Kostritskii, A.N. Alkaev, E.I. Maslennikov, E.M. Paderin, D.V. Apraksin, F. Laurell, "Proton exchanged $LiNbO_3$ and $LiTaO_3$ optical waveguides and integrated optic devices," *Microelectronic Engineering* **69**, 228-236 (2003).
5. V.V. Atuchin, I. Savatinova, C.C. Ziling, "High refractive index metastable phases in proton exchanged H:$LiTaO_3$ optical waveguides," *Mat. Lett.* **46,** 189-192 (2000).
6. V.G. Pankin, V.U. Pchelkin, V.V. Shashkin, "On the use of WKB method for determination of the refractive index profile in plane diffusion waveguides," *Kvant. Elektron.* **4,** 1497-1502 (1997), in Russian.
7. P.K. Tien, "Light waves in thin films and integrated optics," *Appl. Opt.* **10,** 2395-2413 (1971).
8. J. Rams, F. Agullo-Rueda, J.M. Cabrera, "Structure of high index proton exchanged $LiNbO_3$ waveguides with undegraded nonlinear optical coefficients," *Appl. Phys. Lett.* **71.** 3356-3358 (1997).
9. N.V. Sidorov, M.N. Palatnikov, V.T. Kalinnikov, "Raman spectra and peculiarities of the structure of $LiNbO_3$ crystals," *Optics and Spectroscopy* **82,** 38-45, (1997), in Russian.

HIGHLY EFFICIENT BROAD-BAND WAVEGUIDE GRATING COUPLER WITH A SUB-WAVELENGTH GRATING MIRROR

Pavel Cheben[1], Siegfried Janz[1], Dan-Xia Xu[1], B. Lamontagne[1], A. Delâge[1] and Stoyan Tanev[2]
[1]*Institute for Microstructural Sciences, National Research Council of Canada, Ottawa, Ontario, Canada K1A 0R6;* [2]*Vitesse Re-Skilling Canada, 349 Terry Fox Drive, Kanata, ON K2K 2V6*

Abstract: A new diffractive device for light coupling between a planar optical waveguide and free space is proposed. The device utilizes a second order waveguide grating to diffract the fundamental waveguide mode into two free propagating beams and a sub-wavelength grating (SWG) mirror to combine the two free propagating beams into a single beam. The Finite Difference Time Domain (FDTD) simulations show that the SWG mirror improves the coupling efficiency of the waveguide fundamental mode into the single out-coupled beam from about 30% to 92%. A high efficiency (>80%) is predicted for a broad wavelength range of 1520 - 1600 nm. The proposed device is compact (~ 80 μm in length), it eliminates the need for blazing the waveguide grating, and it is simple to fabricate using standard CMOS processes.

Key words: silicon-on-insulator (SOI), sub-wavelength waveguide grating, grating mirror, finite-difference time-domain (FDTD), numerical simulations, CMOS

1. INTRODUCTION

A major problem in the design and fabrication of integrated optics devices is the efficient coupling between compact planar waveguides and the outside macroscopic world. This problem has been identified from the earliest years of integrated optics.[1] Many original solutions have been found since, but the coupling still remains a challenge particularly for waveguides of sub-micrometer dimensions[2] made in high index contrast materials such

S. Janz et al. (eds.), Frontiers in Planar Lightwave Circuit Technology, 235–243.

III-V semiconductors, silicon oxynitride, and silicon-on-insulator (SOI). Originally proposed by Dakss et al.[3] and Kogelnik and Sosnowski[4] in 1970, waveguide gratings are often used as waveguide to free-space couplers and their fundamentals are well established.[5-8] Some of the most relevant waveguide grating coupler applications include surface emitting horizontal cavity semiconductor lasers,[9,10] coupling between planar waveguides and optical fibers,[11-12] beam shaping outcoupling elements and waveguide holograms,[13-15] and photonic crystal based coupling structures.[16-19]

An out-coupling waveguide grating is typically a second order grating with first-order diffraction designed to produce two beams diffracted into the superstrate (typically air) and the substrate, respectively. To maximize its efficiency it is required to increase the directionality of the out-coupling by enhancing one of these diffracted beams and suppressing the other. Suppressing one of the first order diffracted beams can be achieved by using blazed (asymmetric) grating profiles,[20-21] typically fabricated by reactive ion beam etching with the ion beam incident at an angle to the waveguide plane normal. However, the fabrication of precise grating profiles is quite demanding and is generally associated with a strong dependence of the directionality and the efficiency of the coupler on various grating parameters such as its depth, blaze angle, and duty cycle.[15,22] One of the possible ways to obviate the need for blazing is to use a simple rectangular (symmetric) grating and redirect one of the free propagating beams by a multilayer or metallic mirror[11,23,24] or a photonic crystal structure,[16-17] so that both diffracted beams add in phase. However, the fabrication tolerances for making the multilayer and photonic bandgap structures are demanding.

2. DEVICE GEOMETRY

We propose a simple waveguide coupler comprising a waveguide grating and a sub-wavelength grating mirror, the former diffracting a guided wave into two free propagating waves and the latter recombining the two free-propagating waves. The main advantage of this device is that no blazing is required for the waveguide grating and a single layer of a transparent (non-metallic) material is sufficient for the mirror effect. The schematic of the device is shown in Figure 1. A grating with a period $\Lambda_g = \lambda_0/n_{eff}$ is formed by partially etching the silicon waveguide core (where n_{eff} is the effective index of the fundamental mode in the grating section, and λ_0 is the free-space wavelength). The first order diffraction at a wavelength λ (in the proximity of λ_0) yields two out-of-plane diffraction orders propagating in approximately opposite directions out of the waveguide plane, with the respective angles in the air α_a and in the substrate α_s with respect to the waveguide plane normal:

$$\alpha_{a,s} = \arcsin[(n_{eff}k - K)/(n_{a,s}k)],$$

where $K = 2\pi/\Lambda_g$ is the modulus of the grating vector, $k = 2\pi/\lambda$ is the free-space propagation constant, and n_a and n_s are the refractive indices of the air and the substrate, respectively.

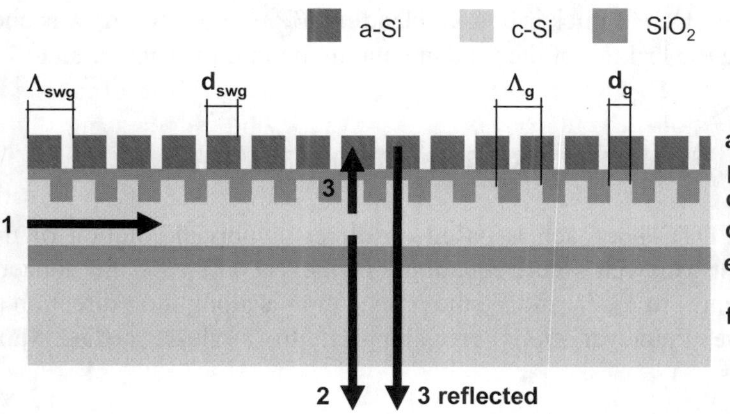

Figure 1. A schematic of the waveguide grating coupler with a sub-wavelength grating mirror. A second-order grating is etched in the single crystal silicon (c-Si) waveguide core. The first order diffraction of the input field (1) yields two out-of-plane diffraction orders (2 and 3) that are recombined in-phase by reflecting the upward propagating order by the amorphous silicon (a-Si) grating mirror. a: a-Si sub-wavelength grating mirror; b: SiO_2 spacer; c: waveguide grating; d: Si waveguide core; e: SiO_2 bottom cladding; f: Si substrate.

These two diffraction orders are recombined by reflecting the upward propagating order by a sub-wavelength grating (SWG) mirror. Our SWG mirror was inspired by a free-space mirror proposed by Mateus et al.[25] In the implementation shown here, the mirror comprises a periodic array of amorphous silicon pillars deposited on a low refractive index SiO_2 spacer, the latter also filling the waveguide grating trenches. It is assumed that a-Si has the same refractive index, $n \sim 3.46$, as single crystal Si at $\lambda \sim 1.5$ μm) The grating mirror acts as a phase mask with all the diffraction orders (including the zero order) being suppressed. For light incident normally on our grating mirror, the scalar grating equation is: $\sin\alpha_a = m\lambda/\Lambda_{swg}$, where m is the diffraction order, all the diffraction orders being evanescent (suppressed) for $\Lambda_{swg} < \lambda$, hence the term sub-wavelength grating The mirror function can be intuitively explained by the fact that the portion of light transmitted through each a-Si pillar is phase shifted π or odd multiples thereof with respect to the light transmitted in between, the two parts canceling each other. This results in a suppressed transmission through the SWG layer, the latter hence acting as a mirror. The thickness of the SiO_2 spacer is chosen in a way that the resulting co-propagating beams add in-phase modulo 2π. Also, it is advantageous to eliminate any partial reflections of the two co-propagating beams as they cross the core-cladding and the cladding-substrate interfaces. This can be achieved by choosing the bottom cladding of a quarter-wavelength thickness, or odd multiples thereof, such that the Fresnel reflections at the two interfaces interferometrically cancel each other. Here a thicker bottom cladding ($3\times\lambda/4 \sim 0.78$ μm) was chosen to minimize the leaking of the fundamental mode into the substrate.

3. THE FDTD SIMULATION METHOD

The FDTD approach is based on direct numerical solution of the time dependent Maxwell's curl equations. In the 2D TM case the nonzero field components are E_x, H_y and E_z, the propagation is along the z direction and the transverse field variations are along x. In lossless media, Maxwell's equations

$$\varepsilon_0\varepsilon_r \partial E_x / \partial t = -\partial H_y / \partial z$$

$$\mu_0 \partial H_y / \partial t = \partial E_x / \partial z - \partial E_z / \partial x$$

$$\varepsilon_0\varepsilon_r \partial E_z / \partial t = \partial H_y / \partial x$$

are discretized by means of central difference approximations for the numerical derivatives in space and time.[18] $\varepsilon = \varepsilon_0\varepsilon_r$ is the dielectric permittivity and μ_0 is the magnetic permeability of the vacuum. The sampling in space is on a sub-wavelength scale. Typically, 10 to 20 steps per wavelength are needed. The sampling in time is selected to ensure numerical stability of the algorithm. The time step is determined by the Courant limit: $\Delta t \le 1/\left(c\sqrt{1/(\Delta x)^2 + 1/(\Delta z)^2}\right)$. The numerical scheme is excited by an input cw field equivalent to the fundamental mode profile of the initial segment of the waveguide. The optical wave propagates until it reaches a stationary state within the computational domain. A Discrete Fourier Transform (DFT) procedure is applied in the last time period of the simulation to provide the complex values of the field components indicated above. The z- and x-components of the Poynting vector are used to give an estimate for the power flow in both longitudinal and transverse directions.

4. SIMULATION RESULTS

To study the output coupling efficiency of the device we have performed an analysis based on two-dimensional Finite Difference Time Domain (FDTD) simulations. Figure 2 shows some of the results. The input field is TM polarized waveguide fundamental mode (λ = 1.55 μm) and the waveguide is along the z axis. The simulation window dimensions are 80 μm (horizontal) by 4 um (vertical). The mesh size is 0.022 μm in both dimensions. The simulation ran for a total 40,000 time steps each of dt = 9.8×10^{-17} s. The waveguide grating is 0.4 μm deep etched in a 1 μm thick silicon waveguide core. The SWG mirror is formed in a 0.45 μm thick a-Si layer. Duty ratios of the SWG mirror and the waveguide grating are $r_{swg} = d_{swg} / \Lambda_{swg} = 0.74$ and $r_g = d_g / \Lambda_g = 0.5$, respectively. The periods of the respective gratings are $\Lambda_{swg} = 0.7$ μm and $\Lambda_g = 0.46$ μm. The thickness, the period and the duty ratio of the sub-wavelength grating have been chosen to maximize device bandwidth, taking into account the calculations in ref. **Error! Bookmark not defined.**. The SiO_2 spacer thickness is 0.25 μm, the substrate is silicon, and the superstrate is the air.

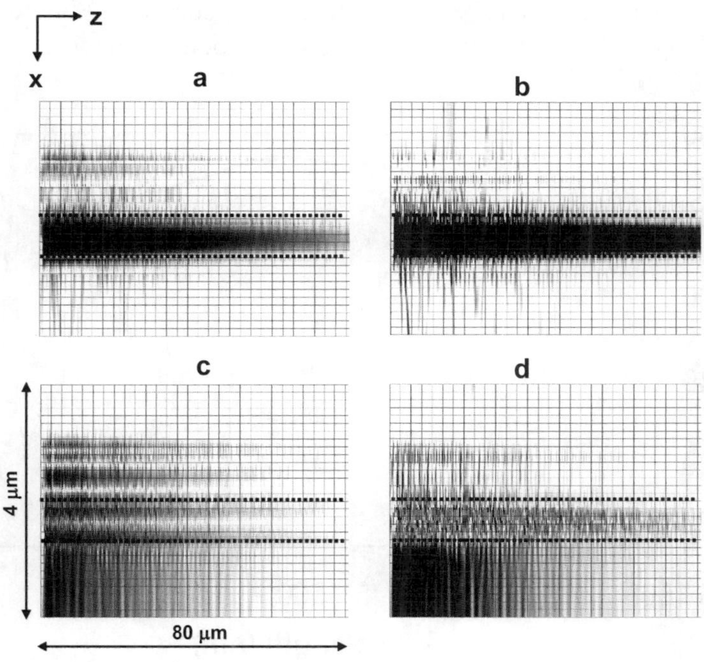

Figure 2. The FDTD simulation results for the electric field components E_x (a) and E_z (c) and the Poynting vector component S_z (b) and S_x (d). The waveguide is along z axis, the dashed lines contouring the non-etched 0.6 μm thick region of the core.

The panels a and b in Figure 2 show the electric field component E_x and the Poynting vector component S_z, respectively. The out-of-plane beam propagation (in +x direction) is evident from E_z and S_x evolution (Figure 2, panels c and d), where the Poynting vector component $S_x = -\mathrm{Re}(E_z H_y^*)/2$. This design of the one-dimensional grating mirror is effective for out-of-plane propagating waves perpendicular to the SWG grooves (i.e. for TM-like polarized waveguide modes).

It has been demonstrated[25] that the efficiency of a SWG grating is maximum for a free propagating wave with the electric field vector perpendicular to the grating grooves. To use the SWG mirror effectively for the TE-like polarized guided mode would require SWG grooves oriented parallel to the waveguide direction z, that would demand a 3D FDTD simulation. Based on these arguments, it appears possible to optimize the mirror efficiency for both TE- and TM-like polarizations by arranging the a-Si pillars in a two dimensional array.

Figure 3 shows the spectral dependence of coupler efficiency with (curve a) and without (curve b) the SWG mirror, respectively, for TM polarization. The coupling efficiency was calculated as P_x/P_0 , where P_0 is the power in the waveguide at z = 0 (see Figure 2), and P_x is power in the out-of-plane propagating wave calculated by integrating the S_x component of the Poynting vector along the bottom edge of the computation window.

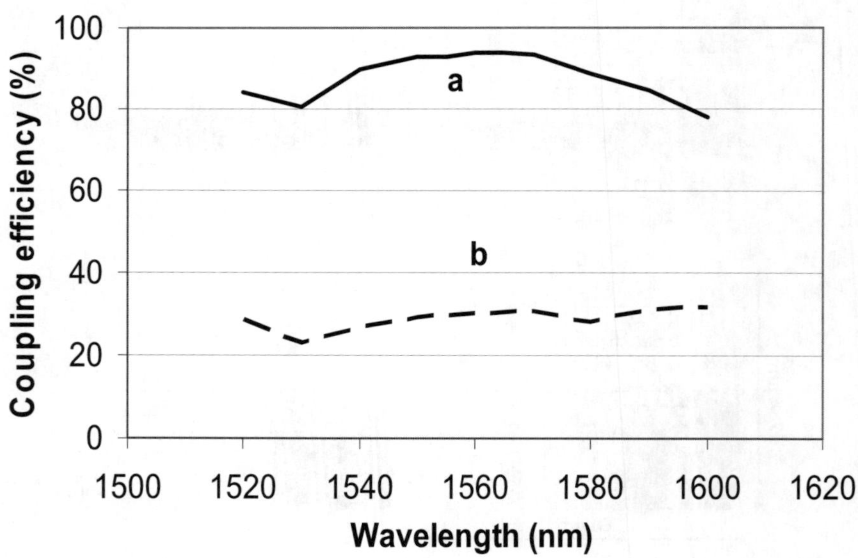

Figure 3. Coupler efficiency for a waveguide grating with (a) and without (b) the sub-wavelength grating mirror.

The calculation of the coupling efficiency is an automatic post-processing feature of the FDTD software used to perform the simulations.[26] It is observed that the grating mirror improves the coupling efficiency from 30% to 92% near 1550 nm. The coupling efficiency larger than 80% is achieved over a broad wavelength range of 1520 - 1600 nm. The central wavelength can be adjusted by scaling of device dimensions, i.e. the periods of the two gratings and thicknesses of the layers. From the fabrication point of view, the tolerance on the oxide spacer thickness appears critical, as an accurate thickness is required for in-phase superposition of the two diffracted beams. According to our FDTD calculations, the oxide spacer thickness tolerance of ±0.05 μm (i.e. ±20%) yields an efficiency of >92% near the central wavelength.

The device shown in Figure 1 provides coupling between the fundamental waveguide mode and the free propagating beam in the downward direction, hence in a substrate emitting geometry. Similar coupling function can also be achieved in the upward direction (surface emitting), for example making use of a double SOI platform, as it is shown in Figure 4. Here, the waveguide grating is formed in the first Si layer and the SWG mirror in the second Si layer of the double SOI wafer. In this geometry, the calculated coupling efficiency was found to be similar to that of the downward coupling device.

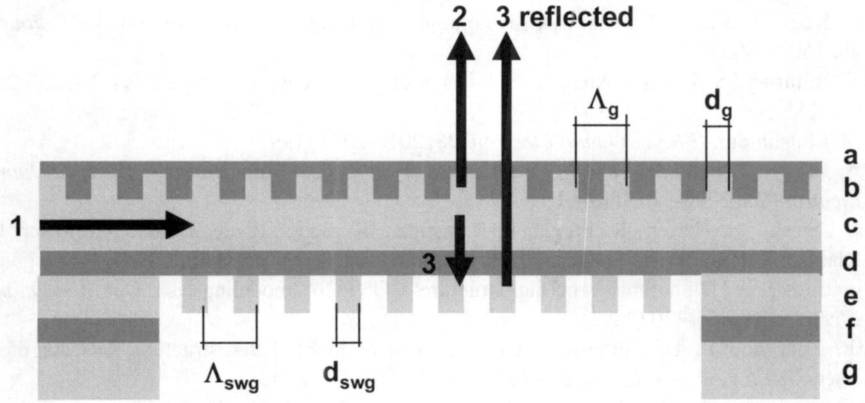

Figure 4. A schematic of a surface emitting waveguide coupler with a sub-wavelength grating mirror formed in a double SOI platform. a: cladding, SiO_2; b: waveguide grating; c: waveguide core, 1st Si layer of SOI; d: cladding, 1st buried oxide (box) layer of the SOI; e: sub-wavelength grating mirror, 2nd Si layer of the SOI; f: 2nd box layer of the SOI; g: the Si substrate.

5. CONCLUSIONS

In the coupler examples shown here the light guided in a planar waveguide is diffracted into a free-propagating beam, but the principle described here also applies for light coupling from an optical fiber to a planar waveguide device. Important advantages of the proposed coupler are that it is simple to fabricate using standard CMOS process, it can be made compact in size, and deposition of multilayer dielectrics or metals is avoided. No blazing of waveguide grating is required, so that simple grating etching processes can be used. The SWG mirror can also be used with blazed waveguide gratings to further enhance the coupling efficiency. The Si-based materials have been chosen here to benefit from the compatibility with the standard CMOS process, but the coupler scheme may also be useful to other, particularly high index contrast waveguide platforms.

REFERENCES

1. T. Tamir, *Integrated Optics* (Springer Verlag, Berlin, 1975).
2. S. Janz, P. Cheben, A. Delâge, B. Lamontagne, M.-J. Picard, D.-X. Xu, and W. N. Ye, *Proc. Materials Research Society Symposium*, Vol. 832 (in press).
3. M. L. Dakss, L. Kuhn, P. F. Heindrich and B. A. Scott, Grating coupler for excitation of optical guided waves in thin films, *Appl. Phys. Lett.* **16,** 523-525 (1970).
4. H. Kogelnik and T. P. Sosnowski, Holographic thin film couplers, *Bell Sys. Tech. Journ.* **49,** 1602-1608 (1970).
5. T. Tamir and S. T. Peng, Analysis and design of grating couplers, *Appl. Phys.* **14,** 235-254 (1977).
6. M. Matsumoto, *IEEE J. Quant. Electron.* **28,** 2016-2023 (1992).
7. A. Hardy, D. F. Welch, and W. Streifer, Analysis of 2nd-order gratings, *IEEE J. Quant. Electron.* **25,** 2096-2105 (1989).
8. J. Čtyroký, S. Helfert, R. Pregla, P. Bienstman, R. Baets, R. de Ridder, R. Stoffer, G. Klaasse, J. Petráček, P. Lalanne, J.-P. Hugonin, and R. M. de la Rue, Bragg waveguide gratings as a 1D photonic band gap structure: COST 268 modeling task, *Opt. and Quant. Electron.* **34,** 455-470 (2002).
9. P. Zory and L. D. Comerfold, Grating-coupled double-heterostructure AlGaAs diode lasers, *IEEE J. Quant. Electron* **QE11,** 451-457 (1975).
10. G. A. Evans, N. W. Carlson, J. M. Hammer, and J. K. Buttler, in: *Surface Emitting Semiconductor Lasers and Arrays*, G. A. Evans and J. M. Hammer, Eds., (Academic Press, San Diego, 1993), chapter 6.
11. D. Taillaert, W. Bogaerts, P. Bienstman, T. F. Krauss, P. V. Daele, I. Moerman, S. Verstuyft, K. D. Mesel, and R. Baets, An out-of-plane grating coupler for efficient butt-coupling between compact planar waveguides and single-mode fibers, *IEEE J. Quant. Electron.* **38,** 949-955 (2002).
12. G. Z. Masanovic, V. M. N. Passaro, and G. T. Reed, Coupling optical fibers to thin semiconductor waveguides, *SPIE* **4997**, 171-180 (2003).
13. T. Suhara, H. Nishihara, and J. Koyama, Waveguide holograms: A new approach to hologram integration, *Opt. Commun.* **19,** 353-358 (1976).

14. T. Suhara and H. Nishihara, Integrated-optic components and devices using periodic structures, *IEEE J. Quant. Electron.* **22,** 845-867 (1986).
15. A. Larsson, N. Eriksson, S. Kristjansson, P. Modh, M. Uemukai, T. Suhara, and H. Nishihara, Grating coupled surface emitters: integrated lasers, amplifiers, and beam shaping outcouplers, *SPIE* **3626,** 190-200 (1999).
16. R. W. Ziolkowski and T. Liang, Design and characterization of a grating-assisted coupler enhanced by a photonic-band-gap structure for effective wavelength-division demultiplexing, *Opt. Lett.* **22,** 1033-1035 (1997).
17. S. Tanev, D. Feng, V. Tzolov, and Z. J. Jakubczyk, Finite-difference time-domain modeling of complex integrated optics structures, *Technical Digest*, Integrated Photonics Research Conference Edition (Optical Society of America, Washington, 1999), pp. 202-204.
18. S. Tanev et al., Advances in the development of simulation tools for integrated optics devices: FDTD, BPM, and mode-solving techniques, Proc. SPIE **4277,** 1-20 (2001).
19. A. Mekis, A. Dodabalapur, R. E. Slusher, and J. D. Joannopoulos, *Opt. Lett.* **25,** 942-944 (2000).
20. S. T. Peng and T. Tamir, Directional blazing of waves guided by asymmetrical dielectric gratings, *Opt. Commun.* **11,** 405-409 (1974).
21. M. Hagberg, N. Eriksson, T. Kjellberg, and A. G. Larsson, Demonstration of blazing effect in detuned second order gratings, *Electron. Lett.* **30,** 570-571 (1994).
22. T. W. Ang, G. T. Reed, A. Vonsovici, A. G. R. Evans, P. R. Routley, and M. R. Josey, Effects of grating heights on highly efficient Unibond SOI waveguide grating coupler, *IEEE Photon. Tech. Lett.* **12,** 59-61 (2000).
23. D. Mehuys, A. Hardy, D. F. Welch, R. G. Waarts, and R. Parke, Analysis of detuned second-order grating output couplers with an integrated superlattice reflector, *IEEE Photon. Tech. Lett.* **3,** 342-344 (1991).
24. N. Eriksson, M. Hagberg, and A. Larson, *IEEE J. Quant. Electron.* **32,** 1596-1605 (1996).
25. C. F. R. Mateus, M. C. Y. Huang, Y. Deng, A. R. Neureuther, and C. J. Chang-Hasnain, Ultrabroadband mirror using low-index cladded subwavelength grating, *IEEE Photon. Tech. Lett.* **16,** 518-520 (2004).
26. The FDTD simulations were performed by OptiFDTD software from Optiwave Systems, Ottawa, ON, Canada, www.optiwave.com

INTEGRATED OPTICS DESIGN: SOFTWARE TOOLS AND DIVERSIFIED APPLICATIONS

Christoph Wächter

Fraunhofer Institute for Applied Optics and Precision Engineering, Albert-Einstein-Strasse 7, 07745 Jena, Germany

Abstract: The design of integrated optics (IO) devices requires dedicated numerical algorithms and related software. Today's commercial tools cover a wide range of the needs, and manifold tasks can be solved this way. A short survey on specific IO-design tools and other design software that is related to particular aspects of IO is given. The basic physics and math behind the graphical user interface is discussed in some detail for a few of the popular schemes for mode solving and field propagation. Aspects of system design and the multi-physics character of IO-microsystems are considered, and the potential complexity of IO-design is demonstrated for a micro-optical coupling of a fibre and a photonic crystal waveguide.

Key words: integrated optics, design software, mode solver, field propagation, multi-physics, photonic crystal coupling

1. INTRODUCTION

Numerics has been used intensively in the field of integrated optics (IO) since its early days, simply due to the fact that even the basic example of the slab waveguide requires the solution of a transcendental equation in order to calculate the propagation constants of the slab- guided modes. Of course, the focus was directed to analytical methods, primarily, as long as the power of a desktop computer did allow for a few coupled equations and special functions only e.g. to describe the nonlinear directional coupler in a coupled mode theory (CMT) picture[1]. During the years, lots of analytic and semi-analytic approaches to solve the wave equation have been developed in order

S. Janz et al. (eds.), Frontiers in Planar Lightwave Circuit Technology, 245–280.

to calculate eigenmodes and to study the evolution of the electromagnetic field within an integrated optical device. Nonlinear material properties, as they are utilized for ultra long-haul soliton based data transport nowadays, and anisotropic media and high index contrast which require full-vectorial schemes, were incorporated in the design procedures. From time to time IO-theory could borrow ideas from related disciplines like acoustics, optics of the atmosphere, and micro-wave theory. So, e.g. the beam propagation method[2] (BPM), the method of lines[3] (MoL) and the finite difference time domain method[4] (FDTD) became popular in IO, as well. This, of course, was directly related to the incredible dynamic of change of hardware power, which made CPU speed of multiple GHz and memory of multiple Gbyte accessible even at the PC level.

Today, there an established software tool set does exist for the primary task, the calculation of modes and the description of field propagation. Approaches based on the finite element method (FEM) and finite differences (FD) are popular since long and can be applied to complex problems[5,6]. The wave matching method[7], Green functions approaches[8], and many more schemes are used. But, as a matter of fact, the more dominant numerical methods are, the more the user has to scrutinize the results from the physical point of view. Recent mathematical methods, which can control accuracy absolutely – at least if the problem is well posed, help the design engineer with this[9].

Real devices may suffer from batch dependent material losses, anisotropy, inhomogeneities due to fabrication tolerances etc., which intensifies the need for parametric models to change over from a description of the electromagnetic field within a FD- or a FE-grid to a more complex device and system level description. This is accomplished with the modelling of the evolution of modal amplitudes, generally, which usually is applied in those specific modelling tools which have been developed for point-to-point communication system design including modulation schemes, signal regeneration, pulse- and soliton propagation, etc. And, there are waveguide related topics like laser modelling, grating design, and even the free-space coupling via micro-optics which belong to the multitude of tasks the IO-design engineer is faced with.

Thus, an efficient design of IO devices does not require reliable software and an adequate understanding of the underlying physics and numerics, only. Moreover, problem complexity requires to use dedicated software tools and corresponding interface routines, and the inventiveness of the design engineer is in demand when gaps are to bridge. To this end, this paper will memorize some of the popular software tools and numerical methods behind, and will direct the readers attention to several diversified applications.

In section 2, different categories of software tools are listed – with no claim of completeness, of course, and some comments on the design flow, in general, are given. In section 3, physics and math of different modelling approaches are elucidated in some detail, and section 4 concentrates on requirements of device design with respect to specific applications. In section 5, a complex example related to the fibre-chip coupling of high numerical aperture (NA) photonic crystal (PhC) waveguides illustrates the need for schemes, which allow for a multi-scale modelling of electromagnetic field propagation.

2. SOFTWARE TOOLS FOR IO-DESIGN

It is not surprising that the giant boom in telecommunications at the end of the last century did contribute to the development and commercialisation of IO-software tools, massively. Various software companies have been founded which did enter an emerging market, and at R&D facilities and institutes lots of proprietary, (and casually freeware) tools have been established.

For a snapshot what software is applied in IO-design, with no claim of completeness, of course, it is advisable to define several categories. The first, naturally, is the category of mode calculation and field propagation.

There, in alphabetical order, commercial tools are provided by e.g. Apollo Photonics[10], Concept to Volume[11] (C2V), JCM-Wave[12], Optiwave Systems[13], Photon Design[14], and RSOFT Design Group[15]. The tool set of Apollo Photonics covers Material Design, Waveguide Design, Device Design, Circuit Design, and FOMS, a Fibre Optic Mode Solver, where device and circuit design are intended to develop complex waveguide circuitry. C2V, which did start as BBV more than a decade ago, comes with its OlympIOs platform with Prometheus, TempSelene, and StressSelene, where ODIN is a related tool for mask-layout. As indicated by the names, extra attention is payed to multi-physics aspects of IO, which take into account e.g. temperature-dependent refractive index changes and stress-induced birefringence. J(ames) C(lerk) M(axwell)-Wave, a spin-off from the Zuse-Institut in Berlin, is a newcomer in the field. Its FEM-based mode-solver relies on adaptive grid refinement techniques with special emphasis on reliable error estimation techniques. Optiwave Systems offers OptiBPM and OptiFDTD, including mode solver capabilities, and OptiFIBER. In contrast to the preceding tools, FIMMWAVE offered by PhotonDesign is based on the wave matching method, and the field propagation tool, FIMMPROP, is based on eigenmode expansion. CrystalWave addresses the light propagation in photonic crystals. BeamPROP and FullWAVE are the tools by RSOFT, again FD-based, and with state of-the-art capabilities for the design of intricate IO-devices.

A next category comprises tools which are IO-related via the topic or kindred ship, covering waveguide-gratings, -lasers etc. Apollo Photonics has FOGS, a fiber optics grating solver, and ALDS, an advanced laser diode simulator for the design of active elements in III-V's, including Fabry-Perot, distributed feedback, and distributed Bragg Reflector configurations. Optiwave Systems has an CMT-based tool for fiber and waveguide grating design, OptiGrating, and OptiAmplifier is a tool for Erbium-doped and Raman fiber amplifiers. OptiHS covers the electronical and optical properties of semiconductor heterostructures. CLADISS-2D, offered by Photon Design, is a design tool for laser diode and travelling wave amplifiers, which in its origin stems from the University of Gent. It includes a wide variety of geometries and electronical effects, as well. RSOFT offers tools for grating and laser modelling, GratingMOD, and LaserMOD, too. BandSOLVE is a stand-alone tool for the calculation of band-structures in PhCs, and DiffractMOD models diffraction provoked by 3-dimensional periodic structures. Virtual Photonics Inc.[16] (VPI) has its VPIcomponentMaker tools based on the transmission line laser model, which address fixed and tuneable lasers as well as semiconductor optical amplifier (SOA) signal processing and wavelength conversion.

A third category of software tools is dedicated to the design of optical communication systems. Necessarily, those tools include parametric models of all the components in the system, lasers, modulators, waveguide circuits for wavelength division multiplex (WDM), the fibre with its intrinsic nonlinearity, dispersive effects, mixed electrical and optical signals, etc. This at the end allows for e.g. bit-error-rate (BER) calculation, system analysis, and system optimisation, potentially. Optiwave Systems' product in this field is OptiSystem. A tool with not too much market penetration, PHOTOSS[17], goes back to initial work at the university of Dortmund, Germany. RSOFT has OptSIM and LinkSIM, and a special tool for multimode communication system design named ModSYS. VPItransitionMaker are tools offered by VPI, which address either WDM or cable access systems.

Specific tools do exist for the laser design, too, where the focus is directed more to semiconductor device and process simulation, potentially. Crosslight Software Inc.[18] delivers PICS3D for the 3D simulation of semiconductor optoelectronic devices, Integrated System Engineering[19] has DESSIS-Laser in its program, an advanced semiconductor laser simulator, and there are many tools for the design of vertical cavity semiconductor lasers (VCSL), a special topic which is omitted for brevity.

As well, software developers with background in FEM and multi-physics did extend their activities towards electromagnetics and optics. ANSYS[20], with its multiphysics and high-frequency electromagnetics capabilities stepped into optical and radio-frequency passives, i.e. fibre optics and micro-stripline waveguides, to some extent. Computational Fluid Dynamics Research Corporation[21] got active in FDTD modelling, its CFD-Maxwell is a full vector solver on arbitrary grids, and there are several more related activities.

But, concerning IO-design, there are much more commercial software tools which are of relevance. If high index contrast and deep gratings are considered, CMT-approaches are questionable. Wave matching, or rigorously coupled wave analysis (RCWA), equivalently, then is an appropriate ansatz. Thus, tools like Grating Solver[22] or UNIGIT[23] can be taken into account profitably for waveguide geometries as well. Thin film software is essential if anti-reflective (AR) coating is considered, a must for e.g. efficient coupling to waveguides in high index materials. Tools like Essential McLeod[24] or Film Wizard[25] are widely used, and slab waveguide problems may be enclosed in thin film analysis, too, e.g. in TRAMAX[26]. If more general problems including scattering are addressed the multiple multipole method may be considered, there a tool Max-1[27] is offered. The coupling from and to the waveguide may include micro-optical components, e.g. in fibre-switch fabrics. There, commercial ray-trace and wave propagation software like CODE V[28], ZEMAX[29], GLAD[30] or VirtualLab[31], to name a few only, can be successfully applied when appropriate interfaces to BPM, FDTD or other guided-wave software do exist. Furthermore, waveguide fabrication requires the transfer of the design data to the wafer in exactly the same manner as micro-electronics or silicon micro-mechanics. Thus, corresponding mask-layout tools, e.g. DesignWorkshop-2000[32], Mask-Generation-System-6[33] or C2V's[11] layout editor ODIN may be applied to generate appropriate data sets for waveguides, electrodes, marks, etc., as needed for passive and active waveguide circuitry.

Beside commercial software, there a limited number of free programs for the IO-design and related topics does exist. The Centre for Ultrahigh Bandwidth Devices for Optical Systems in Sydney presents CUDOS-MOF, multipole-based utilities for micro-structured fibres. For the analysis of PhC-bandstructures, MIT-Photonic-bands is a plane-wave-expansion based tool of the ab-initio-group at MIT[35]. From the Applied Analysis and Mathematical Physics-Group of the University of Twente e.g. an implementation of the wave matching method[36] is available. From Zuse-Institut in Berlin one can download HelmPole[37], a finite element solver for scattering problems on unbounded domains in an implementation based on perfectly matched layers (PML). CAMFR[38], which stands for cavity-modeling framework, is a free vectorial eigenmode expansion tool with PML from INTEC, Gent.

Last not least, there is a great variety of proprietary IO-design software that has been developed at research facilities and university institutes. E.g., at the TU Delft the S-matrix oriented software Photonic CAD[39] was established several years ago, an innovative IO design framework based on Hewlett Packard's Microwave Design System. Actual academic work e.g. addresses modes of bent waveguides, BMS-3D[40] is a quite new bend mode solver of the IRE, Prague, or FDTD-schemes with non-uniform grids[41], a topic of special importance to improve computational efficiency when multi-scale feature sizes are requested, to name a few of recent tasks, only.

The considerable number of software tools that may be applied in the IO-design is what is supplied to help the IO design engineer to master his specific task. The individual tools have their individual scope, naturally, have different capability characteristics, and, have different prices. And, the demands are different, as well. They may go from a R&D-design to a design for manufacture, can cover a quite different device complexity, and can require a device sketch or complete mask data, finally.

Supply and demand is interrelated, consequently, and for various standard tasks, which belong to the generalized design flow, figure 1, software assistance is readily available.

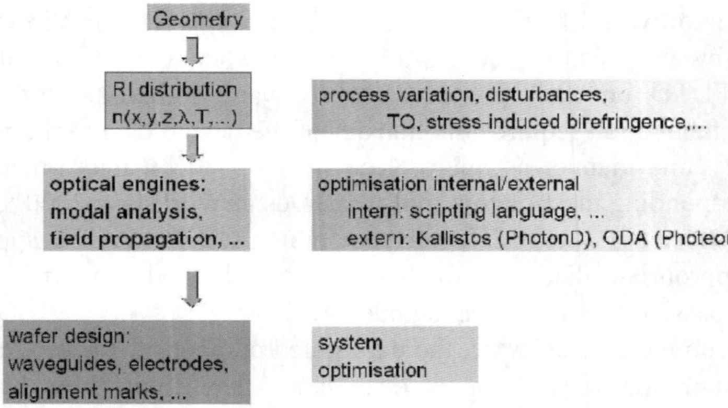

Figure 1. Design flow.

So, influences of process variations, which e.g. might affect the layer height achieved by the deposition process, the waveguide width achieved by the etch process, etc., enter the refractive index distribution. In order to either estimate the impact on the optical characteristics, or to optimise the design, the established IO-design software tool-sets offer multi-dimensional parameter sweeps, scripting capabilities, models for electro-optically or thermo-optically induced refractive index changes, etc. Moreover, for optimisation external modules that rely on appropriate strategies can be used, which then rule the optical engines. For instance, the tool Kallistos of Photon Design can optimise FIMMWAVE device geometries, and ODA – Optical Design Automation by Photeon Technologies[42], acts on its individual BPM, ray-trace and FDTD tools. As well, the implementation of specific genetic algorithms that refer to commercial IO-design software has been reported[43].

Once the device geometry is designed, technological aspects have to be considered, concerning adjustment of waveguides and fibres, waveguides and electrodes, concerning control structures etc. Finally, the complete design data are exported to one of the standard data formats that are accepted by the mask-shop, GDSII or DXF, usually. This again is assisted by all the established IO-design software tool-sets.

But, some aspects of the design are covered less-usually, and some are still challenging. So, if multi-physics effects are considered, internal models included in the IO-design environment cover the dominant aspects. For an intended electro-optical (EO) and thermo-optical (TO) device operation this works well, but perturbations by non-ideal material properties or the like are neglected easily. Concerning mechano-optical properties, stress-induced birefringence is rarely considered, there C2V has a pioneering position. What is rare, too, are interfaces to process simulation, project management, and yield analysis. This of course is somewhat beyond the IO-design, but belongs to an overall development process, too. What still remains challenging is the design and analysis of compact high index-contrast devices, e.g. the estimation of PhC-waveguide losses within the light-cone, the calculation of branching losses of PhC-waveguides, or the 3D fibre to PhC-chip coupling.

Moreover, what is still unrealistic is an automated all-in-one design tool suitable for every purpose. Thus, the design engineer first of all has to know what the requirements of the specific design task are, and, what software capabilities are needed, consequently. This determines, what tool or what set of individual tools is appropriate, basically. If missing, interface routines between tools have to be established. This may refer to the electromagnetic field itself, simply, or may e.g. cover the extraction of a parametric model of the temperature-induced refractive index change out of a series of FEM runs, if the design of a TO-switch is considered.

3. IO-DESIGN – SOME PHYSICS AND MATH

Modal analysis and field propagation techniques are the basic means in the design process for IO-devices if the spatially resolved optical field is considered. The numerical burden to be managed depends on the geometry under consideration and the approximations used, i.e. there quite different levels of complexity can occur. What might be left to the designer is to find a reasonable compromise of accuracy and design efficiency. Similarly, when complex waveguide circuitry is pondered, a change of the abstraction level can be required to retain design efficiency by the use of parametric models, which are based on results of modal analysis and propagation techniques, consequentially. So, the following sections will consider mode solvers, field propagation, and more general aspects with respect to system design.

3.1 Mode Solvers

Starting from Maxwell's equations with the ansatz of monochromatic, z-propagating fields, $\vec{E},\vec{H} \sim \exp(i(k_z z - \omega t))$, transversal and longitudinal components get decoupled if the refractive index distribution is z-independent. Two physically equivalently meaningful equations for the transversal electric and the transversal magnetic field,

$$\left\lfloor \Delta_t + k_0^2\,(\hat{\varepsilon} - \hat{\beta}_{eff}^2)\right\rfloor \vec{E}_t = -\vec{\nabla}_t\left(\vec{\nabla}_t\left(\ln\,\hat{\varepsilon}\right)\cdot \vec{E}_t\right) \tag{1a}$$

$$\left\lfloor \Delta_t + k_0^2\,(\hat{\varepsilon} - \hat{\beta}_{eff}^2)\right\rfloor \vec{H}_t = -\vec{\nabla}_t\left(\ln\hat{\varepsilon}\right)\times\vec{\nabla}_t\times\vec{H}_t \quad , \tag{1b}$$

result, and its solutions are the modes supported by the geometry. The vacuum wave number k_0 relates effective refractive index of the guided mode, β_{eff} or n_{eff}, equivalently, and the propagation constant

$$\hat{k}_z = \hat{\beta} = k_0 \hat{\beta}_{eff} \quad , \; k_0 = 2\pi/\lambda \, .$$

The tick mark of the propagation constant indicates that it might be complex valued, either due to a complex valued dielectric constant or refractive index, $\hat{n}(x,y) = sqrt(\hat{\varepsilon}(x,y))$, equivalently, or due to a leaky wave geometry. For brevity, in the following tick marks are omitted.

The effort to solve Eqs.(1) evidently depends on the refractive index profile. For isotropic media in a one-dimensional refractive index profile the modes are either transversal-electric (TE) or transversal-magnetic (TM), thus the problem to be solved is a scalar one. If additionally the profile consists of individual layers with constant refractive index, Eq.(1) simplifies to the Helmholtz-equation, and the solution functions are well known. Thus, by taking into account the relevant boundary conditions at interfaces, semi-analytical approaches like the Transfer-Matrix-Method (TMM) can be used. For two-dimensional refractive index profiles, different approaches can be pursued. If the refractive index contrast is quite limited, scalar approximations like the effective index method (EIM) may work sufficiently well for isotropic media[44]. Scalar and semi-vectorial approaches can be improved by e.g. the weighted index method[45], perturbation and variational methods[46], or the spectral index method[47]. Direct numerical solutions of Eq.(1), including anisotropic material properties, like WMM, FDM and FEM became more and more popular with recent hardware improvements.

In the following the attention will be directed to TMM and EIM as numerically inexpensive methods, and WMM, FDM, and FEM as fully-vectorial mode solvers. A few examples will be used to illustrate notable peculiarities with the mode calculation.

3.1.1 Slab waveguides

The well-known TMM allows tackling arbitrary loss-less and lossy multi-layer configurations. Continuous profiles can be approximated by a corresponding layer sequence, simply, thus TMM is a quite general tool.

For TE-polarization, $\vec{E} = (0, E_y, 0)$, and

$$\left[\frac{d^2}{dx^2} + k_0^2\left(\varepsilon(x) - \beta_{eff}^2\right)\right] E_y(x) = 0 \tag{2}$$

is to be solved. The corresponding x- and z-components of the *H*-field follow from Maxwells curl-equation. TMM starts with the decomposition of the fields into its parts propagating in +x and –x direction, where j indicates the layer number, cf. figure 2,

$$E_j = E_j^+ + E_j^- \qquad H_j = H_j^+ + H_j^- . \tag{3}$$

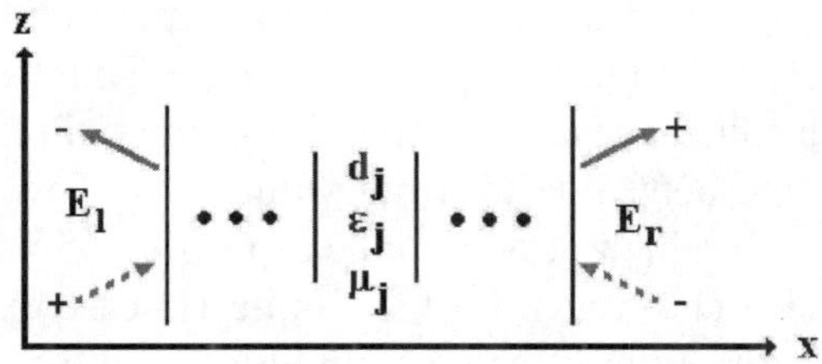

Figure 2. Layer stack geometry.

The x-component of the wave number follows from

$$k_0^2 \varepsilon_j = k_0^2 n_j^2 = k_{x\,j}^2 + k_z^2 . \tag{4}$$

The tangential field components within each layer are related by

$$H_{z\,j} = (\omega\, \mu_0)^{-1} k_{x\,j}\, E_{y\,j} = \gamma_j^{\,TE}\, E_{y\,j} \tag{5}$$

With the relation of the ± field components across a layer

$$t_j = \exp(\pm\, i\, k_{x\,j}\, d_j) \tag{6}$$

and the continuity of the tangential fields at any interface, the electric fields at the left and the right of the layer sequence are directly related by a 2 x 2 matrix. A guided mode is a stationary state with no power flow into the layer stack, neither from substrate, nor from cladding, i.e. the dispersion relation is

$$\begin{pmatrix} 0 \\ E^- \end{pmatrix}_l = \begin{pmatrix} a & b \\ c & d \end{pmatrix} \begin{pmatrix} E^+ \\ 0 \end{pmatrix}_r \quad \Leftrightarrow \quad a = 0 \tag{7}$$

What has to be considered explicitly is the uniqueness of k_x . This is assured by the branch cut

$$\Re(k_x) + \Im(k_x) > 0\,, \tag{8}$$

which ensures the proper sign of the Poynting vector when reflection and transmission are considered, and, for guided modes, the field decay in substrate and cladding. Thus, $a(\beta^2\ )$ is an analytic function, and any appropriate root finding method can be applied to Eq. (7) [49,50].

A potential complication within TMM can be anticipated directly from Eq.(6) concerning the field calculation. The accuracy of the root search is limited by numerics, e.g. to double precision in β^2. Thus, a residual value of *a* does exist, which can cause an improper field growth in a region where the field should be evanescent, due to a big value of d. This e.g. may happen when a simple, real-valued graded index profile, which guides several modes, is considered. Thus, even for the simple 1D-case examples, attention has to be turned to potential effects of numerics, and, as a general remark, numerical stability and accuracy is a quite general topic in IO-design.

An interesting configuration is a waveguide with a thin, high-index overlay. This is shown in figure 3 for a 4-layer configuration operating at

$\lambda = 0.633\mu m$, where a monomode glass-waveguide with $\Delta n = 0.005$ carries a film of a medium with $\Re(n_{ovl}) = 4.0$ in an air-covered system. If the configuration is loss-free, the increase of the optical thickness causes the cut-off of a second mode, and the mode of the glass-waveguide shifts towards the high-index medium. But, when the overlay is lossy, the behaviour is quite different. There, after the cut-off of the second mode, the mode in the glass-waveguide is the mode with the lowest $\Re(n_{eff})$, and around the cut-off of the second mode $\Im(n_{eff})$ has a Lorentzian-shape. This gives evidence to the fact that if leaky modes are considered, simple orthogonality does not longer apply. Although known since long[51], this particular resonance feature is still of interest[52], e.g. for impedance matched vertically integrated photodetectors.

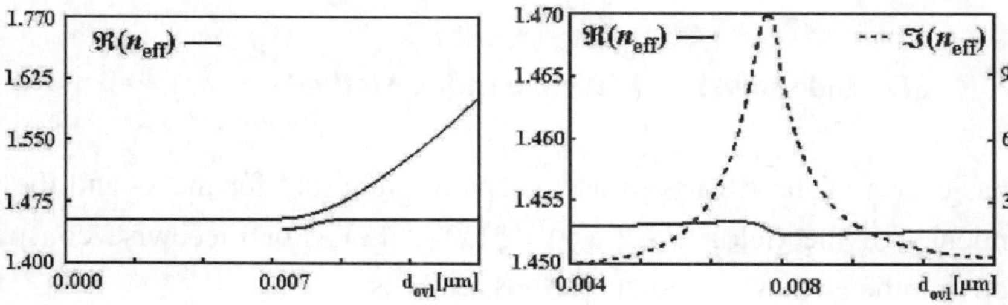

Figure 3. n_{eff} vs. overlay thickness, left: $\Im(n_{ovl}) = 0$, right: $\Im(n_{ovl}) = 0.2$.

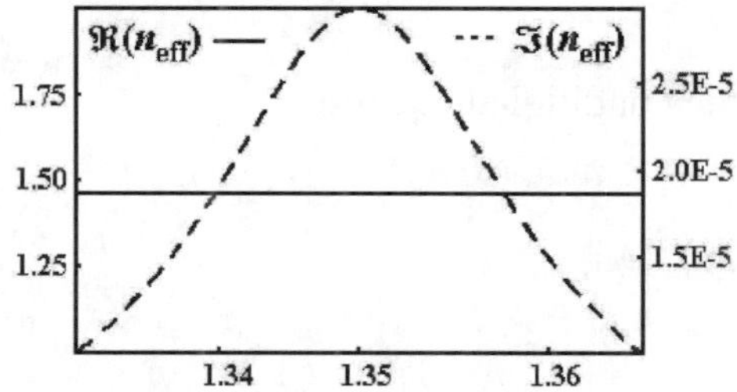

Figure 4. Dispersion curve vs. cladding index for SPP-configuration at λ=0.6μm, substrate n_s=1.45, waveguide thickness 4μm, Δn=0.01, buffer index 1.40, buffer thickness 0.5μm, Au thickness 50nm.

TMM handles thin metallic films as well, as they are used in IO-sensors based on surface-plasmon-polaritons (SPP). SPPs appear at the dielectric-metal interface for TM polarization, exclusively. The sensor principle is to have a waveguide mode and the SPP close to resonance, and screen the resonance vs. angle or vs. wavelength to detect refractive index changes of the cladding. Figure 4 shows the resonance of the absorption vs. the refractive index to be probed, which indicates a resolution of some 10^{-4} for this elementary configuration.

The intention of the discussions of those examples is to punctuate that leaky wave configurations and numerical problems even in basic dielectric slab waveguides may require some in depth knowledge about the mathematical and physical backgrounds. This can be even more important if numerical algorithms are regarded, which are less strict than TMM, as it applies e.g. to finite difference approaches and mode discretization schemes, where the number of points or modes, respectively, enters the quality of the solution. And, as well, the thickness of the substrate and cladding considered within the configuration as well as the boundary conditions (BC) along the periphery of the computational window may affect both mode index and field.

3.1.2 2D Mode solvers – Effective Index Method

The basic idea of the EIM is to use a separation ansatz for the x- and the y-dependency of the field, $\Psi^{\mu\nu}(x,y)=\xi(x)\cdot\phi(y)$. For piecewise constant refractive indices the wave equation thus reads as

$$\left[\frac{\partial^2}{\partial x^2}+\frac{\partial^2}{\partial y^2}+k_0^2\varepsilon(x,y)\right]\Psi^{\mu\nu}(x,y)=\beta_{\mu\nu}^2\,\Psi^{\mu\nu}(x,y), \tag{9}$$

which leads to the conditional equation

$$\left[\frac{\partial^2}{\partial y^2}+k_0^2\varepsilon(x,y)\right]\phi=\beta_\nu^2\,\phi\quad. \tag{10}$$

The x-dependent separation constant β_ν is the y-component of the wave-vector, and acts as an effective index profile, cf. figure 5,

$$\left[\frac{\partial^2}{\partial x^2}+\beta_\nu^2\right]\xi=\left[\frac{\partial^2}{\partial x^2}+k_0^2\,n_{eff}^2(x)\right]\xi=\beta_{\mu\nu}^2\xi\quad. \tag{11}$$

Thus, in both transversal directions a 1D-mode search, cf. Eq.(7), does apply subsequently, which is easily performed by TMM. From Fig. 5 it is intuitively clear that TE and TM is reversed for the mode calculations in y and x. Taking this into account, polarization effects are incorporated into EIM to some extent.

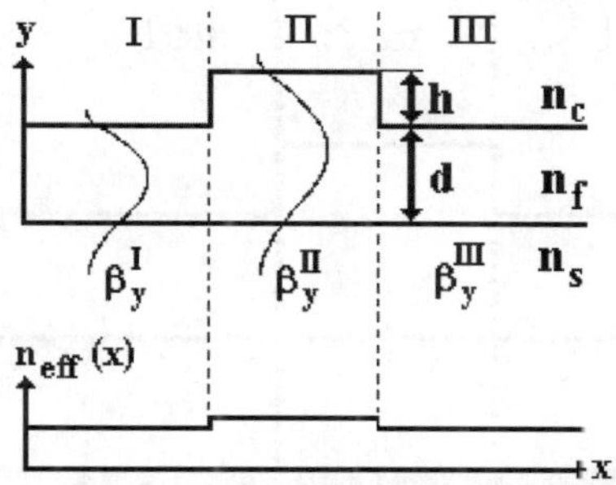

Figure 5. Rib waveguide and corresponding EIM-scheme.

What makes EIM susceptible to be mistaken are configurations with regions, where no guided mode can be determined. If more than an estimate of the propagation constant is required, e.g. in order to calculate the half-beat length of a buried channel waveguide directional coupler configuration from the mode indices of the symmetric and anti-symmetric supermodes, EIM should be superseded by a more accurate method.

The reduction of the mode search in the step-like or smoothly varying two-dimensional index profile to two consecutive inexpansive calculations for a one-dimensional index profile is the main advantage of EIM, which makes it a 'work horse' for limited index contrast. But, what is always to be considered is the fact, that the separation ansatz used is an approximate one only. If highly precise mode indices are essential e.g. in the design of interferometric devices, more rigorous calculations are inevitable. Nevertheless, a useful first guess for effective mode indices is readily obtained by EIM for various configurations. And, what is to be appreciated explicitly, the reduction of dimension can speed up propagation schemes by orders of magnitudes. This may greatly help with the design of area-intensive waveguide circuitry.

3.1.3 2D Mode solvers - Wave Matching Method

The primary ansatz of the WMM[7], which is familiar as the transversal resonance method in the microwave community, is to enclose the cross-section of the waveguide between horizontal metallic walls, figure 6.

By this, in all the different vertical slices complete discrete sets of 1D-TE- and TM-modes with predefined maximum order can easily be obtained semi-analytically.

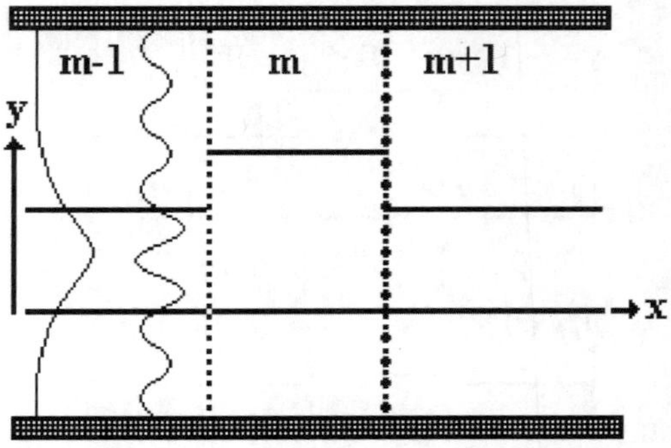

Figure 6. Sketch of WMM-geometry, matching line indicated by bullets.

The field in each vertical slice m thus can be represented by a weighted sum of products of y-dependent functions, the slab modes for each polarization (2 sketched within slice m-1), and corresponding x-dependent functions,

$$\varphi^{(m)} = \sum_{k=1}^{\infty} u_k^{(m)}(x)\ \varphi_k^{(m)}(y) \quad . \tag{12}$$

If a specific k_z -value is used as a trial value for a 2D-waveguide-mode, the x-component of the wave-vector which applies to slab-mode k in slice m follows from k_k ,

$$k_{xk}^{(m)} = \sqrt{k_k^{(m)\,2} - k_z^2} \quad . \tag{13}$$

Thus, the x-dependency $\sim \exp(i(k_x x))$ of the field, Eq.(12), and its derivative is given within the slice in terms of (complex) sine and cosine-functions with argument $(k_x x)$, which relates the fields across any slice. Field continuity determines the coupling of modal amplitudes at slice interfaces, which is taken into account by means of appropriate coupling matrices – this resembles TMM above. By applying appropriate BC which may either be magnetic, electric or even PML[53] ones at the left and at the right of the configuration, mode matching at an inner interface establishes a transverse resonance condition, which is fulfilled for appropriate k_z , i.e. for the effective refractive indices of guided modes.

WMM represents a stable mode solving technique, where, similar to TMM, continuous geometrical profiles and gradient index media require a staircase-approximation. Accuracy is affected if boundaries are too close to the guiding structure, or if the number of 1D modes is too low, which both is easy to check. So, WMM is a stable and accurate mode-solving technique. It

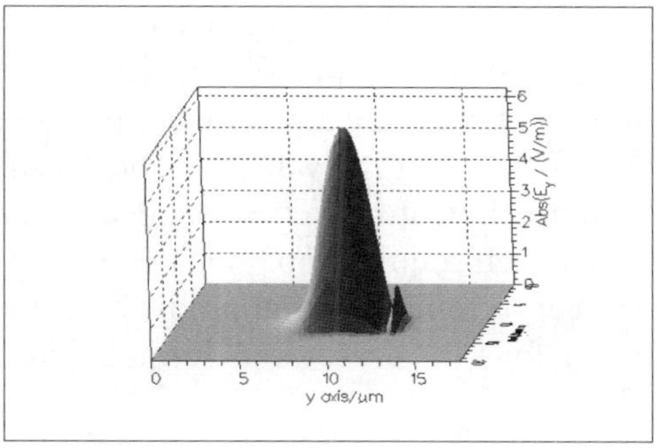

Figure 7. SPP rib waveguide configuration, E_y mode field.

can be applied to lossy configurations, and anisotropic material properties can be considered to some extent, too. One noteworthy peculiarity is the fact that the matching line should be situated at x-positions where non-negligible field components of the modes of interest can be expected, i.e. a zero would hide a solution.

As an example, one and the same layer sequence of the SPP-configuration is reused, now in a rib configuration of 5μm width, figure 7. For a surrounding medium of $n_a = 1.35$, with 250 1D-modes and an appropriate computational domain an effective refractive index of $n_{eff} = 1.4575628 + i\ 2.899\ 10^{-5}$ is found.

3.1.4 2D Mode solvers – Finite Difference Methods

FDM is one of the methods where the whole problem is fully discretized in both transversal coordinates and no suppositions concerning a special geometry are made. For a standard uniform mesh grid-points (x_i , y_k), are located at x_i =(i-1)Δx and y_i =(k-1)Δy, respectively. Derivatives are approximated by finite differences by e.g.

$$\frac{\partial^2}{\partial x^2} H^x_{i,k} = \frac{H^x_{i+1,k} - 2H^x_{i,k} + H^x_{i-1,k}}{\Delta x^2} \tag{14}$$

for the individual field components, and the Laplace-operator in (1b) thus gives the typical five-point-stencil, see figure 8. For the $\vec{H}$ -field all field components are continuous along interfaces, which makes the *H*-formulation the preferred one for various numerical approaches. With piecewise constant refractive indices from equation (1b) the 2D-Helmholtz-equation,

$$\left\lfloor \Delta_t + k_0^2\ \hat{\varepsilon} - \hat{\beta}^2 \right\rfloor \vec{H}_t = 0 \quad , \tag{15}$$

follows for the transversal magnetic field. With the divergence equation $\vec{\nabla} \cdot \vec{H} = 0$ and with $\vec{\nabla} \times \vec{H} = i\omega\, \varepsilon_0 \varepsilon\, \vec{E}$, z-components and respective boundary conditions at any interface can be expressed in terms of the transversal *H*-field, and vectorial modes of isotropic and an-isotropic configurations can be accessed. The FD-discretization of the wave equation results in an algebraic equation system whose coefficient matrix is sparse and banded, and, in case of a scalar approximation, self-adjoint, too.

Compared to the scalar or semi-vectorial problem the matrix-dimension is increased by a factor of four for a vectorial formulation, and the matrix is non-Hermitean,

$$\beta^2 \begin{pmatrix} \overline{H}_x \\ \overline{H}_y \end{pmatrix} = \begin{pmatrix} \overline{\overline{D}}_{xx} & \overline{\overline{D}}_{xy} \\ \overline{\overline{D}}_{yx} & \overline{\overline{D}}_{yy} \end{pmatrix} \begin{pmatrix} \overline{H}_x \\ \overline{H}_y \end{pmatrix} \quad . \tag{16}$$

The number of grid-points by far exceeds the number of physical meaningful modes, i.e. iterative matrix solvers can be used advantageously, and thus the numerical burden is kept within reasonable limits.

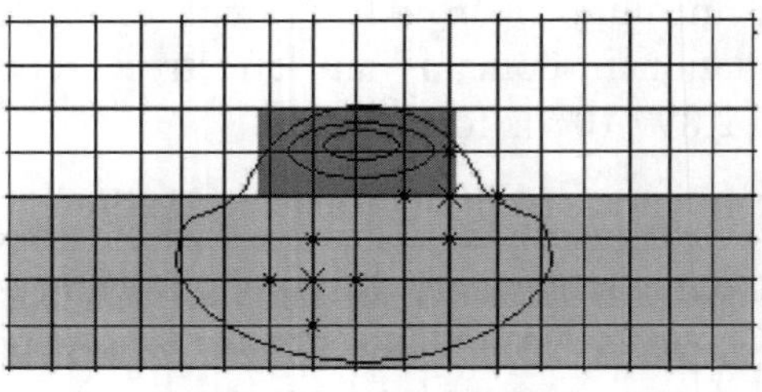

Figure 8. Finite difference grid with 5-point stencil.

Usually, after the geometry of a problem is defined, an appropriate spatial resolution is chosen, and some accuracy requirements for eigenvalues and fields are defined as well. But, it should be noticed explicitly that for a fixed grid the accuracy requirement imposed on a solver routine is a measure of its convergence rate, i.e. of accuracy of the numerical solution only. If the discretized model does not image the physical problem correctly, even a perfect numerical solution is not the solution of the physical problem. This includes, similar to WMM, the potential influence of the dimension of the computational window as sketched in Figure 8. If the decay of the mode field is too slow, mode field and effective refractive index will be affected by the window and the BC applied there. Especially near cut-off this would require a massively increased transverse cross-section for conventional BC, von Neumann or Dirichlet ones, which would dramatically increase the workload. If transparent BCs are incorporated into the solver routine[54], the overall effort is reduced considerably. Although PML is employed for mode solvers successfully as well, there is still some doubt concerning its accuracy in case of leaky and evanescent modes[55].

What is to be noted with respect to accuracy, too, is the fact that a refined discretization gives a better approximation to the physical problem. But, excessive run-time increase and accumulating round-off errors limit this approach. If the $\bar{\nabla}_t\,(\ln\hat{\varepsilon})$ term in Eq.(1) is discretized directly, this internally requires to smooth the discontinuities of the dielectric constant at interfaces to prevent from indefinite expressions. This, of course changes the geometry under consideration, again the more the coarser the mesh is. If Eq.(15) is discretized with implicit consideration of boundary conditions, figure 8 demonstrates intuitively that material interfaces affect the inherent accuracy of the FD-approximation. Improved finite-difference equations operating at a nine-point stencil have been reported which greatly improve accuracy whereas a moderate total effort in mesh density is retained[56,57]. What should be noted here is the fact that similar improvements have been reported for boundary element approaches recently[58].

Typically, problems may occur with highly degenerate and closely spaced modes, which may be hard to resolve. In the first case, independent of the mode solving scheme, a slight break of symmetry usually helps, in the latter case the design engineer is well advised to use a general purpose mode solver where the initial guess for mode field and/or mode index can be accessed to direct the mode search.

For FD schemes, especially if constant grid spacing is used, SPP-configurations can get problematic. A metallic layer of say 20 nm thickness enforces a resolution of a few nm only, thus the number of transverse grid points can get excessive if fibre-matched waveguides determine the overall computational window. Multi-grid algorithms[59] reduce the cost of mode solving advantageously[60].

3.1.5 2D Mode solvers – Finite Element Methods

Similar to FD- in FEM-schemes the cross-section under consideration is divided in a set of little areas, rectangles or triangles, preferentially. Now, within the finite elements, the field is defined not only by nodal amplitudes, but expressed within the whole element in terms of low-order polynomial functions. If e.g. the full magnetic field is considered, Maxwells curl equations give

$$rot\left(\varepsilon^{-1}\, rot\,\vec{H}\right) - k_0^{\,2}\vec{H} = 0\,. \qquad (17)$$

In case of real ε with arbitrary anisotropy this is equivalent to the functional

$$F(\vec{H}) = \int_{\Omega} \left(\left(rot\, \vec{H}^* \right) \left(\varepsilon^{-1}\, rot\, \vec{H} \right) - k_0^2\, \vec{H}\, \vec{H}^* \right) d\Omega \ . \tag{18}$$

The variation of Eq.(18) results in a generalized eigenvalue problem,

$$\overline{\overline{A}}\, \overline{H} - k_0^2 \overline{\overline{B}}\, \overline{H} = \overline{0} \ , \tag{19}$$

with solutions for $\overline{H}$ which correspond to the vectorial mode fields. Spurious modes occur, if $\vec{\nabla} \cdot \vec{H} = 0$ is not considered explicitly. The divergence condition can be taken into account explicitly by adding a penalty-term[61] to Eq.(18), or by the use of special finite elements, so called edge- or Whitney elements[62], which are inherently free of divergence. Formulations restricted to the transversal field $\vec{H}_t$ result in fewer nodal amplitudes, but a matrix inversion is needed to eliminate H_z and matrix sparsity of the eigenvalue problem is violated[5].

Recent mathematical methods combine fully vectorial formulations

$$\begin{bmatrix} \nabla_\perp \times \varepsilon^{-1} \nabla_\perp \times + k_z^2 \varepsilon^{-1} & -ik_z \varepsilon^{-1} \nabla_\perp \\ -ik_z \nabla_\perp \cdot \varepsilon^{-1} & -\nabla_\perp \cdot \varepsilon^{-1} \nabla_\perp \end{bmatrix} \begin{bmatrix} \vec{H}_\perp \\ H_z \end{bmatrix} = \omega^2 \mu \begin{bmatrix} \vec{H}_\perp \\ H_z \end{bmatrix} \tag{20}$$

which result in a quadratic eigenvalue problem for k_z, with mesh refinement and reliable error estimation techniques[9,63]. With appropriate BC and well adapted solver routines, complex problems can be tackled. Coming back to the SPP-configuration of figure 7, with JCMWave[64] an effective mode index of β = 1.4575627 + i 2.896 10^{-5} results with an assured accuracy of 10^{-6}. The comparison with the result of the WMM shows excellent coincidence, and the solution can be considered trustworthy.

What is to be noticed, finally, is the fact that different FEM-schemes may impose restrictions to the dielectric constant, e.g. they may require loss-less materials or a specific structure of the ε-tensor. Sometimes, FEM-problems are posed the way that eigen-frequencies for specified k_z are computed, which is less common to the optical design process. Nevertheless, a special advantage of FE-based grids is the fact, that almost any complicated geometry can be included easily into standard triangulation schemes.

3.2 Field Propagation Schemes

For z-invariant structures the spatial field evolution is simply ruled by

$$H(x,y,z)=\sum a_{\mu\nu}^{+}\vec{H}_{\mu\nu}^{+}(x,y)\exp\left(i\beta_{\mu\nu}\,z\right) \tag{21}$$

if the modal amplitudes are known at $z=0$, and only forward propagating fields are considered for the moment. Discontinuities in z perturb the unidirectional propagation and cause scattering in forward and backward propagating modes, and modal amplitudes for reflected and transmitted waves are related by means of the continuity requirements of the tangential fields[65]. Usually, radiation modes are necessary to model waveguide discontinuities correctly, and evanescent modes may have considerable influence, too. Thus, the calculation of adequate mode-sets can be quite expansive for 2D waveguide-cross-sections. But, if only a few modes are relevant in a specific device like a multi-mode-interference (MMI) coupler, MoL-BPM[66] and eigenmode-expansion[67] (EME) are very efficient for device analysis, simply due to the fact that in each z-invariant section propagation is just a complex factor applied to the individual amplitude. If the number of different sections is limited, even a periodic section sequence can be incorporated into the scattering matrix evaluation with reasonable effort. What is to be pointed out explicitly again is the fact that EME is bi-directional in nature, which allows tackling waveguide resonator problems, where e.g. BPM fails.

Probably most popular for the modelling of field propagation in IO-design is BPM, which applies to z-variant geometries, too. The method is derived from the Helmholtz-equation, here for brevity depicted in scalar notation for one transverse direction, only,

$$\left[\frac{\partial^2}{\partial x^2}+\frac{\partial^2}{\partial z^2}+k_0^2 n^2(x,z)\right]E=0 \quad . \tag{22}$$

The common exponential ansatz $E \sim \exp(i\,k_0\,n_0)$ now is used with a mean or reference refractive index n_0, and Eq.(22) is transformed to

$$\frac{\partial}{\partial z}E=\pm i\left(\sqrt{\underline{\underline{P}}+k_0^2 n_0^2}-k_0 n_0\right)E\,. \tag{23}$$

This first order differential equation now governs the evolution of an initial field. For a finite step length the propagation is approximated by

$$E(z_0+\Delta z)=\exp\left(i\left\lfloor \underline{\underline{O}}\right\rfloor\Delta z\right)E(z_0) \tag{24}$$

From Eq.(23) the main limitation of BPM is apparent – one algebraic sign of the root which corresponds to one propagation direction has to be chosen. To evaluate Eq.(23) the square root is expanded, which in first order,

$$\frac{\partial}{\partial z}E = \frac{1}{2ik_0 n_0}\left[\frac{\partial^2}{\partial x^2} + k_0^2\left(n^2(x,z) - n_0^2\right)\right]E, \tag{25}$$

gives the paraxial approximation of the wave equation, i.e. the Fresnel-equation. This approximation is valid only for limited index contrast and limited angular spectrum, of course. Quite different solution techniques can be applied to Eq.(25), where Fast-Fourier-transform (FFT) based split-step methods[2] have been the first, historically. In order to model high index contrast appropriately, higher order expansions of Eq.(23) are required. Adequate techniques for wide-angle BPM have been developed for FFT- and FD-based schemes[68,69].

Appropriate propagation step length, accuracy of the propagator expansion, refractive index contrast, and reference refractive index are interrelated, thus BPM-results should be validated for different step length. The spatial sampling in the transversal dimensions sensibly affects accuracy, too. All the relevant geometrical aspects of the device must be reflected by the sampling sufficiently. Whereas regular grids are used for FFT- and most of the FD-based BPM implementations, FEM-schemes offer more flexibility in grid definition, but at the price of a somewhat more intricate code. Semi- and full-vectorial analysis in 3D is state of the art, as well as the consideration of lossy, anisotropic or nonlinear media[70-72].

The extension of the classical BPM towards bi-directional propagation is of special interest for long[73]. Combinations with EME[74] and e.g. complex operators representations[75], have been tried for this.

Within standard BPM-schemes almost any geometry *n(x,y,z)* can be considered except for abrupt z-dependent changes of the refractive index. Fortunately, IO-devices usually are intended to have low loss, good transmission and low back-reflection, which is congruent with the above limitation. Thus, BPM got such popular over the years

One special topic for field propagation techniques in general is the minimization of the effect of the transversal boundaries. Uncared, they correspond to abrupt changes of the refractive index distribution, and back-reflections from the boundary into the computational domain do occur. After the obvious ansatz of absorbing BC, TBC[54] and PML[53] indicate the major improvements so far, which eliminate the problem almost completely.

A rigorous treatment of field propagation is accessible for IO-design, too, when the FDTD-method[76] is applied to optical waveguide elements like micro-cavity resonators[77]. Similar to Eq.(23) Maxwell's curl equations

$$\frac{\partial}{\partial t}\mu_0 \vec{H} = -\nabla \times \vec{E} \ , \ \frac{\partial}{\partial t}\varepsilon_0 \varepsilon \, E = \nabla \times \vec{H} \tag{26}$$

constitute an initial value problem in time, which can be solved directly in a finite difference scheme[4]. Now, FD- discretization applies to all spatial coordinates and time as well. Thus, the field evolution in time can be modelled in its full complexity, i.e. continuous wave and pulse propagation are accessible, and dispersive and nonlinear material properties can be included in the analysis by adequate models for frequency- and field-dependent dielectric function.

The main limitation of FDTD is the need for computer memory and run-time, dominantly. The Courant criterion limits the time step of FDTD in order to ensure numerical stability,

$$\Delta t \leq 1 / \left(c \sqrt{\left(1/\Delta x^2\right) + \left(1/\Delta y^2\right) + \left(1/\Delta z^2\right)} \right) \quad . \tag{27}$$

So, with a spatial resolution of at least λ/20, which applies to the wavelength in the medium, roughly 50000 data are required for the field per cubic wavelength, and time steps are ultimately in the sub-fs range for optical frequencies.

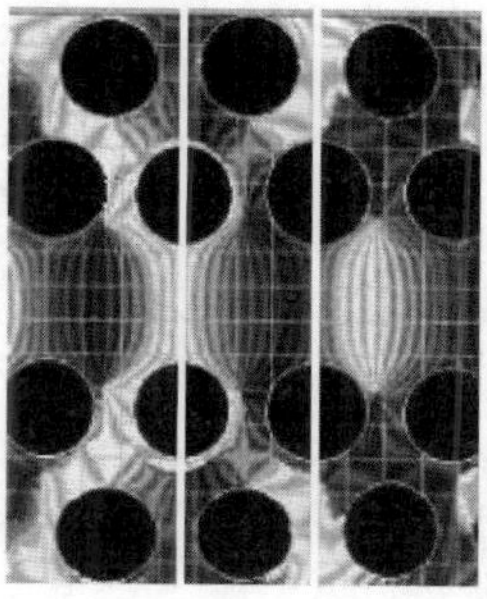

Figure 9. Phase plot for PhC-waveguide, vertical white lines indicate unit cell extension.

If applicable, EIM may help to reduce dimensionality, which decreases the numerical burden by more than an order of magnitude, definitely. Evidently, the use of FDTD in integrated optics design is restricted to small device structures, only. But, FDTD-calculations may offer a deeper insight into details of the spatial field distribution of linear and nonlinear photonic devices which is hardly accessible by other methods.

E.g., figure 9 demonstrates the variation of the phase front along a unit cell of a W1-PhC-waveguide, which is formed by just one missing row of air holes in a silica embedded silicon film. This for instance is an indication that a direct butt-coupling from and to photonic crystal waveguides would be affected by the relative position of the PhC-interface relative to the unit-cell.

3.3 General Aspects of System Design

What has been considered so far are issues of numerics for waveguide analysis, dominantly, which have to be related to the applications which are relevant in a specific IO-design process. E.g. for an adiabatic power splitter in a fibre adapted waveguide geometry BPM is the appropriate tool, and a scalar approach is sufficient for isotropic media, at least. If, at the other hand, an interferometric device is considered, an intimate knowledge of the effective propagation constants is advisable. And, since any deviation from a quadratic core geometry breaks polarisation degeneracy even for isotropic media, semi- or fully-vectorial analysis is mandatory to evaluate polarization dependent device operation. A detailed eigenmode analysis additionally helps to select an adequate field propagation scheme. If EIM looks reasonable, a punctual comparison of the results of a 2D and a 3D propagation model should assure a 2D approach, provided that similar spatial resolutions fit to the computers memory.

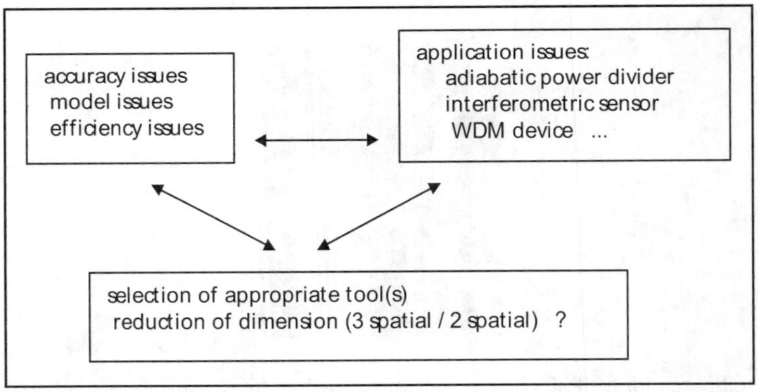

Figure 10. Principal considerations concerning application of design tools.

So, it is one of the basic responsibilities of the design engineer to analyse the requirements and issues of the specific design task in order to facilitate an efficient operation with a sufficient accuracy, figure 10. If e.g. the design considers a WDM system, say an optical-add-drop-(de-)multiplexer

(OADM), there the step from waveguide design to system design is to be included, necessarily, and a change of the abstraction level is required.

If a multitude of individual waveguides is contained in a complex circuitry, then parametric models for basic entities are preferentially used, which apply to the optical signal level, i.e. to complex (mode) amplitudes.

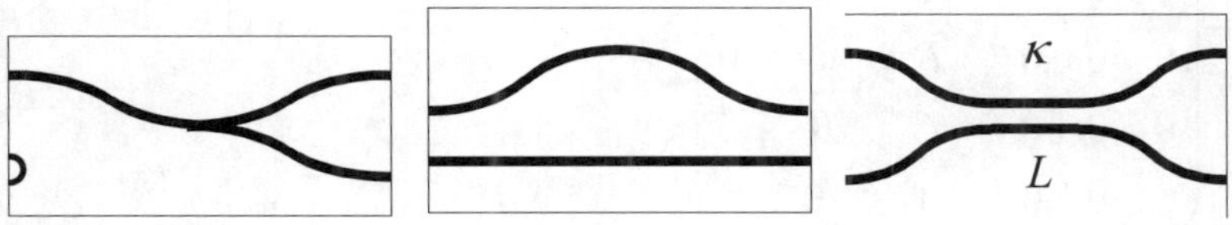

Figure 11. 4-port devices: power splitter, relative phase shifter, directional coupler.

The characteristics of those basic entities like power splitters, combiners, directional couplers etc. can be related to some basic waveguide properties, and the 4-port devices are described by 2 x 2 matrices, Eqs(28),

$$U_{splitter} = \frac{1}{\sqrt{X+1}} \begin{pmatrix} \sqrt{X} & 0 \\ 1 & 0 \end{pmatrix}$$

$$U_{phase\ shift} = \begin{pmatrix} \exp(i\Delta\varphi/2) & 0 \\ 0 & \exp(-i\Delta\varphi/2) \end{pmatrix} \qquad (28)$$

$$U_{DC} = \begin{pmatrix} B & C \\ -C^* & B^* \end{pmatrix} \quad .$$

2-port devices like straight waveguides, amplifiers and attenuators are simply characterised by a complex constant. A waveguide (sub-)system consisting of N waveguides, figure 12, corresponds to a N^2-port.

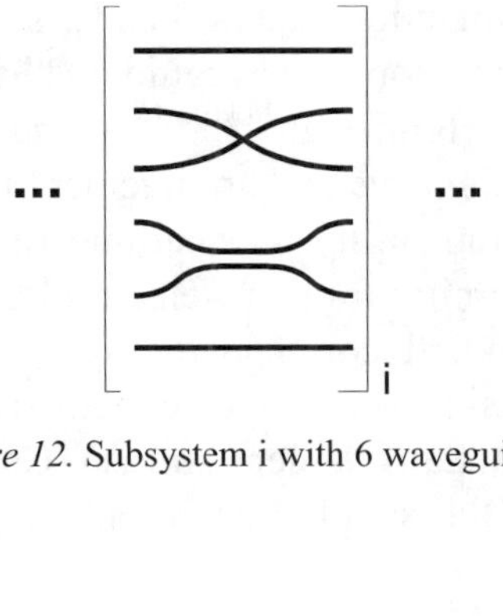

Figure 12. Subsystem i with 6 waveguide paths.

From the individual elements matrices, cf. Eqs.(28), a NxN transfer matrix for a (sub-)system can be constructed[78],

$$\underline{\underline{U}} = \begin{pmatrix} \gamma & 0 & 0 & 0 & 0 & 0 \\ 0 & 0 & i & 0 & 0 & 0 \\ 0 & i & 0 & 0 & 0 & 0 \\ 0 & 0 & 0 & B & C & 0 \\ 0 & 0 & 0 & -C^* & B^* & 0 \\ 0 & 0 & 0 & 0 & 0 & \gamma \end{pmatrix} . \tag{29}$$

If those matrices U are made unitary and are related to their mean phase shift, stacking gives the transfer matrix of the full network,

$$\underline{S_{out}} = \prod_i \underline{\underline{U_i}}\, \underline{S_{in}} , \tag{30}$$

which relates the signal at input and output sides. At this abstraction level e.g. signal modulation may be applied to any relevant waveguide path, and the systems characteristics can be analysed with PSPICE, for instance, or with any of the sophisticated tools mentioned in section 2.

4. SPECIFIC APPLICATIONS AND REQUIREMENTS

Standard IO-devices rely on just one layer containing waveguides, and very few vertically stacked layers are used for some exceptional geometries like ring resonators[79]. In the wafer plane a multitude of waveguides can occur, and the requirements on the positioning of individual sections of a waveguide path are quite strict. What helps with the layout is a library of basic elements, linear waveguides, tapers, bends, splines etc., which allow for an relatively easy composition of waveguide paths and complete devices. The commercial tools for IO-design[10,11,13,15] contain quite general CAD systems with sets of basic elements and arrangement facilities by hierarchical structures and user programmability. Concerning the individual waveguide plane, the layout is quite similar to micro-electronics, and export to formats like GDSII and DXF is standard, correspondingly.

For standard design tasks which may contain pretty high numbers of individual waveguides like star couplers and AWGs, standard solutions do exist in parametric form. But, standard solutions may fail. For DWDM the typical banana AWG design[80] works well, but for CWDM other solutions[81] have to be used or found. The AWG is a typical example where design steps at a higher abstraction level are required as mentioned above. Typically, in the AWG-design, star couplers for splitting, combining and some amplitude shaping, potentially, are treated by BPM. But the individual phase shifts in the array waveguides of the AWG are handled by corresponding complex factors for the mode amplitudes. To be correct, this requires accurate knowledge of the effective refractive index of the guided mode including potential polarization effects, naturally.

Fig. 13 shows an example where data for an irregular structured multilayer thin-film-filter (TFF) are imported into a waveguide design. A FDTD-simulation carried out in order to check the influence of the internal resonator-like structure of the TFF onto the waveguide device shows a significant transversal shift for oblique incidence. This hinders symmetrical multi-port designs, but may be used for specific WDM functionality[82].

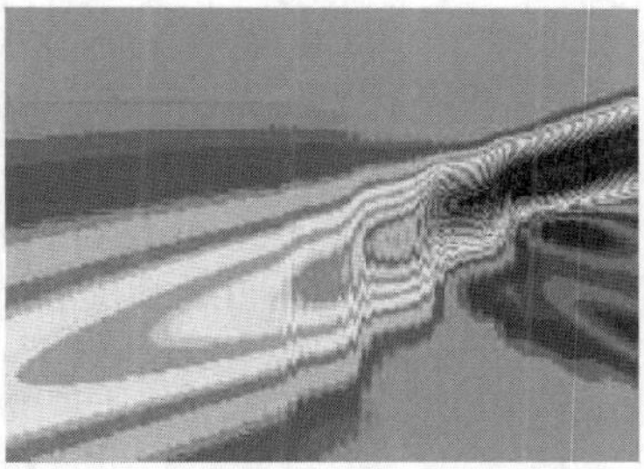

Figure 13. Waveguide mode impinging from the right onto a multi-cavity TFF.

In IO there a great variety of materials is used, and their optical constants may be affected e.g. by film deposition technologies. What is thus required is the access to data for material dispersion with relation to technological parameter as well, either as Sellmeier or related formula, or as tabulated values. Additionally, refractive indices respond to temperature, which may be intended for device operation in case of a TO-switch, or unintended in field use. The temperature dependence of the refractive index can be attributed to the individual material, simply, but the influence of heater electrodes needs special consideration. If an IO design-tool comes with inherent TO or EO capabilities, those effects are taken into account in the optical design directly.

Just another approach is to calculate e.g. the temperature distribution within a separate program like ANSYS™, figure 14, and to transfer the temperature induced index change to the IO design. If the refractive index change due to the heater can be represented in a parametric model of a

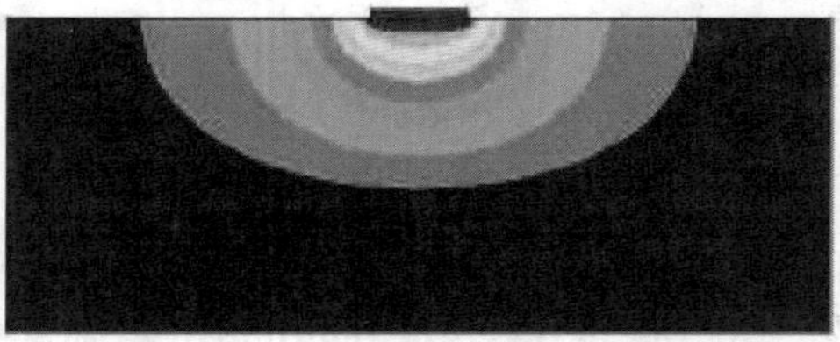

Figure 14. Temperature distribution beneath a heater electrode.

fictive waveguide type which adds the TO-induced index change to the initial waveguide layout, a great design flexibility is achieved.

Equation (31) shows an instance of a temperature profile function which incorporates the thickness of a specific polymer t_{poly}, the width of the heater electrode w_{heat}, and the heat generation rate H_v as parameter, and constants c_1 and c_2 are fitted from calculated temperature profiles,

$$T(x, y, t_{poly}, w_{heat}, H_v) = T_{heat}(t_{poly}, w_{heat}, H_v) \times \exp\left(-\left(x/(c_1 \cdot t_{poly}\right)^2\right) \cdot \exp(-y/c_2) \quad (31)$$

If the resulting refractive index profile due to the 'electrode waveguide' is coded in a DLL, this gives a very efficient z-dependent refractive index update in a BPM-run. Such an approach gains attractiveness, e.g. when quite different cross sections are necessary or intended for the modelling in the thermal and the optical domain.

In order to optimise individual structures in the IO-design the definition of parameter-loops and scanning routines is required. In combination with this, post-processing of BPM-runs to evaluate channel-specific mode amplitudes, system transmission etc. is necessary, as well, which is assisted by commercial IO-design tools, naturally. This facilitates the generation of parametric models for any optical sub-system, cf. figures 11 and 12, which at the end is a prerequisite for an efficient system design.

Design in integrated optics always is intimately connected with technology. Thus, if the experiment indicates deviations of actual and design data, clarification of facts is needed, where again parametric models ease the work. What beside signal level measurements turns out to be quite useful are detailed measurements of the refractive index distribution, e.g. by the method of the refracted near-field[83]. If the measured index profile is transferred to the IO-design tool, propagation constant, field profile and related characteristics can be analysed in detail. Based on this, design or technology parameter can be updated accordingly.

If the coupling from and to the waveguide is considered, design-tool interfaces are naturally defined via the spatially discretized optical field.

What is required, accordingly, are field import and export for propagation tools, and field export from the mode solver, too, in order to be able to evaluate overlap integrals, q.v. the subsequent section, too.

5. MICRO-OPTICAL COUPLING OF FIBRE AND PHOTONIC CRYSTAL WAVEGUIDE

The fibre-chip coupling of high numerical aperture (NA) waveguides including photonic crystal (PhC) waveguides is a demanding problem, mainly due to the apparent mismatch of mode field diameters (MFD) which can exceed a ratio of 1:20. For the necessary MFD adaptation several strategies can be tracked. Tapering of waveguide cross-sections is widely used for waveguides in III-V's[84], and this approach is used for PhC-waveguides as well[85]. If monolithic taper integration conflicts technological constraints e.g. due to the material used, an increased assembly effort for hybrid integration has to be accepted. Apt for hybrid integration are either individual 3D-tapers[87], or focussing micro-optical elements[88]. From the point of view of IO-design the coupling with micro-optical elements is of special interest, since it requires a multi-scale modelling of electromagnetic field propagation as well as the consideration of the Bloch-wave nature of the guided mode of the PhC-waveguide.

Spot size reduction of an optical beam is generally limited if focussing in air is considered due to the NA limit of unity. If the angular spectrum of the waveguide mode field comes close to or exceeds this limit, two ways can be considered. The first one is to widen the mode field by a 2D-taper in such a way that it comes close to a diameter which is available with focus formation in air. The second one is to focus in a high index lens material, and to attach the lens to the IO-chip, finally, just similar to what is achieved with a solid immersion lens[89]. The principal geometry is shown in figure 15.

The light emanating from the fibre goes through an glass wafer of appropriate thickness and is collimated by a refractive spherical or slightly aspherical surface. Next, the collimated beam enters a high speed Si-micro-lens and gets focussed. The focus is situated directly at the back surface of the wafer, thus, if the lens has close contact to the subsequent chip, the waveguide mode is efficiently excited. Whereas the field diameter in the free-space sections in glass, air, and at the entry surface into the silicon is much greater than the wavelength, close to focus it shall get as narrow as possible, intentionally.

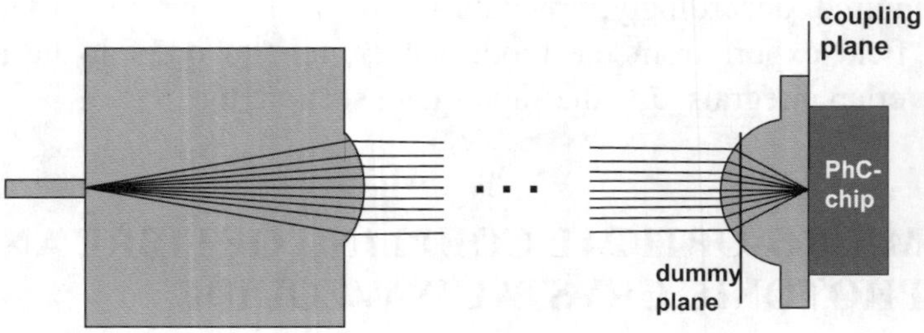

Figure 15. Fibre-chip coupling scheme.

Thus, an efficient ray-trace analysis can be used for the design of the refractive surfaces, but if the field comes close to focus wave propagation schemes are required. What is mandatory for high index contrast waveguides and PhC-waveguides in particular is a fully-vectorial field analysis just from the beginning, i.e. polarization ray-tracing is a must. The ray-trace procedure is used until an artificial dummy-plane is reached. In this plane, from the ray data the electric field is obtained by means of the intensity law of geometrical optics and the well-known relation between electric and magnetic fields[90]. Due to the fact that a plane wave description holds when the field diameter is still quite extended, the corresponding curl-equation can be evaluated without derivatives,

$$\mathrm{rot}\left(\vec{E}\right) = i\omega\mu_0\vec{H} \quad \Rightarrow \quad \vec{k}\times\vec{E} = \omega\mu_0\vec{H}\,, \tag{32}$$

taking into account the plane waves direction given by the ray-trace engine. What is to be noticed explicitly is the fact, that large longitudinal field components do occur simply due to the noticeable inclination of the outer rays towards the optical axis. To propagate the optical fields up to the focus a fully vectorial model is needed, again. In the dummy plane the field dimension is much too big to fit to a FDTD-scheme, thus another method is required. Since the distance from the dummy to the coupling plane is homogeneous, i.e. it is free-space propagation, the angular spectrum propagation[90] (ASP) can be applied. Thus, all field components in the dummy plane are Fourier-transformed, propagated, and transformed back which gives the field in the coupling plane, as depicted for the electrical field,

$$\begin{pmatrix} E_x \\ E_y \\ E_z \end{pmatrix}(x, y, z_{dum}) \ \underrightarrow{FT} \ \exp\left(i\sqrt{k^2 - k_x^2 - k_y^2}\ \Delta z \right)$$
$$\underrightarrow{FT^{-1}} \begin{pmatrix} E_x \\ E_y \\ E_z \end{pmatrix}(x, y, z_{\text{coup}}) \qquad (33)$$

Finally, FDTD may be used to model the coupling of the focal field into the PhC-waveguide, potentially with the presence of an air or glue gap. Even such a simulation procedure with adapted numerical methods for each part of the propagation requires a considerable computation time. To speed up the simulation process for system optimisation remarkably, the FDTD-simulation can be replaced by a formula for the coupling efficiency to a conventional high-index or a PhC-waveguide[91],

$$T = \frac{1}{\iint\limits_{(\text{cp})} dxdy \left[\vec{E}_{\text{Foc}}^{(t)^*} \times \vec{H}_{\text{W}}^{(t)} + \vec{E}_{\text{W}}^{(t)} \times \vec{H}_{\text{Foc}}^{(t)^*} \right]_z} \times$$
$$\left\{ \left| \iint\limits_{(\text{cp})} dxdy \left[\vec{E}_{\text{Foc}}^{(t)^*} \times \vec{H}_{\text{W}}^{(t)} + \vec{E}_{\text{W}}^{(t)} \times \vec{H}_{\text{Foc}}^{(t)^*} \right]_z \right|^2 - \left| \iint\limits_{(\text{cp})} dxdy \left[\vec{E}_{\text{Foc}}^{(t)} \times \vec{H}_{\text{W}}^{(t)} - \vec{E}_{\text{W}}^{(t)} \times \vec{H}_{\text{Foc}}^{(t)} \right]_z \right|^2 \right\} \qquad (34)$$

which goes back to orthogonality and reciprocity for Bloch-waves[92]. In Eq.(34) subscripts Foc and W indicate transversal focal and waveguide fields, respectively, and the integration includes the whole coupling plane. The fields at both sides of the coupling plane are normalized by

$$\left| \int\limits_{-\infty}^{+\infty} dx \int\limits_{-\infty}^{+\infty} dy \left[\vec{E}^{(\text{t})} \times \vec{H}^{(\text{t})^*} + \vec{E}^{(\text{t})^*} \times \vec{H}^{(\text{t})} \right]_z \right| = 1 \ , \qquad (35)$$

i.e. to unity energy flux in z-direction. Equation (34) is focused on dominating modes or fields, respectively, and doesn't take into account all possible coupling losses, e.g. due to edge modes at a PhC interface. The validity of this approximation should be checked at least for the final result of an optimisation by FDTD. Nevertheless, Eq.(34) implies the main features of the coupling process which are of relevance. The integration of the product of the focal and modal fields over the coupling plane reflects the criterion of MFD adaptation for maximum coupling efficiencies, and the cross product of electric and magnetic fields at the left and the right of the coupling plane reflects the impedance dependent coupling efficiency. Furthermore, Eq.(34) is phase sensitive, thus it takes into account potential coupling losses due to defocus or due to a nontrivial phase distribution of the Bloch mode of a PhC-waveguide. And, by the normalization chosen, a mismatch of group velocity is related to a mismatch of the absolute value of the electromagnetic fields in Eq.(34), directly.

As an example a PhC-waveguide configuration is considered, where a silicon film embedded within silica, i.e. an insulator-on-silicon-on-insulator (IOSOI) geometry, provides guiding in the vertical direction, and a missing row of air holes in the Γ–K direction of a triangular lattice ensures the guiding in the horizontal direction, figure 16.

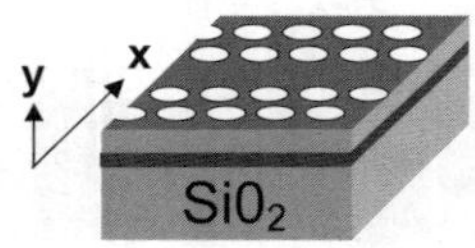

Figure 16. IOSOI PhC-waveguide geometry.

The period of the W1-PhC-waveguide under consideration amounts to $a = 530$ nm, the hole radius is $0.366a$, and the Si-film is 210 nm thick. For the operating wavelength of 1.55 μm a MFD of the fibre of 10.3 μm is assumed. The lens surfaces are described by

$$z = \frac{1}{R} \frac{r^2}{1+\sqrt{1-r^2/R^2}} + \alpha_2 r^2 + \alpha_4 r^4 \qquad (36)$$

with lens radius R and asphere parameters α_2 and α_4, whereas r is the radial coordinate in the transversal plane. For lens parameter R = -1.77 mm, α_2 = -0.46, α_4 = -2.01 for the collimating glass lens, and R = 0.13 mm, α_2 = 0.16, α_4 = 2.05 for the silicon lens a full-width-half-maximum focal spot size of about $\rho_f \sim 0.38$ μm with negligible diffraction rings is obtained for the Poynting vector in the propagation direction, S_z, figure 17.

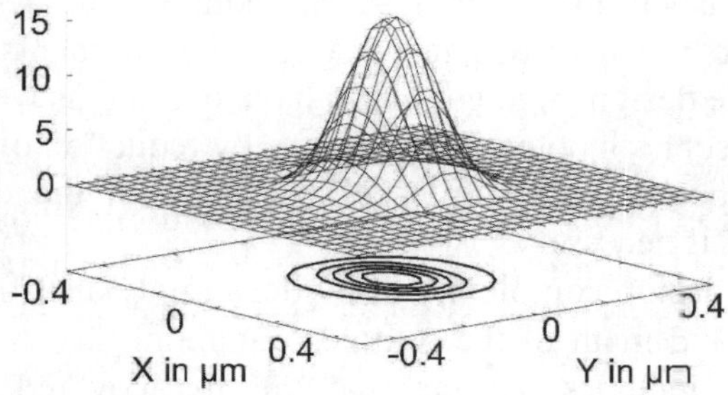

Figure 17. z-component of time-averaged Poynting vector in the focal plane.

The PhC-waveguide mode used to evaluate the generalized coupling integral, Eq.(34), is determined by an adaptive FEM-solver[93], and the resulting coupling efficiency is ~80%, if an appropriate AR-coating of the lenses is assumed. This marvellous result is double-checked by a transfer of the fully-vectorial field to FDTD shortly in front of the focus, where the transversal cross section is narrow enough so that a FDTD-model fits into a PC's memory, figure 18.

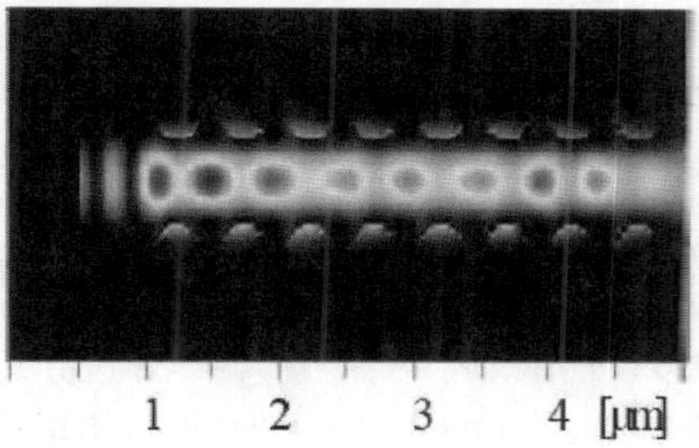

Figure 18. Coupling to PhC-waveguide, E_x , x-z-plot at half the film height.

The intensity transmitted to the waveguide can be related to the Poynting vector of the incident free-space radiation and that of the PhC-mode a few microns away from the interface, which confirms the 80% coupling efficieny.

6. CONCLUSIONS

Today, various capable software tools and bundles facilitate IO-design and the power of modern PCs allows tackling quite complex design problems. Due to the diversity of IO-applications, the diversity of customer needs and the diversity of algorithms quite different software tools do exist. In order to select and apply adequate programs for a certain design task correctly, it is indispensable to analyse the requirements due to an intended application and to translate this to capability requirements concerning the software. What the design engineer has to have in mind as well are efficiency issues, which reflect both potential speed-ups by reduction of dimensionality, and the use of specific tools for specific aspects of the design including interface routines, if necessary.

Naturally, the design complexity in IO does imply a sufficient familiarity with the physics and math of the routines commonly used. This helps, too, when the design includes aspects that are not covered by commercial software and therefore requires filling or bridging gaps with own development work.

ACKNOWLEDGEMENTS

The author wish to thank S. Burger, L. Zschiedrich and F. Schmidt from Zuse-Institut, Berlin, for stimulating discussions on mode solvers and for providing FEM-results for the SPP-configuration and PhC-waveguide modes, and S. Kudaev and D. Michaelis from Fraunhofer-IOF, Jena, for continuous communication on complex aspects of IO design.

REFERENCES

1. S. M. Jensen, The nonlinear coherent coupler, *IEEE Journal Quantum Electronics*, **QE-18** 1579-1583 (1982)
2. M. D. Feit, and J. A. Fleck, Light propagation in graded-index optical fibers, *Applied Optics* **17**, 3990-3998 (1978).
3. R. Pregla, and W. Pascher, The method of lines, in T. Itoh (ed.), *Numerical Technics for Microwave and Millimeter Wave Passive Structures*, Wiley, New York, 381-446 (1989).
4. K. S. Yee, Numerical solution of initial boundary value problems involving Maxwell's equations in isotropic media, *IEEE Transactions on Antennas and Propagation* **14**, 302-307 (1966).
5. M. Koshiba, and K. Inoue, Simple and efficient finite-element analysis of microwave and optical waveguides, *IEEE Transactions Microwave Theory & Technology* **40**, 371-377 (1992).
6. P. Lüsse, et al., Analysis of vectorial mode fields in optical waveguides by a new finite difference method, *Journal of Lightwave Technology* **12**, 487-493 (1994)
7. A. S. Sudbø, Improved formulation of the film mode matching method for mode field calculations in dielectric waveguides, *Pure and Applied Optics* **3**, 381-388 (1993).
8. M. Paulus, and O.J.F. Martin, A fully vectorial technique for scattering and propagation in three-dimensional stratified photonic structures, *Optical and Quantum Electronics* **33**, 315-325 (2001).
9. F. Schmidt, et al., Adaptive multigrid methods for the vectorial Maxwell eigenvalue

problem for optical waveguide design, in W. Jäger et al. (eds.), *Mathematics - Key Technology for the Future: Joint Problems between Universities and Industry*, Springer, 270-292 (2003).

10. Apollo Photonics; http://www.apollophoton.com/apollo/
11. Concept to Volume B.V. (C2V); http://www.c2v.nl/
12. JCMWave GmbH; http://www.jcmwave.de/
13. Optiwave Systems; http://www.optiwave.com/
14. Photon Design; http://www.photond.com/
15. RSOFT Design Group; http://www.rsoftdesign.com/
16. Virtual Photonics; http://www.vpiphotonics.com/photonics_home.php
17. PHOTOSS; http://www.lenge.de/english/
18. Crosslight Softwrae Inc.; http://www.crosslight.com/Product_Overview/prod_overv.html#PICS3D
19. ISE; www.ise.ch (until 2004), now http://www.synopsys.com/products/acmgr/ise/dessis_ds.html
20. ANSYS; http://www.ansys.com/products/multiphysics-hf-emag.asp
21. CFDRC; http://www.cfdrc.com/bizareas/microelec/ph_oe_dev/photonics.html#fullwave
22. Grating Solver; http://www.gsolver.com/
23. Unigit; http://www.unigit.com/
24. Essential McLeod; http://www.thinfilmcenter.com/GMacleod.asp
25. Film Wizard; http://www.sci-soft.com/Film%20Wizard.htm
26. TRAMAX; http://www.photonik.uni-jena.de/, select software, select TRAMAX
27. Max-1 ; http://www.wiley.co.uk/max-1/overview.html
28. CODE V ; http://www.opticalres.com/products_f.html#codev
29. ZEMAX ; http://www.zemax.com/zemax/index.html
30. GLAD ; http://www.aor.com/
31. VirtualLab; http://www.lighttrans.com/
32. DesignWorkshop-2000; http://www.designw.com/PDA-Overview.php
33. MSG-6; http://www.ipms.fraunhofer.de/products/index.shtml?products_07.shtml select MSG
34. CUDOS-MOF; http://www.physics.usyd.edu.au/cudos/mofsoftware/index.html
35. S. G. Johnson, and J. D. Joannopoulos, Block-iterative frequency-domain methods for Maxwell's equations in a planewave basis, *Optics Express* **8**, 173-190 (2001); http://ab-initio.mit.edu/mpb/download.html
36. http://wwwhome.math.utwente.nl/~hammer/Wmm_Manual/index.html
37. HelmPole; http://www.zib.de/nano-optics/hiphocs/
38. CAMFR; http://camfr.sourceforge.net/
39. PhotonicCAD; http://www.dimes.tudelft.nl/1998/s1/c5/c5-6-photoniccad.html
40. BMS3D; http://www.ure.cas.cz/~prkna/msolvers/bms3d.php
41. J. Moloney, et al., AMR FDTD solver for nanophotonic and plasmonic applications, Proc. SPIE **5451**, 97-104 (2004).

42. Photeon; http://www.photeon.com/

43. R. Moosburger, et al., Shaping the digital optical switch using evolution strategies, *IEEE Photonics Technology Letters* **9**, 1184-1186 (1997).

44. R. März, *Integrated Optics: Design and Modeling*, Ch. 4.9, Artech House (1995).

45. P. C. Kendall, et al., Theory for calculating approximate values for the propagation constants of an optical rib waveguide by weighting the refractive indices, *IEE Proceedings A* **134**, 699-702 (1987).

46. W. Huang, H. A. Haus, and H. N. Yoon, Ananlysis of buried-channel waveguides and couplers: scalar solution and polarization correction, *Journal of Lightwave Technology* **8**, 642-648 (1990).

47. C. J. Smartt, et al., Exact polarised rib waveguide analysis, *Electronics Letters* **30**, 1127-1128 (1994).

48. C. Vasallo, Optical mode solvers, *Optical and Quantum Electronics* **29**, 95-114 (1997).

49. R. Smith, S. N. Houde-Walter, and G.W. Forbes, Mode determination for planar waveguides using the four-sheeted dispersion relation, *IEEE Journal Quantum Electronics* **28**, 1520-1526 (1992).

50. C. Wächter, and K. Hehl, General treatment of slab waveguides including lossy materials and arbitrary refractive index profiles, *Physica Status Solidi* **102**, 835-842 (1987).

51. T. E. Batchman, and G. M. McWright, Mode coupling between dielectric and semiconductor planar waveguides, *IEEE Journal Quantum Electronics* **QE-18**, 782-788 (1982).

52. J. M. Hammer, et al., Isolators, polarizers and other optical waveguide devices using a resonant-layer effect, *Journal Lightwave Technology* **22**, 1754-1763 (2004).

53. J. P. Bérenger, A perfectly matched layer for the absorption of electromagnetic waves, *Journal of Computational Physics* **114**, 185-200 (1994).

54. G. R. Hadley, and R. E. Smith, Full vector waveguide modeling using an iterative finite-difference method with transparent boundary conditions, *Journal of Lightwave Technology*, **13**, 465-469 (1995).

55. T. Tischler, and W. Heinrich, The perfectly matched layer as lateral boundary in finite difference transmission line analysis, *Int. Microwave Symposium Digest* **1**, 121-124 (2000).

56. G. R. Hadley, High-accuracy finite-difference equations for dielectric waveguide analysis I: uniform regions and dielectric interfaces, *Journal of Lightwave Technology* **20**, 1210-1218 (2002).

57. G. R. Hadley, High-accuracy finite-difference equations for dielectric waveguide analysis II: dielectric corners, *Journal of Lightwave Technology* **20**, 1219-1231 (2002).

58. T. Lu, and D. Yevick, Comparative Evaluation of a Novel Series Approximation for Electromagnetic Fields at Dielectric Corners With Boundary Element Method Applications, *Journal of Lightwave Technology* **22**, 1426-1432 (2004).

59. W. Hackbusch, *Multi-grid methods and applications*, Springer, Berlin (1985)

60. P. Lüsse, K. Ramm, H.-G. Unger and J. Schüle, Comparison of a vectorial and new

semivectorial finite difference approach for optical waveguides, *Optical and Quantum Electronics* **29**, 115-120 (1997).

61. B. M. A. Rahman, and J. B. Davies, Penalty function improvement of waveguide solution by finite element, *IEEE Transactions Microwave Theory and Techniques* **32**, 922-928 (1984).
62. A. Bossavit, Whitney forms: a class of finite elements for three-dimensional computations in electromagnetism, *IEE Proceedings A* **135**, 493-500 (1988) .
63. F. Schmidt, and L. Zschiedrich, Adaptive numerical methods for problems of integrated optics, *Proc. SPIE* **4987**, 83-94 (2003).
64. L. Zschiedrich, JCMWave GmbH, Putzbrunn, Germany; http://www.jcmwave.de/
65. W. Biehlig, et al., Light propagation in a planar dielectric slab waveguide with step discontinuities, *Optical and Quantum Electronics* **18**, 219-228 (1986).
66. R. Pregla, MoL-BPM Method of Lines Based Beam Propagation Method, in W. P. Huang (Ed.), *Methods for Modeling and Simulation of Guided-Wave Optoelectronic Devices, Progress in Electromagnetic Research* **11** (Elsevier, New York, 1995).
67. D. F. G. Gallagher, and T. Felici, Eigenmode expansion methods for optical propagation in photonics – pros and cons, *Proc. SPIE* **4987**, 69-82 (2003) .
68. L. Thylen, and C. M. Lee, Beam propagation method based on matrix diagonalisation, *Journal of the Optical Society of America A* **9**, 142-146 (1992).
69. G. R. Hadley, Wide angle beam propagation using Padé approximant operators, *Optics Letters* **17**, 1426-1428 (1992).
70. H. Hoekstra, On beam propagation methods for modelling in integrated optics, *Optical and Quantum Electronics* **29**, 157-171 (1997).
71. J. P. da Silva, H. E. Hernández-Figueroa, and A. M. F. Frasson, Improved vectorial finite-element BPM analysis for transverse anisotropic media, *Journal of Lightwave Technology* **21**, 567-576 (2003).
72. T. Fujisawa, and M. Koshiba, Full-vector finite-element beam propagation method for three-dimensional nonlinear optical waveguides, *Journal of Lightwave Technology* **20**, 1876-1884 (2002).
73. P. Kaczmarski, and P. Lagasse, Bidirectional beam propagation method, *Electronics Letters* **24**, 675-676 (1988).
74. K. Hayashi, M. Koshiba, Y. Tsuji, S. Yoneta and R. Kaji, Combination of beam propagation method and mode expansion propagation method for bidirectional optical beam propagation analysis, *Journal of Lightwave Technology* **16**, 2040-2045 (1998).
75. H. Rao, M. J. Steel, R. Scarmozzino and R. M. Osgood, Jr., Complex propagators for evanescent waves in bidirectional beam propagation method, *Journal of Lightwave Technology* **18**, 1155-1160 (2000).
76. K. L. Shlager, and J. B. Schneider, A survey of the finite-difference time domain literature, in A. Taflove (Ed.), *Advances in computational electrodynamics: the finite difference time domain method* (Artech House, 1998), pp. 1- 62.
77. S. C. Hagness, Microcavity resonators, *ibid.*, pp. 561-609.
78. R. März, *Integrated Optics: Design and Modeling*, Ch. 6.5 (Artech House, 1995).
79. S. Suzuki, Y. Kokubun, M. Nakazawa, T. Yamamoto and S. T. Chu, Ultrashort

optical pulse transmission characteristics of vertically coupled microring resonator add/drop filter, *Journal of Lightwave Technology* **19**, 266-271 (2001).

80. M. K. Smit, Integrated Optics in silicon-based aluminium oxide, *PhD Thesis*, Ch. 6 (Delft University of Technology, 1991).

81. R. Adar, et al., Broadband array multiplexers made with silica waveguides on silicon, *Journal of Lightwave Technology* **11**, 212-219 (1993).

82. M. Gerken, and D. A. B. Miller, Multilayer thin-film stacks with step-like spatial beam shifting, *Journal Lightwave Technology* **22**, 612-618 (2004).

83. K. I. White, Practical application of the refracted-near-field technique for the measurement of optical fibre refractive index profiles, *Optic and Quantum Electronics* **11**, 185-196 (1979).

84. I. Moerman, P. P. Van Daele, and P. M .Demeester, A review on fabrication technologies for the monolithic integration of tapers with III-V semiconductor devices, *IEEE Journal of Selected Topics in Quantum Electronics* **3**, 1308-1320 (1997).

85. Th. D. Happ, M. Kamp, and A. Forchel, Photonic crystal tapers for ultracompact mode conversion, *Optics Letters* **26**, 1102-1104 (2001).

86. N. Moll, and G.-L. Bona, Comparison of three-dimensional photonic crystal slab waveguides with two-dimensional photonic crystal waveguides: efficient butt coupling into these photonic crystal waveguides, *Journal of Applied Physics* **93**, 4986-4991 (2003).

87. J. J. Fijol, E. E. Fike, P. B. Keating, D. Gilbody, J. Leblanc, S. A. Jacobson, W. J. Kessler, M. B. Frish, Fabrication of silicon-on-insulator adiabatic tapers for low loss optical interconnection of photonic devices, *Proc. SPIE* **4997**, 157-170 (2003).

88. C. Wächter, et al., Fibre-chip coupling with silicon micro-lenses, to appear in *Proc. SPIE* **5728** (2005).

89. I. Ichimura, S. Hayashi, and G. S. Kino, High–density optical recording using a solid immersion lens, *Applied Optics* **36**, 4339-4348 (1997).

90. M. Born, and E. Wolf, *Principles of Optics*, (Cambridge University Press, Cambridge, 1999).

91. D. Michaelis, U. Peschel, C. Wächter, A. Bräuer, Coupling coefficients of photonic crystal waveguides, *Proc. SPIE* **4987**, 114-125 (2003).

92. D. Michaelis, et al., "Reciprocity theorem and perturbation theory for photonic crystal waveguides", Phys. Rev. E **68**, 065601(R) 1-4 (2003)

93. S. Burger, R. Klose, A. Schädle, F. Schmidt, and L. Zschiedrich, Adaptive FEM solver for the computation of electromagnetic eigenmodes in 3D photonic crystal structures, *Proc. SCEE04 - Scientific Computing in Electrical Engineering*, Springer Verlag (to be published)

THREE-DIMENSIONAL INTEGRATED OPTICS: MMI COUPLERS FOR MULTILEVEL PHASARS

Ayman Yehia Hamouda and Diaa Khalil
Ain Shams University, Faculty of Engineering, Electronics and Communications department, 1 Elsarayat st., Abbassia, Cairo, Egypt. Email: aymanyehia@ieee.org , diaa.khalil@ieee.org

Abstract: In this work, we report the use of two-dimensional multimode interference (2D MMI) couplers for the construction of three-dimensional multimode interference phased array (3D MMI PHASAR) demultiplexers. The 2D MMI coupler is thus studied in details and the results are used in the design of a 3D MMI PHASAR. The analysis of the 3D MMI PHASAR shows that there is an appreciable performance enhancement and size reduction compared to the planar MMI PHASAR

Key words: integrated optics, DWDM, two-dimensional waveguides, 3D integrated optics, wavelength demultiplexing, PHASAR

1. INTRODUCTION

Planar integrated optics technology is a very reliable technology that has been used in the production of many photcnic circuits for different applications. However, this planar technology has its own limitations either on the level of the circuit size or the signal routing. To overcome these limitations, three-dimensional (3D) integrated optical circuits are required. However, 3D optical circuits[1-4] are still technologically complicated and need more development. One of the promising solutions to build 3D optical circuits is the use of multi-layer technology.

In this case, the lightwave devices could be stacked in different layers resulting in an appreciable decrease in the overall device area. Also, the signal routing in three-dimensions avoids the problem of waveguide

S. Janz et al. (eds.), Frontiers in Planar Lightwave Circuit Technology, 281–287.

crossing. Two-dimensional multimode interference (2D MMI) couplers[5,6] are capable of coupling the input signal into a large number of outputs in different levels with a very good uniformity, bandwidth and fabrication tolerance. This makes them very suitable for the multilevel coupling function in three-dimensional integrated optics.

Three-dimensional wavelength demultiplexing has also been recently introduced[7] as the multilevel extension of the planar AWG. It, however, suffers from difficulties in its packaging. On the other hand, the 3D MMI PHASAR[8] is compatible with the planar waveguide fabrication technologies and has no special packaging problems.

2. TWO-DIMENSIONAL MULTIMODE INTERFERENCE (2D MMI) COUPLERS

The 2D MMI structures are multimode waveguides in both the transverse directions. The modal interference along the propagation causes self and multiple images in the whole transverse plane. Thus, to have $N_x \times N_y$ images we need to have

$$L = \frac{3L_{\pi x}}{S_x N_x} = \frac{3L_{\pi y}}{S_y N_y}$$

where S_x and S_y are the excitation symmetry factors in the x and y directions respectively and they are equal to 1 for general excitation, 3 for paired excitation and 4 for symmetric excitation. $L_{\pi x}$ and $L_{\pi y}$ are the beating lengths of the 2D MMI in the x and y directions respectively. The beating lengths are given by

$$L_{\pi x} = \frac{\pi}{\beta_{00} - \beta_{10}} = \frac{4n_g W_{xeff}^2}{3\lambda} \quad \text{and} \quad L_{\pi y} = \frac{\pi}{\beta_{00} - \beta_{01}} = \frac{4n_g W_{yeff}^2}{3\lambda}$$

where β_{00} , β_{10} , β_{01} are the propagation constants of the fundamental mode, the mode of order (1,0) and the mode of order (0,1) respectively. W_{xeff} and W_{yeff} are the effective width and height of the 2D MMI with a core refractive index of n_g at the free space wavelength λ. Figure 1(a) shows the 3×3 multiple images formed in a $80 \times 80\ \mu m^2$ 2D MMI hollow waveguide fabricated in silicon using deep reactive ion etching (DRIE)[9]. The guide itself is shown in Figure 1 (b).

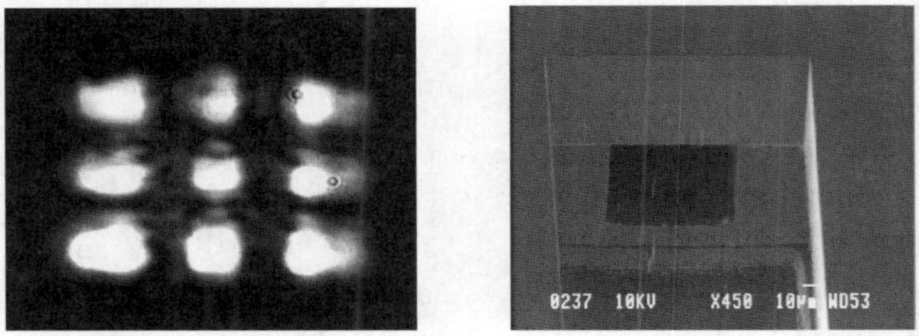

(a) (b)

Figure 1. (a) The 3 × 3 multiple images formed in a 2D MMI hollow waveguide (b) SEM of the 80 × 80 μm^2 silicon hollow waveguide showing the guide cross-section

According to the previous results, 2D MMI structures could be used as 3D power splitters/combiners. It is capable of splitting the input signal into a large number of output ports with a good power splitting uniformity and a reasonably small insertion loss. Figure 2(a) shows a schematic diagram of a 1 × 65 2D MMI power splitter, which splits the input signal into 5 rows and 13 columns as shown in the 3D finite difference beam propagation method (3D FDBPM) simulation demonstrated in Figure 2(b). This 2D MMI has a 133 × 82 μm^2 cross-sectional area and a length of 5546 μm. Figure 3(a) shows the image uniformity produced by the 2D MMI in comparison with a planar 1 × 65 MMI power splitter having a width of 490 μm and a length of 14673 μm. We can notice the improved uniformity of the 2D MMI as well as the appreciable area reduction (approximately 10 times reduction in area). The spectral responses of both power splitters are shown in Figure 3(b). We can see that the 2D MMI has a larger bandwidth than its planar counterpart.

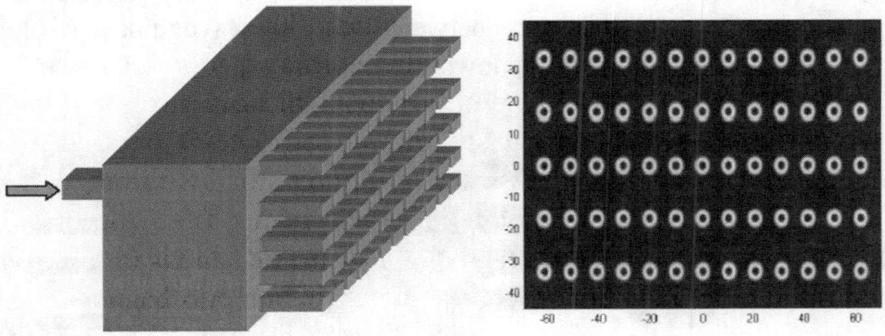

(a) (b)

Figure 2. (a) A schematic diagram of a 2D MMI 1 × 65 power splitter, and (b) a 3D FD-BPM simulation of the output intensity distribution showing 1 × 65 power splitting.

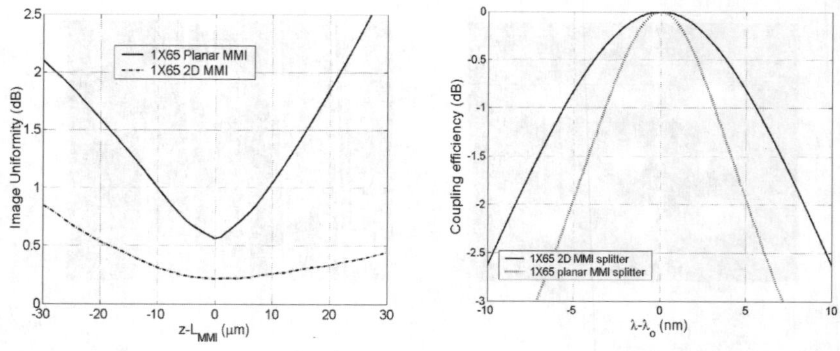

(a) (b)

Figure 3. (a) A comparison of uniformity for the 1 × 65 planar and 2D MMI power splitters. (b) A comparison of the bandwidth of the 1 × 65 planar and 2D MMI power splitters.

3. 3D MMI PHASAR

To demonstrate the importance of the 2D MMI in 3D integrated optics, we use it for the constructions of a 3D MMI PHASAR demultiplexer. The schematic diagram of the 3D MMI PHASAR[8] demultiplexer is shown in Figure 4. It is the 3D generalization of the planar MMI PHASAR[10]. And it is basically an interferometric device, where the input signal is divided using a 2D MMI splitter into N ($N_x \times N_y$) parts, and then different signal portions acquire a wavelength dependant relative phase difference and finally recombine in a 2D MMI combiner. Thus, the single mode interconnecting waveguides are the dispersive element in the PHASAR, where the dispersion is achieved by the path difference between different waveguides. A block diagram of the MMI PHASAR is shown in Figure 5.

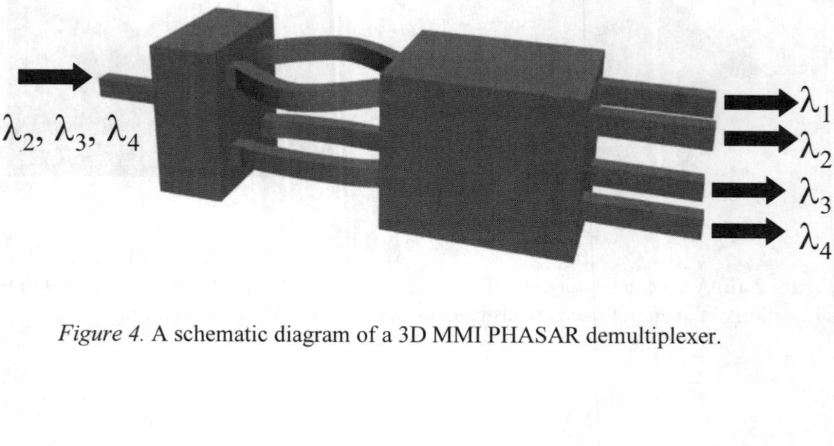

Figure 4. A schematic diagram of a 3D MMI PHASAR demultiplexer.

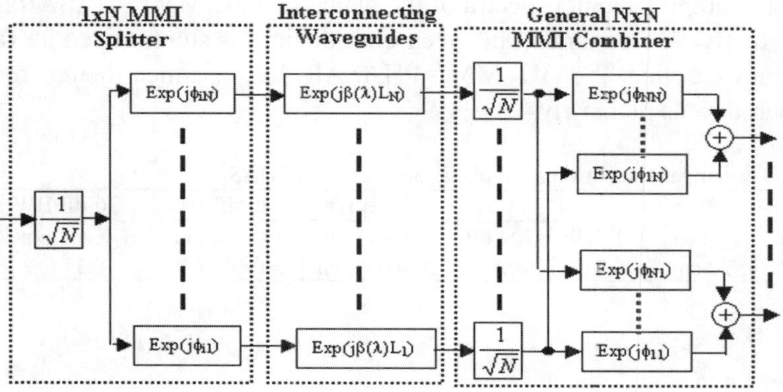

Figure 5. A block diagram of a 3D MMI PHASAR demultiplexer.

To evaluate the performance of the 3D MMI PHASAR we compare it with a planar MMI PHASAR with the same specifications. For this purpose the PHASAR is designed for the multiplexing of 65 DWDM channels with a channel separation of 0.2 nm (25 GHz). We used the planar and the 3D power splitters analyzed in the previous section in the splitting and combining in both designs. The longest path length for the interconnecting waveguides of the 3D MMI PHASAR is much smaller than its planar rival. This is the main advantage of using the third dimension. Modal analysis is used to simulate the spectra response of both devices. Figures 6(a) and 6(b)

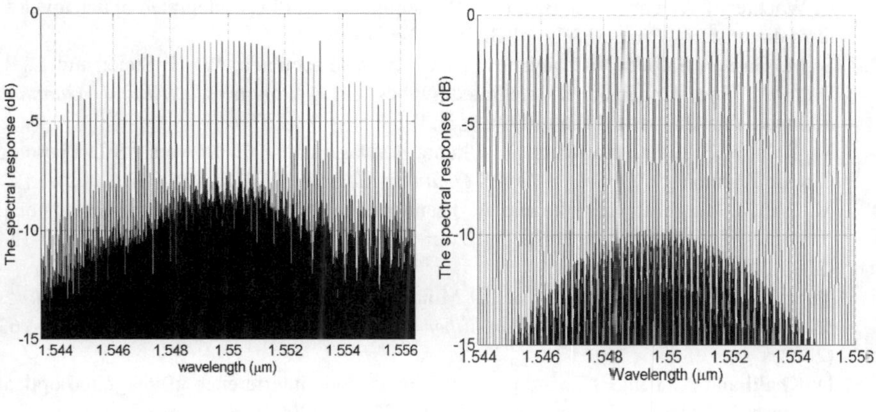

Figure 6. (a) The spectral response of the planar 65 channel MMI PHASAR. (b) The spectral response of the 3D 65 channel MMI PHASAR.

show the output channel spectra of the planar and 3D MMI PHASARs, respectively. A comparison between both devices is summarized in Table 1. We can see that the 3D MMI PHASAR has a much better response compared to its planar rival.

Table 1. A comparison between the planar and 3D MMI PHASAR

	Area	IL	Flatness	3-dB BW	20-dB BW	30-dB BW	XT
Planar	330 cm^2	1.3 dB	4 dB	0.015 nm	0.056 nm	0.067 nm	16 dB
3D	20 cm^2	0.7 dB	0.5 dB	0.065 nm	0.142 nm	0.33 nm	45 dB

4. CONCLUSION

In this work, we presented the 2D MMI couplers as a basic component for 3D integrated optics. Two basic applications of the 2D MMI, the power splitter and 3D MMI PHASAR, are studied and compared to their planar equivalent. The simulations show an appreciable performance enhancement in both cases. These promising devices open the door to multilevel integrated optical circuits with higher packing density and better performance.

REFERENCES

1. C. Wächter, Th. Hennig, Th. Bauer, A. Brauer and W. Karthe, "Integrated optics towards third dimension," *Proc. SPIE*, Vol. 3278, 102-111 (1998).
2. S. M. Garner, S. Lee, V. Chuyanov, A. Chen, A. Yacoubian, W. H. Steier and L. R. Dalton, "Three-Dimensional Integrated Optics Using Polymers," *IEEE J. Quantum Electron.*, Vol.35, 1146-1155 (1999).
3. M. Raburn, B.Liu.,K. Rauscher, Y. Okuno, N. Dagli and J. E. Bowers, "3-D Photonic Circuit Technology," *J. Select. Topics Quantum Electron.,* Vol. 8, 935-942 (2002).
4. M. Will, J. Burghoff, S. Nolte and A. Tuennermann, "Fabrication of three-dimensional photonics devices using femtosecond laser pulses," *Proc. of SPIE*, Vol. 4978, 147-154 (2003).
5. Yehia and D. Khalil, "Analysis of 2D Multimode interference structures," *Proceedings of 2002 IEEE/LEOS Workshop on Fiber and Optical Passive Components*, 57- 63 (2002).
6. D. Khalil and A. Yehia, "Two-dimensional multimode interference in integrated optical structures," *J. Opt. A: Pure and Appl. Opt.* **6**, 137–145 (2004).
7. J. Yang, X. Jiang, M. Wang, Y. Wang, "wo-dimensional wavelength demultiplexing employing multilevel arrayed waveguides," *Optics Express* **12**, 1084-1089 (2004).

8. A.Yehia and D. Khalil, "Design of a compact three-dimensional multimode interference phased array structures (3D MMI PHASAR) for DWDM applications," *IEEE J. Sel. Top. Quant. Electron.* (in press, 2005).
9. A.Yehia, K. Madkour, H Maaty and D. Khalil, "Multiple-Imaging in 2D MMI Silicon Hollow Waveguides," *IEEE Phot. Technol. Lett.* **16**, 2072-2074 (2004).
10. M. R. Paiam, and R. I. MacDonald, "Design of phased array wavelength division multiplexers using multimode interference couplers," *Appl. Opt.* **36**, 5097-5108 (1997).